Sexuality and Aging

Jennifer Hillman

Sexuality and Aging

Clinical Perspectives

 Springer

Jennifer Hillman
Psychology Department
Pennsylvania State University
Berks College
Reading, PA, USA

ISBN 978-1-4614-3398-9   ISBN 978-1-4614-3399-6 (eBook)
DOI 10.1007/978-1-4614-3399-6
Springer New York Heidelberg Dordrecht London

Library of Congress Control Number: 2012933767

Printed on acid-free paper

Springer is part of Springer Science+Business Media (www.springer.com)

Dedicated to my patients, clients, and students

Preface

Sometime around the late 1970s my mother, a high school home economics teacher, was asked to teach sex education. Until then, no such thing was taught in the district. We lived in a small town in Southeastern Pennsylvania that was socially conservative, to say the least. In my mother's own experience as a teacher there were many students who came to her for help when facing challenging situations about boyfriends, girlfriends, alcohol use, birth control, and what to do when birth control failed. Although my mother was uncomfortable with the idea of teaching sex education initially, she knew that it was desperately needed. Once she gathered her teaching materials, she experimented with her delivery on my younger brother and myself. I remember sitting with my mouth open, being shocked to hear my mother talk about erections, wet dreams, menstruation, gynecological exams, and how to use tampons, condoms, and other such things while she was pointing out details on colorful anatomically correct charts. I also remember that these lessons had a huge impact upon me and allowed me to make wise choices about various things when the time came. Probably most importantly, I learned that it was OK to talk about sex.

While in college, I vividly remember the campus nurse and de facto sex educator talking to us in at a large assembly about a deadly new sexually transmitted disease called AIDS. The nurse managed to do this in such a way that we left her talk not feeling afraid, but empowered and informed. Soon after, I volunteered as a peer health educator. I went around campus speaking to various sororities and other primarily female groups about the free annual gynecological exams available at the college health center. I explained what a gynecological exam was, how important they were, and that you didn't lose your virginity by having one. I was shocked how many girls never had an exam because they were afraid, because their parents never talked to them about it, or because their parents even forbid it. The nurse lent me a collection of speculums, applicators for pap tests, and tubes of lubricants so that I could show the girls what kind of equipment was used. I also made it fun to pass around these materials as part of a health "show and tell" game. I also informed the girls that you had the right to talk with the doctor with all of your clothes on before the exam and to ask the doctor explain and get permission for everything he was going to do.

During my graduate work in clinical psychology, I found my niche working with older adult clients. I was always very close with my grandparents who were socially and physically active well into their 80's. I also remember my grandparents being very affectionate with one another and chuckling about needing their alone time. As a result, I viewed older age as a time for family, friendship, travel, hobbies, volunteer work, and general happiness. So, when I encountered older clients who needed assistance with even the simplest of tasks or who were alone or depressed, I knew I wanted to help them. What soon became clear among my class in graduate school was that I was the only one who wanted to work with older adults. Fortunately my mentor, George Stricker, who conducted research on grandparenting, helped connect me with the North Shore-Long Island Jewish Health System and their geriatric training program. I am indebted to Rick Zweig, Greg Hinrichsen, and Eileen Rosendahl for giving me such a wonderful experience there and later to Mike Bibbo and Tom Skoloda for continuing my outstanding training in geropsychology at the Coatesville VA. The negative social stigma about aging fascinated me, and I was conducting research on this topic as well.

An overarching issue that quickly emerged in my research and clinical work was that health care providers, mental health providers, family members, friends, and older adults themselves were hesitant to talk about sexuality and aging. For example, when a 73-year-old client in a geriatric day treatment program reported that she was raped 2 years ago, most of the health care professionals on the unit did not want to "retraumatize an old woman" by even suggesting that she get tested for HIV and other STDs. In various nursing home settings, I encountered adult children of residents became enraged, upset, or extremely embarrassed when they learned that their single, cognitively intact mother or father had started a romantic relationship with another resident. Older clients recovering from breast or prostate cancer often reported feeling ashamed or shunned when they broached the topic of sex with their health care providers. Probably most vividly I remember conducting an intake session in a locked inpatient unit with a 68-year-old client who sat impassively with his wrists wrapped in bandages after a suicide attempt. According to my client, who was unable to achieve orgasm while taking Prozac, his suicide attempt was fueled by his psychiatrist's response to his complains about sexual side effects. His psychiatrist told him, "I don't care if you have side effects from your medication. You are old and you are single. You don't need to have sex." When I turned to the literature for guidance about issues of sexuality and aging, only limited information was available.

A few years later, with encouragement from George Stricker I published a book on elderly sexuality with Kluwer in 2000 that incorporated both clinical and research findings. In the midst of my research program, I was fortunate to publish the article, "Sexual issues and aging within the context of work with older adult patients" in 2008 in the APA journal, *Professional Psychology: Research, and Practice.* What I was most pleased to learn in 2010 from Jeffrey Barnett, one of the journal's editors, was that my article was the most frequently downloaded article from the journal within the last decade. In other words, people clearly wanted information about sexuality and aging, particularly from a clinical perspective. Buoyed by this finding, I contacted my generous editor, Sharon Panulla at Springer. She agreed to let me revise my text for a new release in 2011. What I soon realized, however, was that my book did not only need to be revised, but essentially rewritten to incorporate new research findings, clinical cases, and emergent and vital areas of interest.

Just some of these emergent areas of interest included LGBT issues; cross-cultural and cross-national findings; Viagra and other PDE-inhibitors; sexuality at midlife; sexuality in palliative and hospice care; legal rights and policies related to sexuality in nursing home and assisted living; the assessment of sexual consent capacity for cognitively impaired elders; age-related risk factors for HIV/AIDS and other STDs; the role of the Internet upon dating, sex education, and pornography use; sexual side effects from popular psychotropic and other prescription medications; and updates for prostate, breast, and gynecological cancers. The role of baby boomers in influencing media and other social perspectives also needed to be highlighted. Even the title of the book was changed to *Sexuality and aging: Clinical perspectives*. It is my hope that this new book will prove useful for anyone, particularly psychologists and other mental health providers, who has an interest in empowering adults of all ages to maintain their sexuality and dignity.

Without my clients, patients, previous supervisors, colleagues, students, friends, and family this text would not have been possible. I need to acknowledge the following clinicians and researchers for all of their generous help and insight including George Stricker, Richard Zweig, Eileen Rosendahl, Tom Skoloda, Mike Bibbo, Patrick Ross, Christine Li, and Donna Chimera. A very special thank-you goes to two of my students at Penn State Berks, Lacey Lott from the Applied Psychology Program and Jamie Brackman from the Professional Writing Program, who delighted me with their interest and high quality of work. I wish them both well and know that they will excel in whatever field they choose. There are many other people I would like to acknowledge for their friendship and emotional support including Stephanie Padgett, Jennifer Rosa, Cheryl Lynn Malloy, Amy Gingrich, Amber Mintz, Trish Luberta, Erin Byrne, Mitzi Wilke, Kristin Jones, Patti Brownmiller, and Tom and Linda Brownback. A special thank-you goes out to my sister-of-the-heart, Daniele Richards, for being the best friend I could possibly imagine. I also am indebted to my husband and children, Sean and Francesca, for all of their love and patience during the many times I was busy writing. As penned by Browning, "Grow old along with me. The best is yet to be."

Reading, PA, USA   Jennifer Hillman

Contents

Chapter 1
An Introduction including Media, Boomer, and Cross-cultural Perspectives

Contrary to popular stereotypes, middle-aged and older adults as a group are more heterogeneous than their younger counterparts. In recognition of this diversity, clinicians should only expect that the sexuality expressed among their older clients will be just as unique and varied.

In prior decades the topic of sexuality and aging was viewed as having no real importance, as a waste of professionals' time, or at its worst, as an oxymoron. Only in recent years have sexuality and aging been addressed seriously and responsibly from a clinical perspective. Empathic, attentive work from clinicians and researchers has allowed the field to gain increased respect as well as a measurable body of knowledge. With the change in our country's demographics, including a dramatic increase in the sheer number of older adults and our society's general tolerance for more open discussion of sexuality in general, a substantial need for clinical expertise in sexuality and aging has become readily apparent.

The Importance of Sexuality and Aging

Middle-aged and older adults represent diverse populations who may engage in a variety of sexual behaviors within the context of various long-term, short-term, and traditional and nontraditional relationships. The sexual orientation of older adults includes lesbian, gay, bisexual, and transgender, as well as heterosexual. Aging adults also engage in any variety of activities including dating, cohabitation, affairs, having protected and unprotected sex, sexual abuse, masturbation, and abstinence. Middle-aged and older women and men also may find themselves negotiating new relationship and sexual situations after the divorce or death of a spouse or long-term partner, seeking new partners through the Internet and other forms of social media,

J. Hillman, *Sexuality and Aging: Clinical Perspectives*,
DOI 10.1007/978-1-4614-3399-6_1, © Springer Science+Business Media New York 2012

or attempting to regain their sexual function after treatment of breast or prostate cancer. These adults may have sexual relationships within the context of community living, assisted living, or full care nursing facilities.

Despite potential differences in physical health and living arrangements, aging adults can and do engage in the same types of sexual behaviors as their younger counterparts, and like younger adults, may be satisfied or dissatisfied with their sex lives. However, because aging adults often encounter mixed societal messages about what constitutes appropriate sexual behavior (e.g., diametrically opposed views featuring either complete abstinence or high frequency and performance) and must cope with the presence of chronic illness and other physiological changes that may occur with aging, an understanding of the multidisciplinary nature of sexuality and aging becomes essential.

The Relevance of this Text

This text is intended to provide up-to-date information and practical advice regarding clinical issues in sexuality and aging. Detailed case examples will be used throughout to illustrate both theoretical constructs and therapeutic techniques, thus providing a unique clinical perspective. Empirical findings will be introduced in a clear, easy to understand fashion, along with a critical assessment of the underlying research methodology. Although this book is geared primarily for mental health students and professionals (e.g., psychologists, social workers, psychiatrists, counselors, clergy), its material is suitable for physical therapists, occupational therapists, nurses, nutritionists, nursing home administrators, geriatricians, retirement home coordinators, and others who work with aging adults. It also is important to note that material within this volume would be of interest to middle-aged and older adults who have questions about sexuality and aging, and to people of various ages who may be family members, partners, caregivers, or friends of such aging adults.

This volume is unique in that the topics presented include many commonly neglected themes in sexuality and aging including:

- Sensuality and sexuality with or without a partner.
- Women's issues such as body image, vaginal dryness, breast cancer, and sexual abuse.
- Men's issues such as erectile dysfunction (ED), prostate cancer, and body image
- Sexuality within the context of disabilities such as Alzheimer's disease.
- Sexuality in long-term care, assisted living, and other institutional settings.
- The impact of prescription and over-the-counter medications on sexual functioning.
- Cross-cultural perspectives.
- Lesbian, gay, bisexual, and transsexual (LGBT) issues.
- HIV/AIDS and other sexually transmitted infections.
- The assessment of sexual consent capacity among older adults with cognitive impairment.

- Changes in dating and cohabitation versus marriage.
- The impact of the Internet and social media.
- Sexuality in palliative, hospice, and other end-of-life care.

This introductory chapter is designed to provide a general overview of sexuality and aging, including societal attitudes and stereotypes, historical approaches, and an introduction to research methodology. Chapters 2 and 3 introduce basic, requisite knowledge of sexuality and aging, as well as suggestions for clinicians in establishing open communication with patients. Additional information will be included to help practitioners manage a variety of attitudes toward sexuality and aging, including potential transference and countertransference.

The next two chapters highlight elderly sexuality within an institutional context such as a nursing home or assisted living facility (Chap. 4) and within the context of disabilities and chronic illness such as depression, diabetes, arthritis, and heart and Alzheimer's disease (Chap. 5). Chapter 6 discusses the typically ignored and overlooked topic of sexually transmitted infections among aging adults, including participation of middle-aged and older adults in high-risk behaviors, the increasing incidence of HIV/AIDS, and HIV-induced dementia. Chapters 7 and 8 are devoted, respectfully, to specific women's and men's issues in sexuality and aging. Chapter 9 presents information related to sexuality in both short- and long-term relationships, and within the context of aging heterosexual, gay, lesbian, and transgender couples. Chapter 10 concludes with information on emergent areas of research in sexuality and aging. This final chapter provides suggestions for coping with the medicalization of sexuality and aging (e.g., how to work effectively in interdisciplinary settings), end-of-life issues related to sexuality (e.g., in hospice), and the impact of the Internet and social media upon new relationships, health education, pornography, sexuality, and aging. All of these chapters provide case examples based upon the author's experience with middle-aged and older clients as a licensed psychologist.

Aging Demographics

Defining Older Adulthood

In order to discuss older adulthood as a developmental stage or specific age cohort, it is necessary to first define the age at which one becomes an older adult. Various arguments can be made to support the notion that older adulthood is not a function of chronological age, but rather a function of physical ability and mental health. Many older adults who exercise, eat properly, maintain fulfilling personal relationships, and demonstrate optimism and resiliency are in better mental and physical health than their younger counterparts who have a more sedentary or isolated lifestyle. Other experts maintain that the wisdom accumulated through life provides older adults with protection from stressful life events, and that an increase in the exploration of opposite-sex roles generates a youthful outlook among many chronologically older adults (Gutmann 1994).

As cohorts age, they tend to regard older adulthood as beginning at later and later ages, For example, consider the new phrase in US popular cultures in which "Sixty is the new 40." However, for the sake of analyzing demographics and making some generalizations, it becomes necessary to set a specific, yet somewhat arbitrary age as a cutoff point for older adulthood to conduct research and provide more effective means of generalizing findings. For ease of categorization in this text, older and elderly adults will be defined as women and men who are 65 years of age or older.

Although 65 years of age has been accepted generally as a defining point for older adulthood, it also is readily apparent to epidemiologists that adults who range from 65 to well over 100 represent a diverse group. Society itself tends to view (i.e., stereotype) a 65-year-old quite differently than a 95-year-old (e.g., Hummert et al. 1995). Although age 65 has been defined historically as the requisite age of retirement by the US Social Security Administration, this age of retirement has already increased and is likely to advance over time as more adults work later into life for both financial reasons and personal satisfaction.

In addition, the current cohort of adults over the age of 85 appears to represent a unique subgroup of older adults, labeled as the oldest-old. These oldest-old adults are identified as a demographic who may require additional clinical attention and consideration because of their advanced age (e.g., Hillman et al. 1997) and greater likelihood of experiencing chronic illness, poverty, and lack of social support (e.g., Pennix et al. 1999).

The Oldest-Old as a Distinct Population

Within the cohort of adults aged 65 and older, the oldest-old segment of the population (i.e., aged 85 years and older) demands special attention as a unique subgroup. This group of the oldest-old is experiencing an even higher rate of growth than their young-old peers. In 1990, 3 million Americans were more than 85 years of age, representing approximately 1.2% of the nation's total population. By the year 2050, the number of oldest-old Americans is expected to reach more than 18 million people—a sixfold increase in their population since 1990.

Related more directly to clinical issues that may present themselves in terms of sexuality and aging, significant gender differences exist in this oldest-old cohort. The vast majority of this oldest-old population is comprised of women. Among US citizens aged 85–89, there are more than 2.5 women for every man. Among individuals aged 95 years and older, women begin to outnumber men by 4–1 (U.S. Census Bureau 2011). These ratios are even more striking when one considers the numbers of single or "available" older men and women. (This statistic also assumes that there are issues of supply and demand among heterosexual elders; it remains virtually unknown what proportion of elderly men and women consider themselves homosexual. Ostensibly, lack of an available homosexual partner would be more of an issue for elderly men than for elderly women.) By age 85, nearly 1 out of 2 men is married, whereas nearly 4 out of 5 women (80%) are widowed. Although men are expected to live almost as long as their female counterparts during the next century, there will continue to be a dramatic shortage of older men compared to older women.

Gender, Racial, and Economic Diversity

Despite the heterogeneity of the older adult population, one fact is clear; this segment of the USA and world population is significantly increasing. Various sources (e.g., Administration on Aging 2011; U.S. Bureau of the Census 2011) show that by 1990, about 1 in 8, or 31.1 million Americans were over the age of 65. By the year 2000, adults over 65 increased in number to more than 39.6 million. Currently, older adults represent approximately 13% of the US population, representing nearly one in eight Americans. By the year 2020, the older adult population is expected to increase to more than 54 million people, representing 1 of every 6 Americans. By 2050, this older adult demographic is expected to double in size to more than 79 million people. In other words, by the year 2050, one in five, or nearly 20%, of all Americans will be older than 65 years of age. This growth rate of the older population is nearly double that of the overall growth rate for the total US population. A longer life span, increased access to improved health care, and the large number of baby boomers aging into older adulthood contribute to this population increase.

The diversity of the older adult population also is increasing in terms of race and ethnicity, income status, level of education, numbers of men and women, and even marital status. By 2050, the numbers of older African-Americans are expected to quadruple from 2 to more than 9 million people, representing a virtual doubling of their number of 8% to more than 15% of the US population. Among American Indians and Eskimos, their proportion of older adults is expected to double from 6 to more than 12%. Among Hispanic and Asian Americans, the ranks of their older adult members are expected to nearly triple from 6 to 16% and 5 to 15%, respectively. The absolute number of Hispanic older adults estimated to be living in 2050 (more than 12 million) will be 11 times the number of Hispanic elders living now. Appreciation for cross-cultural issues remains essential in clinical work with an aging population.

Diversity among older adults also is apparent in their distribution of income. Although significantly more older adults live above the poverty line than in the 1970s, older women continue to have nearly double the chance of living in poverty than older men; 16% of older women versus only 8% of older men live below the poverty line. Adults over the age of 65 with higher levels of education are more likely to have more disposable income and retirement savings. The median net worth of households headed by someone aged 65 and older is more than $100,000, whereas the median net worth of households headed by those under the age of 35 is under $8,000. Adults over 50 own three quarters of the nation's wealth and spend more than one trillion dollars annually on goods and services including 60%

of all health care costs, 51% of all over-the-counter medication, and 74% of all prescription medication. Adults over 50 also represent the fastest growing segment of online users, boasting more than 7 billion dollars of annual Internet purchases (AARP 2010; Bureau of Labor and Statistics 2011). Ethnic disparities also exist; 52% of all African-American and 56% of all Latino retirees feel insecure about their income whereas only 30% of all Whites report such insecurity (Meschede et al. 2010). All of these demographic factors, including access to personal and health resources, are relevant when considering sexuality and aging.

Middle-Aged Adults

Middle-aged adults, identified here somewhat arbitrarily as men and women aged 40–64, also represent a sizable portion of the US population. A large cohort of Americans in midlife can be identified as "middle boomers," born from 1952 to 1958. A 2010 MetLife report indicates that this cohort of middle-aged Americans comprises 10% of the US population. Unlike the current cohort of already retired older adults, middle-aged women and men in the USA need to work even longer to retire with full social security retirement benefits. When HIV and AIDS were recognized in the late 1980s, the youngest members of this middle boomer cohort were already in their 30s. Additional information about baby boomers, as a whole, and factors that will influence their sexuality follows.

Baby Boomers' Unique Characteristics

When working with middle-aged and older clients, it becomes vital to consider that new cohorts of older adults are reaching their prime. These cohorts carry with them different social mores, expectations, and experiences that can shape their clinical presentation. Because of exposure to different historical events and general socioeconomic, political, and societal pressures; for example, a woman born in the USA in 1900 would be expected to come from a very different social culture and may have very different attitudes toward sexuality and aging than a woman born in 1950. Throughout the current millennium, the cohort that has come to represent a major financial, political, and cultural presence is that of the baby boomers. Understandably, their approach to and expectations for their own sexuality can be expected to be quite different from those of the cohorts that came before them.

What defines a baby boomer? Demographers identify baby boomers as those approximately 78 million individuals who were born between the years of 1946 and 1964, during the literal population boom that occurred when large numbers of American GIs returned home after World War II and had children. Many of these boomers grew up during good financial times, and expected to live the proverbial American Dream. Although women often maintained traditional roles as young wives and mothers, as they approached midlife they often attended college, entered the workplace, and established successful careers outside of the home. Men in this cohort typically reaped great benefits from secondary education and enjoyed job security for many years. The men and women from this cohort also experienced life-changing events such as the landing of men on the moon, an increase in life span through advances in medicine, the introduction of modern conveniences in the home such as television and the microwave, and the civil and women's rights movements. Baby boomers also lived through the 1960's sexual revolution, had unprecedented access to relatively effective birth control (i.e., the pill), and had no knowledge (or fear) of life-threatening sexually transmitted infections such as HIV.

Also relevant to clinicians, the numbers of baby boomers seen in therapists' offices can be expected to increase considerably. The first of the baby boomers turned 65 in the year 2011, with nearly half of their demographic expecting to celebrate their 65th birthday by the year 2015. The length of the average American's life span (e.g., 81 years for women and 76 years for men; Kulkarni et al. 2011) will continue to swell the number of baby boomers relative to the rest of the US population. As noted, the middle boomers themselves now account for 10% of the US population.

In addition to this emergent increase in sheer numbers of middle-aged and older adults, baby boomers can be expect to have considerable impact on our country's financial and political makeup. Lobbyist organizations such as AARP and various political groups have begun to harness the power of this large group of voters. Aging boomers express significant political concerns about the viability of social security, Medicare's long-term care benefits, changes in health insurance policies, tax laws affecting retirement planning, and rules governing pension plans, as most can expect to live well into their late 70s or 80s. Both small businesses and large-scale corporations recognize the financial power of middle-aged and older baby boomers as consumers. Contrary to general societal myths that older adults are poverty stricken, older adults and baby boomers in particular, dispel that myth even further. Middle-aged and older adults over the age of 50, who account for approximately 25% of the population, hold both the majority of US assets and its discretionary income (Bureau of Labor and Statistics 2011).

Additional changes in the general population and popular culture have emerged in response to this growing cadre of baby boomers. Throughout the next few decades, we can expect to encounter a media-friendly blitz toward middle-aged and older adults. Fortunately or unfortunately, this increase in positive media portrayals of aging adults will be motivated primarily by corporate entities interested in the baby boomers' enormous spending power. Compared to other older cohorts who survived the Great Depression as youngsters and managed to carve out a living for a typically large family, many baby boomers benefited from the country's economic prosperity. The average boomer raised only 2.5 children and was able to achieve his or her maximum earning potential while maintaining substantial savings. Many baby boomers had pension plans that afforded them a more secure financial future in retirement. As a result, baby boomers as a demographic possess the highest levels of disposable income relative to any other age cohort.

Advertisers are aware that the majority of boomers are able to afford expensive prescription drugs (e.g., Viagra, which may cost up to $20 per dose) and various health care procedures and services. Corporate entities recognize that boomers are a consumer demographic willing to spend enormous amounts of their disposable income on products and services designed to prevent the signs of aging. Such markets include gym memberships, personal trainers, vitamins and holistic medicines, gourmet foods, hair transplants, dietetic services, cosmetics, plastic surgeries, home exercise equipment, and even bras and girdles (now conveniently reintroduced as body shapers by the company Spanx). Other emerging consumer markets for boomers include children's toys (i.e., gifts from doting grandparents), financial planning, life insurance, second homes, retirement communities, home security systems, and upscale spa and vacation packages.

In order to entice baby boomers to buy their particular services and products, advertisers now typically feature models and spokespersons who are older adults themselves (e.g., Sally Field is a spokesperson for Boniva, a prescription drug to treat osteoporosis.) Middle-aged and older adults will continue to be featured more frequently as models in print advertisements and television commercials. According to basic principles of advertising, these aging spokespeople will be selected primarily for their beauty and physical fitness.

What Clinicians Can Expect

When working with baby boomers regarding sexuality and aging, as compared to the cohort of older adults raised during the depression era, practitioners can expect to see some significant differences in clinical problems and presentations. Although the following certainly represent generalizations, and individual factors and accounts must take precedence, a variety of cohort differences can be expected along these dimensions.

- Aging baby boomers are more likely to seek out medical and psychological treatment for sexual problems, as they grew up in a time when it was more acceptable to discuss sexuality among their peers. Compared to their older cohorts, baby boomers are more likely to have experienced and have begun to accept some aspects of a "therapy culture." They also are more likely to have taken psychology courses in college, to have read self-help books, and to expect sexual satisfaction as a part of life. This positivity, if realistic and accompanied by appropriate levels of patient motivation, can only be an asset in treatment. This interest in sexuality also will prompt more adults to seek treatment for sexual dissatisfaction in midlife as well as in late life.
- Baby boomers will present more clinical issues related to underlying family dynamics. Compared to their older cohorts, baby boomers are more likely to cope with divorce proceedings, step families, second or multiple marriages, dating in later life, cohabitation, and poorly defined or overwhelming family caregiving responsibilities. These changes in family structure, away from a core

nuclear family, have the potential to introduce significant challenges within the context of intimate relationships. Sexual disturbances may emerge as an outward manifestation of underlying interpersonal problems.

- Higher rates of eating disorders and body image disturbances will be observed among both men and women in this baby boom cohort. One culprit for this expected increase in pathology is the internalization of artificially created media messages and images. The general directive in current US culture is to find the fountain of youth "even if it kills you." Clinicians must be as willing to assess and diagnose the presence of an eating disorder or a problem in body image among their older, as well as their young adult, patients.

- Issues related to sexuality within the context of institutional settings will become more prominent in clinical practice. Although the actual percentage of older adults who live full time in nursing homes or other institutions is relatively small, as the numbers of baby boomers increase the numbers of older adults in long-term care will increase. Expectations for the maintenance of sexual activity in assisted living and other minimal care settings also will remain high among this age cohort. Clinicians may find themselves serving as outside consultants or full-time staff members of such institutions.

- The medicalization of elderly sexuality can only be expected to increase. Baby boomers grew up with a reliance on and respect for the medical profession. The boomers also had the benefit of quick fixes for many of life's problems through scientific advancements. Many members of this cohort will seek out seemingly quick and easy medical approaches for treatment of sexual dysfunction. Many already seek out drug therapies that may or may not be the most effective way to address their sexual dysfunction. Clinicians will need to function effectively as interdisciplinary team members and to emphasize education and appropriate communication with medical professionals between themselves and their patients.

- Aging boomers are more likely to show greater acceptance of alternative relationships in late life within the context of dating, cohabitation, interracial relationships, and both lifelong and emergent homosexual relationships. Although many members of this generation may continue to hold negativistic, stereotypical views toward nontraditional relationships, clinicians are more likely to encounter patients dealing more openly with such issues in their own and in others' lives.

- The impact of less traditional and rigid gender roles on sexuality will become evident. For example, female baby boomers, when compared to their own mothers, are significantly more likely to have earned a college education and to have pursued a career outside of the home. These female boomers may feel less obligated to fulfill a traditional wife's role, which includes directives to place her husband's sexual needs before her own and to wait for her partner to make sexual advances. For aging male boomers, life after retirement often means having more flexibility to experiment with different sex roles. These men may assume more caregiving responsibilities (for either an ailing spouse or a young grandchild), seek greater variety in their leisure activities (e.g., gourmet cooking), and adopt more domestic chores about the home. Although such increases in role differentiation are seen as positive changes from a general psychological perspective, many older adults may not change as readily as their partner, or may feel threatened by their own changes in personal priorities and interests. These psychological pressures can certainly assert themselves within the context of sexuality and aging.

- More boomers are more likely to relate earlier experiences of sexual trauma in therapy, and will be more likely to recognize the possible connection between these earlier events and current sexual dysfunction or dissatisfaction. Individuals from this cohort are more likely to admit to such sexual abuse, unlike prior generations who often felt that such traumatic experiences, for the sake of propriety or as a product of shame, were to be kept as a family secret. Clinicians also should be aware that older adults can experience rape or incest at *any* age and that in late life, older adults themselves may represent the victims *or* perpetrators of sexual abuse.

- The influence of the Internet and related technologies upon Baby Boomers cannot be underestimated. Statistics suggest that, on average, older adults spend more of their disposable income and time on the computer per day, and view mobile Web-based applications as more useful (Yang and Jolly 2008) than their younger adult peers. Clinicians can be invaluable in helping aging baby boomers evaluate the accuracy and legitimacy of information they receive about sexuality and aging over unregulated sites. The use of the Internet, via chat rooms and other social networking sites, also can allow older adults to maintain personal connections if homebound, or to seek out others who experience similar problems, resulting in a beneficial support network. The use of the Internet also is likely to allow older adults greater anonymity in their sexual dealings, including the selection of online partners and access to pornography. (The limited information available about older adults' use of the Internet related to sexuality will be addressed specifically in Chap. 10.)

Intimacy, Sexuality, Sensuality

In any discussion of sexuality, the concepts of intimacy, sexuality, and sensuality must be reviewed. Because various researchers and theorists have defined these terms differently, the meanings ascribed to them here should be regarded as generalizations and aids in nomenclature rather than theoretically driven constructs. The following categorizations are offered for the following:

Intimacy will be defined as the quality of the interpersonal relationship among two people in a romantic interpersonal relationship, who may or may not be actively engaged in sexual activity. Attachment style, prior family dynamics, sexual identity issues, and self-esteem may all contribute to the level of intimacy experienced (or desired) by an individual. In practical terms, intimacy could be manifested by a subjective feeling of love or satisfaction when in the partner's presence or when thinking about the partner, the degree of appropriate self-disclosure between partners, and the willingness or ability to value the partner's needs and desires as well as one's own. For the purposes of this text, *intimacy* will be used to refer exclusively to emotional intimacy (i.e., interpersonal satisfaction and subjective feelings of closeness).

Sensuality can be defined as the experience of pleasure from one's senses leading to an increased awareness of and appreciation for one's own body. Such pleasure may be generated via sexual activity specifically, but also from any activation of the sensory organs. It is essential to note that sensual pleasure can be experienced with or without another person, and that expressions of sensuality are vast and quite individualized. Examples of sensual activities may include taking a hot bath or shower, noticing the breeze against one's face, having a massage, listening to music, lighting candles, getting one's hair done, eating a wonderful meal, molding or shaping clay, dressing up in beautiful clothing, splashing in puddles, lying in a feather bed, wearing silky underwear, singing in a resonant choral group, holding hands, using fragrant body lotions, dancing, engaging in foreplay, feeling muscles warm and loosen during exercise, or appreciating artwork. While sensual activities may induce sexual excitement, the inherent goal of the activity is not sexual intercourse or climax.

Sexuality will be defined here as a *broadly based* term that indicates any combination of sexual behavior, sensual activity, emotional intimacy, or sense of sexual identity. Any individual's wish to engage in any of these activities also may be considered an aspect of sexuality. Sexuality may involve sexual activity with the explicit goal of achieving pleasure (e.g., hugging, kissing) or orgasm (e.g., petting, oral sex, intercourse), sensual activity with or without the explicit goal of sexual pleasure (e.g., hugging, dancing, wearing body lotion to feel attractive or feminine), or the experience of emotional intimacy within the context of a romantic relationship. Thus, sexuality is commonly associated with a variety of issues and concepts including body image, self-stimulation, love, libido, intercourse, homophobia, relationship satisfaction, marital satisfaction, desires for sexual and sensual experiences, and participation in high-risk behaviors. It also is important to note that sexuality encompasses thoughts, feelings, and behaviors that may lead to positive or negative feelings (e.g., consider body image, masturbation, and sexual abuse).

Historical Context

It is useful to view contemporary social attitudes toward sexuality and aging through the lens of historical attitudes. Consistent with current, generally dismissive attitudes toward elderly sexuality, little has been recorded about sexuality and aging in the art or literature of biblical and medieval times. However, the overwhelming majority of these reports are negative (e.g., Covey 1989). During these periods in history, sexual relations among older people were viewed as evil, immoral, perverse, inappropriate, impossible, or pathetically comical, at best.

During the Middle Ages in Europe, the church played a central role in shaping beliefs about sexuality among older adults. At the core of these beliefs was the prohibition that sexual intercourse was designed for procreation only, among people of all ages. St. Augustine wrote that celibacy was a human ideal to strive for, and St. Albertus Magnus and St. Thomas Aquinas professed similar views that sex was for reproductive purposes only. This doctrine promoted great hostility toward older adults who engaged in any type of sexual behavior, as they engaged in a "sin against nature" (Bullough 1976).

These religious prohibitions also were mirrored by general beliefs among the populous that sexuality was reserved for younger adults who could reproduce and multiply. Older adults were seen as entering an "age of life" in which decay, decline, and repose were an inescapable and unavoidable part of life (Burrow 1986), even by preeminent scholars and scientists. In his *Masterpiece,* Aristotle wrote that sexual activity ceased for women at menopause and for men after their fifth decade of life (Stone 1977). Other medieval physiologists professed that women had a stronger sex drive than men, but believed that this powerful drive ceased immediately on menopause. Thus, sexual activity among older adults was viewed as unnatural, inappropriate, and even disgusting. Consistent with this view, religious notions held that if older adults engaged in physical intimacy, they literally chained themselves to the flesh and impeded their ascension into heaven (Burrow 1986).

Despite strong prohibitions against elderly sexuality in the Middle Ages, facets of popular culture suggest that this phenomenon was present but hidden, as represented in plays and famous literature. Chaucer provided an account of elderly sexuality in his *Canterbury Tales.* In the "Merchant's Tale," he describes a 60-year-old knight who wants to marry a young bride in order to satiate his sexual appetite. The implication is that older women simply do not have the ability (or attractiveness) to satisfy a man's needs: "I'll have no woman 30 years of age. That's only fodder … straw for a cage." Chaucer also prompts his older knight to blatantly disregard church prohibitions about elderly sexuality: "A man is not a sinner with his wife, he cannot hurt himself with his own knife." However, on his wedding night, the noble drinks large quantities of an aphrodisiac and falls asleep without consummating the marriage. When he becomes blind a few years later, the knight's young bride has an affair with one of his young servants. In a cruel twist of fate, the older knight regains his eyesight to the sight of his bride and servant making love in a field. As a further insult, his young bride convinces the older knight that he was just imagining things, because surely he has become senile at his advanced age. Although designed to be humorous, this story illustrates medieval beliefs that older men are not virile, are not attractive to younger women, are physically ill, and appear foolish if they attempt to engage in sexual relations. The story's other implications are that older women are inherently unattractive and unable to obtain partners for sexual enjoyment.

It can be assumed that although older adults did indeed show interest and engage in sexual relations during biblical and medieval times, these practices were regarded as hapless, humorous, and even dangerous by the general population. A double standard prevailed regarding male and female elderly sexuality, namely, that men's participation in sexual activity was seen as humorous or as a foolish possibility, whereas older women's participation in sex was viewed as unnatural and evil (Covey 1989). For example, although older men were thought to have virtually no capacity for sexual relations, those who were able to have active sex lives were believed to gain social status and even an increase in their life span. In contrast, an older woman was thought to have sex in her later years only if she were able to trick a man into going to bed with her, a feat so abhorrent that it required the aid of witchcraft. This evilness associated with female elderly sexuality persisted well into the fifteenth century, fueled by the *Hammer of Witches,* a popular text indicating that witchcraft was responsible for both carnal lust among older women (Stone 1977) and impotence among older men (Bullough 1976).

Cross-cultural Perspectives

An appreciation of cross-cultural perspectives on sexuality and aging is essential. Unfortunately, few quantitative or qualitative accounts exist to detail attitudes and practices about sexuality and aging in other cultures. What little information we do have, however, presents a picture that is strikingly different from that espoused by our industrialized North American, youth-oriented culture. An ancient Turkish proverb illustrates the general positivity espoused by the majority of traditional and pre-industrial cultures: "young love is from earth, while late love is from heaven." An understanding of cross-cultural issues is vital (Hillman 2011) as aging adults from China represent one of the largest demographics on the planet, and nearly 33% of all Americans now identify themselves as a member of an ethnic or cultural minority (Shrestha and Heisler 2011).

In a groundbreaking study of more than 106 cultures (Winn and Newton 1982), less than 3% of those cultures were found to have societal sanctions or prohibitions against older people having sex. An analysis of the data gathered by the anthropologists, sociologists, and psychologists studying these cultures revealed that 70% and more than 84% of the societies reported sexual activity among its older male and female members, respectively. In many Eastern and Middle Eastern cultures, men and women commonly engaged in sexual relations well beyond the age of 100 and 80, respectively. African cultures maintained that impotence was not a normal function of old age, but an unnatural loss of ability resulting from illness or witchcraft. In the majority of these traditional cultures, menopause was not associated with either more or less sexual activity among older women; it simply represented a "point in a woman's life." In certain African and Asiatic cultures, an older women's physical attractiveness appeared unrelated to her sexual status; toothless, older women were considered as sexually desirable as younger women. Thus, sexual activity among older men and women in traditional societies is common, and apparently readily accepted throughout most of the world.

An additional difference noted between these traditional societies and our own was that although a double standard appeared to operate with regard to sexuality and aging, it appeared to be in the opposite direction. Specifically, older women were more likely to engage in sexual relations than older men, and older women were often described (in more than one-fourth of the cultures) as becoming less sexually inhibited and more sexually aggressive with advancing age. In certain South American and Eastern cultures, older women were designated teachers for sexually inexperienced young men. Older women also were described as commonly taking younger men for husbands or sex partners, ostensibly because there were few male partners available of their own age. Other older women in South American, Eastern, and North American Indian cultures were described as dressing more seductively, baring their breasts more often in public, and delighting in off-color jokes once past the age of 60.

An African Perspective

It also is important to note that not all contemporary reports of cross-cultural elderly sexuality are positive or accepting. Issues of gender, power, and status are typically linked with sexual expression, and the strength of this link can vary significantly among cultures. Nyanzi (2011) provides a detailed ethnographic analysis of "widow-inheritance," a Ugandan tradition in which a widow is expected to marry her brother-in-law or another relative after the conclusion of final funeral rites. According to Nyanzi, this traditional practice was designed to allow aging women to reassume sexual activity and be taken care of financially, as well as to maintain the integrity and status of the patriarchal family clan.

Currently, a significant proportion of older widows are resisting this traditional practice due to fears of infection or reinfection with HIV and AIDS (which has reached epidemic proportions in Uganda) as these remarriages come with a clear expectation that the new bride will be sexually available for her new partner, and a general lack of personal choice about one's life partner. For older widows in Uganda who do not engage in this practice of widow-inheritance, the social consequences can be severe. If an older widow does not remarry, her adult children are expected to monitor and oversee her sexual behavior. Because traditional Ugandan beliefs indicate that sexuality is acceptable primarily for procreation, most adult children actively scold, chastise, or actively prohibit their widowed mothers from dating and having other sexual relationships.

As noted by one of Nyanzi's (2011) participants, an adult daughter of a widowed, older mother who wished to remarry someone other than her brother-in-law, "Imagine how mad I got…to have another man, ah ah no! I told her off in no uncertain terms that this was not going to happen. I was crude. I said to her, 'Mama, do you really want to have another wrinkled body climbing on top of you…?'" Many adult children insist that their widowed mothers move in with them, which also allows them to monitor their whereabouts and limit their privacy. Loneliness and a loss in social status appear to be the typical result for these displaced widows. As more aging Ugandan women chafe against the tradition of widow-inheritance, it remains unclear if and how these cultural values will change.

Asian Perspectives

Other contemporary studies examine the cultural issues and beliefs surrounding sexuality and aging within Asian cultures. As noted previously, although older adults from China and other Asian countries represent one of the largest demographics in the world, most of what is available in the literature is written in English and is based upon older adults of Asian descent living in American and Eastern European nations (e.g., Corona et al. 2010). Findings from some of the few quantitative studies available indicates that both women and men from China, Japan, and Thailand placed less importance upon sex, report lower levels of satisfaction (Laumann et al. 2006), and have sex with less frequency (Cain et al. 2003) than women and men from Western, industrialized nations.

McCurry (2008) of sexual behavior in Japan suggests that one quarter of all married couples (of all ages) did not have sex with each other within the last year, and that more than one-third of married Japanese couples over the age of 50 stopped engaging in sex completely. Other ethnographic studies (Moore 2010) suggest that among older Japanese married couples, a wife's anger and resentment about her husband's infidelity in prior years account for the couple's significant decline in sexual activity. For even well-educated Japanese men over the age of 65, participation and interest in sex was associated with increased vitality in life and passion at work. Various myths about sexual functioning and desire were espoused, in which many Japanese older men blamed the lack of meat in their diet for a decline in sexual desire.

Sexual power and privilege are distributed differently between Japanese men and women, and particularly so in older aged cohorts. In Japanese society, it is often expected that husbands will seek sexual partners outside of marriage, and that wives will remain true to their husbands. It also is important to note that even when the older Japanese wives reported that they had sex with some frequency with their husbands when they were younger, the wives' ability to communicate any sexual interest or desire was limited solely to indirect measures (Moore 2010). Because it is considered taboo and immodest for Japanese women to discuss or show any interest in sexual behavior, the wives in the study indicated that they would prepare certain foods for their husbands (e.g., broiled abalone and surf clams) as a nonverbal signal that they were interested in having sex. These foods are not considered aphrodisiacs per se in Japanese culture, but rather as symbols of fertility. With this apparent need for nonverbal, indirect communication about sex, it is obvious that clinicians should be aware of such cultural factors when attempting to communicate with certain older Japanese clients about their sexuality.

Information about the sexuality of older adults in China is limited, but cultural traditions appear to be derived from a combination of Confucian, Taoist, and Buddhist beliefs that place value upon sexual behavior primarily for procreation (Moore 2010.) More than one-third of older adult participants in an ethnographic study (Guan 2004) believed that sexual activity among older adults was abnormal and detrimental to one's health. Beliefs exist that men are born with a set number of sperm, and that losing too many through sex can result in fatigue, ED, and even death. For older Chinese who did report that they engaged in sexual activity, having a more positive relationship with their spouse was associated with having sex more often, as well as engaging in other sexual activities such as hugging, kissing, and fondling. (Virtually no information is available about Chinese elders' participation in oral sex and masturbation.) Living arrangements often pose a clear barrier to sexual expression among older Chinese couples. More than 20% of the older Chinese couples sampled did not have their own bedroom as they shared it with their grandchildren. Many of those older couples did not even have their own bed.

Although the small sample makes generalization of the findings questionable, a detailed study of Korean married couples over the age of 75 primarily from urban areas (Youn 2009) suggests that gender inequities are deeply ingrained in traditional Korean culture. In Youn's (2009) study, the majority of the husbands interviewed expected their wives to have sex with them. Unlike the attitudes reported in the aforementioned studies with older Chinese couples, older Korean men in this study associated participation in sex with increased virility, an increased life span, and a way to prevent memory loss. The majority of the husbands believed that they had to have sex at least once a month to avoid ED. Both wives and husbands reported violence in response to refusals to have sex. The violence ranged from verbal threats, to actual beatings, and even to rape (e.g., "My husband…beats me until I undress myself. While covering my face with a quilt and making me stay where I am, he rapes me" p. 235). The older Korean wives reported that they often did not want to have sex with their husbands because they were aware of their husband's prior extramarital affairs. The quality of the husband and wife's relationship appeared to be the best predictor of sexual activity.

Indian and Middle-Eastern Perspectives

In traditional Indian culture, older adults are revered and respected. However, a fatalistic approach to health and aging leads many Indian family members, community members, and even some physicians to believe that physical and cognitive decline is inevitable in advanced age (Chandra 1996). Subsequently, traditional Indian culture posits that sexuality in advanced age is a misnomer. Even though Indian culture celebrates sexuality via the *Kama Sutra* and in many religious beliefs about creation, sexuality is typically viewed as something to be practiced and enjoyed by young and middle-aged married adults.

A report from an Indian researcher describes symptoms of older adults with dementia that are considered problem behaviors by younger relatives. Chronic incontinence and inappropriate sexual behaviors are typically the only ones that family members regard as serious enough to require a physician's attention. For many families, having an older relative make sexual advances, make inappropriate sexual remarks, or disrobe or touch themselves in front of others is considered more deserving of medical attention than having a broken hip, pneumonia, or diarrhea. Many family members do not understand that sexual disinhibition is typically outside of one's personal control in advanced stages of dementia, and may resort to physically abusive behavior in order to stop it (Chandra 1996). As noted by one Indian businessman during a trip to the USA, "I know what 'sexuality' means, and I know what 'aging' means, but I have never heard those words together in the same sentence." In essence, in traditional Indian culture sexuality and aging itself can be viewed as a virtual anomaly.

Although many countries and cultures can be thought of as Middle Eastern, virtually no empirical research is available regarding sexuality and aging among those countries' individuals. Results from one of the first large-scale studies of middle-aged and older Arabic speaking men's sexual functioning became available in 2011 (Shaeer and Shaeer 2011). More than 800 men from 17 countries including Egypt, Libya, Morocco, Sudan, Saudi Arabia, Yemen, Palestine, Syria, Iraq, Kuwait, Qatar, and Bahrain served as participants in an online survey of sexual functioning, including the prevalence of ED and its relationship to age, diabetes, and high blood pressure. The online format, used to provide participant anonymity, was used in response to the authors' statement that "Sexuality is a sensitive issue and is more so in the Middle East" (p. 2153). Findings from the study revealed that the presence of ED increased with age, and that high blood pressure and diabetes emerged as clear risk factors. Concerns about the length of one's penis also were associated with ED. A particularly interesting finding from Shaeer and Shaeer's (2011) study is that most of the men in the study falsely believed that PDE-5 inhibitors such as Viagra were addictive, and that their use caused high blood pressure and heart disease. It also is notable that no empirical studies are currently available regarding middle-aged and older Arabic speaking women's sexuality.

It becomes vital for clinicians to carefully assess underlying beliefs and taboos when working with any individual from another country, culture, or ethnicity. Asking questions and maintaining an attitude of respect is typically the first step to understanding.

Evolutionary and other Perspectives

From a historical, psychodynamic perspective, Freud's introduction of the oedipal complex contributes to our understanding of the pervasively negative attitudes toward sexuality and aging espoused in most industrialized, Western cultures. Essentially, the oedipal conflict emerges when a child falls in love and wants to have sex with his opposite-sex parent. The child also may experience a desire to kill off (and compete with) his same-sex parent. Because incest is taboo in our society, these urges must be repressed, and the child ultimately develops revulsion toward sexualized thoughts of the opposite-sex parent in order to avoid this oedipal conflict. This guilt-induced revulsion becomes repressed further and translated into general revulsion at the thought of one's parents, as a couple, engaging in any sexual activity whatsoever. A distaste for sexuality among older adults is believed to be just as powerful and guilt-ridden because it represents the adult child's continued, unconscious need to renounce any sexual activity even loosely associated with parental sexuality (Kernberg 1991). Thus, the societal taboo against sexuality and aging is likely to begin at a very early age.

Internal conflicts regarding sexuality among older adults also can be framed in Darwinian concepts that appear within the context of evolutionary psychology. Darwin observed that in many groups of primates, an older dominant male chased off the younger males so that he could mate with the large number of females in his collective. Sometimes, the younger males worked together in a primal horde (Freud 1913/1946) or independently to kill the older male so that they could spread their now superior genetic material. This Darwinian observation is mirrored in Freud's supposition that during the resolution of the oedipal conflict, young boys must come to terms with their competitive and aggressive feelings toward their governing fathers.

To expand this metaphor further, current demographics suggest that there are many more older females available and few older males. It may be a natural evolutionary response for older males to take a protective stance, and for younger males (i.e., the younger generation) to ward off any sexual competitors. Rather than resorting to physical violence as in the animal kingdom, another approach is for younger adults to render older adults, through the belief system of their society, as asexual beings who are not capable of engaging in sexual activity. Thus, if popular culture views older adults as nonsexual beings, older adults begin to internalize these beliefs and they lose their abilities as sexual competitors (Covey 1989; Pfeiffer 1977). This process guarantees the younger cohort virtually unlimited access to all sexual partners and resources.

Contemporary Portrayals in US Culture

The Implied Value of Youth

In contrast to these generally accepting attitudes among other world cultures, contemporary US societal attitudes toward elderly sexuality are agist and traditionally negative. Butler (1969) first coined the term *ageism* as treating elderly persons differently and typically negatively, based solely on their advanced age. Consistent with this notion, most segments of US society posit that while sexuality is an essential, desirable, enjoyable part of life, sexuality among elderly adults either does not exist or that it is dirty, disgusting, and taboo.

This derision for sexuality and agism among older adults has been documented extensively among college students, adult children of elderly adults, health care providers, and even among older adults themselves (Ehrenfeld et al. 1999; Harries et al. 2007; Kessell 2001; Minichiello et al. 2000). Although pervasive agist stereotypes suggest that older adults are helpless, depressed, and sexless, a consistent message in US society is to seek out and maintain one's youth, or at least its illusion, at virtually any cost. As a result, we see multimillion dollar industries based on cosmetics, plastic surgery, exercise equipment, nutritional supplements, and male performance enhancing drugs including Viagra.

The crux of the problem appears to be that sexuality, as well as other perceived societal benefits, is thought to be reserved for the young, or for those who are willing and able to maintain an appearance of youth. Our culture values youth extensively for it represents independence, excitement, physical prowess, physical attractiveness, and the potential for growth and change. The American Dream implies that a younger person can do better than his or her parent (or grandparent), and the capitalistic component of our society places incredible value on a person's ability to work, earn a living, and secure even greater independence and autonomy. Although most members of our society accept these values and champion a strong work ethic, the implications for older adults can be devastating. Older adults who may be retired, physically disabled, or who no longer fit traditional standards of physical beauty, much less the majority who function happily and well in the community, are saddled with preconceptions that they are a useless burden to others and no longer entitled to interpersonal benefits such as love and sex.

Limited Media Portrayals of Sexuality and Aging

Findings from various research studies suggest that in the media, older adults frequently are portrayed as cognitively impaired, annoying, lonely, stubborn, depressed individuals whose lives revolve around loss and illness. Positive social interactions, much less positive expressions of sexuality, do not appear to play a role in their lives (e.g., Nusbaum and Robinson 1984). In contrast, portrayals of older adults in prime time television programs appear to be somewhat more positive. Qualitative research findings suggest that older adults in these programs tend to be engaged in positive social interactions, particularly within the context of familial settings. They also have been shown, on average, to perform tasks that require some degree of social and interpersonal independence (Dail 1988) and to be stereotyped or typecast as individuals who are healthy, wealthy, agentic, and sexy (Bell 1992). It remains unclear if such positive portrayals of sexy, powerful aging adults on television are more likely to stem from strategic marketing campaigns designed to capture the growing aging adult market (e.g., Schewe and Balazs 1992) or from the generation of more positive and accepting societal attitudes regarding older adult sexuality.

This increase in positive representations of sexuality and aging would seem to be a good omen, regardless of its actual source. However, taking a more than cursory look at this increase in positive media portrayals reveals some subtle but pervasive stereotypes. First, aging adults tend to be portrayed in only one specific way; if they are not healthy, wealthy, and sexy, they do not seem to exist on prime time television. The general absence of realistically diverse portrayals of older adults also appears in commercials (Moore and Cadeau 1985). This lack of diversity and virtual sexual invisibility, which certainly is not representative of the heterogeneity of the older adult population, can impact quite negatively on older adults (Armstrong 2006) who may have health problems, live in relative financial distress, or feel dissatisfied with their own sex lives.

Second, the context in which elderly sexuality is portrayed on television and in popular movies is circumscribed, stereotypical, and gender biased (Bell 1992; Bildtgard 2000). Although prime time shows employ physically attractive characters who may flirt with members of the opposite sex, they rarely if ever engage in actual sexual behavior (e.g., kissing, lounging in bed). The older male characters in many shows who gain admiring glances from women seem to gain favor only from younger women, not women from their own age cohort. These male protagonists also are depicted as more "safe" (i.e., gentle and romantic) than virile and sexually dynamic. Virtually no consistent love interests or male characters are present in the prime time shows that feature older women characters (e.g., Betty White's character on *Hot in Cleveland*). It is as though older women on television exist contentedly in a chaste sexual vacuum, and only engage in sexual activity off screen, in the viewer's imagination if at all. Thus, although elderly sexuality (via sexy older men) may appear on popular television, its context is limited, stereotypical, and not representative of elderly sexuality within the general population. It also is essential to note that virtually all depictions of sexuality among older adults in the media are between White, heterosexual individuals. No representation whatsoever is offered for the significant numbers of older minority group and LGBT members of our country.

Typically, older adults in the media are portrayed as either androgynous, physically weak, unattractive, asexual, ineffectual beings or as youthful, attractive, highly sexualized specimens of physical perfection (e.g., Bildtgard 2000). Although the latter view initially appears positive in that aging adults are viewed as independent, healthy, and sexual, both of these depictions represent unrealistic extremes and represent a form of agism. These biased images in popular culture show successful aging only as a function of "aging prevention," rather than an acceptance of aging and the successful integration of the potentially positive and negative changes that it can bring. Such polarized views do not allow middle-aged and older adults in our culture to perceive themselves and others along a realistic continuum of physical health, attractiveness, and sexual function and expression.

Unfortunately, aging adults in general US culture appear destined to fail; although some aspects of sexual expression appear to be tolerated (or expected) among middle-aged adults, current cultural norms continue to insist that sexual behavior in late life is a goal that is virtually unachievable, and only with the greatest of investment of time, resources, and sacrifice to sustain the appearance of youth and its related standards of physical attractiveness and uninterrupted levels of sexual performance. Older adults themselves have noted their virtual "invisibility" on prime time television (Healey and Ross 2002).

Extreme Representations in Middle Adulthood

Portrayals of sexuality among middle-aged individuals in the media, including television, movies, and advertisements, appear significantly more often when compared to those of older adults. Even though such portrayals appear with greater frequency, an analysis of their content reveals that the type and quality of the sexuality displayed by middle-aged men and women is limited and significantly circumscribed.

Few empirical studies exist, but a recent content analysis of middle-aged women's sexuality in the more than 4,000 US movies released between 2000 and 2007 reveals that middle-aged female characters engaged in sexual activity in only 13 of those films (Weitz 2010). Within these few films, the middle-aged sexually active female characters were generally thin, White, middle-class women who were in committed heterosexual relationships with men of similar age. Although sexual desire among the female characters was often associated with humor rather than passion, some themes emerged in which middle-aged women were able to seek out and obtain pleasurable sexually intimate relationships. Similarly, a content analysis of US magazines targeted at 40- to 50-year-old women, including *O: The Oprah Magazine, Good Housekeeping, and Redbook,* revealed that sexuality at mid-life was portrayed as a relatively sterile concept associated with "emotional work" and a woman's responsibility for maintenance of long-term, heterosexual relationships.

Also within the last decade, a new term has been coined for a specific type of sexuality displayed by middle-aged women who actively seek out younger men for sex partners: the cougar. Although no empirically based analyses of this pop culture phenomenon have been published in the literature to date, lay references indicate that a "cougar" is a highly attractive, youthful, typically financially independent, and college-educated woman who actively pursues both short- and long-term sexual relationships with men 20 years her junior (Putterbaugh 2010; Reyes 2010). Prime examples of cougars in the mass media include Kim Cattrall's portrayal of the beautiful, successful, sexually aggressive, and promiscuous, 50-year-old Samantha on the television show *Sex and the City,* Courtney Cox's 40-plus-year-old lead character depicted as a hapless but sexy "cradle robber" on *Cougar Town,* the television show *The Cougar* which featured a modified dating game with 40-year-old women and 20-year-old suitors, the Internet based *cougared.com* dating site, and even *More* (geared toward a 40 plus female audience) magazine's "Cougar Tips" on their smart phone application.

This cougar role appears exclusively within the context of White heterosexual relationships and presents with both positive and negative elements. Although characters in this ascribed role appear to be empowered and admired, they often serve simultaneously as a vehicle for comic relief (Nusbaum 2009) and may be portrayed as desperate. The one available content analysis of sexuality among middle-aged female characters in recent US movies revealed that of the "cougars" portrayed in 7 of the 13 films portraying sexually active middle-aged women, none of characters sustained their relationship through course of the film, and five of those characters were portrayed unflatteringly as generally "unhappy" or "drunk" (Weitz 2010). Additional analyses of popular films suggest that the cougar role is often associated with discord and disdain among various family members (Tally 2006.)

Although the term cougar was coined in the last decade (e.g., Gibson 2001), the role of an older, seductive, sexually experienced woman is archetypal (Nusbaum 2009). As noted previously, cross-cultural analyses reveal that in many cultures, older women are specifically chosen or valued for their sexual experience. The wife of Bath from the Canterbury Tales represents an early historical reference to a sexual relationship between a younger man and older woman, and in more contemporary terms, the seductive middle-aged Mrs. Robinson from the 1967 movie, *The Graduate*, represents the prototypical cougar. It also is important to note that in this role, the middle-aged woman typically displays no fear of pregnancy or sexually transmitted disease, is able to support herself financially, and manifests no desire for a long-term relationship. In essence, the cougar appears to represent the "female" version of the stereotypically hypersexual male typically portrayed in the media.

Perhaps a more important question to ask is whether the highly visible social role of the cougar is representative of actual women in the USA. Although the media touts Demi Moore, with her 15-years-younger spouse, Ashton Kucher, as a prototypical cougar, it is unclear what proportion of US marriages is between men and women who are at least 10 years older. Consistent with traditional male gender norms, it remains more common for an older man to marry a younger woman (i.e., a May–December romance), but an AARP survey (2003) suggests that approximately one-third of heterosexual women over the age of 40 report dating younger men and prefer it over dating older men. Because various aspects of the cougar role may be considered positive or empowering for women, including the ability to maintain high levels of self-confidence, initiate and enjoy sexual activity, and demonstrate financial independence, research is needed to examine the reality of this potential trend among both middle-aged and younger adults. Certainly, for middle-aged women who do not fit the ascribed cougar image, either in terms of financial independence or appearance of beauty and youthfulness, disappointment may result if this expectation becomes internalized as typical behavior.

The sexuality of middle-aged men in mainstream media has received surprisingly little empirical study. Consistent with patterns of sexuality observed among older adults in the media, middle-aged men typically engage in sexual behavior with greater frequency than their female counterparts (AARP 1999; 2003). In fact, aging men are more likely to appear in mainstream media as primary characters than aging women. What is interesting about this gender bias is that it portrays a reality that is opposite of what is actually true; with advancing age, aging women significantly outnumber aging men. In essence, middle-aged male sexuality is tolerated or even expected, whereas female middle-aged sexuality is "muted." Another notable gender difference in the media is that with advancing age, men continue to command respect and demonstrate sexual prowess, whereas aging women literally lose the ability to be seen as agents of sexual desire and attraction (Carpenter et al. 2006).

One area aging male sexuality that has received some scholarly attention is that of the influence of print and electronic advertisements for male performance enhancing drugs such as Viagra and Cialis. According to these advertisements, the only kind of sexual behavior deemed healthy, normal, and valued is that of penetrative, heterosexual intercourse (Croissant 2006). Women in commercials for such medications appear happy and satisfied only when their male partner has a large, fully functional phallus (e.g., a large golf club in one advertisement) at his disposal. Interestingly, initial commercials for Viagra featured older men such as Bob Dole.

However, when Viagra's parent company, Pfizer, realized that most older men did not achieve the expected results with this medication, their marketing campaign was changed significantly to feature younger men (Vares and Braun 2006).

Viagra marketing materials are even altered to appeal to potential consumers in different countries and cultures (Csberg and Johnson 2009). Although some US advertisements for Cialis, another male-enhancement drug, display a man and a woman soaking in (two separate) bathtubs, little is done to suggest that ED is associated with anything other than a medical problem, and that taking one pill provides an immediate answer or "quick fix" for any sexual difficulties. The influence of such a pervasive mass media "script" for male sexuality and aging cannot be underestimated; Viagra is now one of the most widely recognized brands in the world, along with Coca-Cola (Harris 2003). Some of the specific ways in which public perceptions of ED and Viagra influence aging men and their partners will be addressed specifically in Chap. 8. In sum, with the exception of the ability to make it more socially acceptable for men to discuss ED, these mass media messages involving a "quick fix" for sexual dysfunction typically have a negative influence. It also remains unclear how Viagra and other performance-enhancing drugs are viewed and used by aging minority group and LGBT community members who are not featured in any of these media appeals.

As noted, representations of LGBT elders are absent in mainstream media, and the majority of lesbian and gay characters appear on cable versus network television (Fisher et al. 2007; Netzley 2010). However, a number of gay and lesbian middle-aged characters have been featured on prime time shows (e.g., *Will and Grace; ER*) and in popular films. However, a content analysis of prime time US television shows revealed that less than 8% of the characters were gay or lesbian (Netzley 2010). With general estimates indicating that approximately 10% of the general population can be categorized as gay or lesbian, this finding suggests that LGBT individuals are certainly underrepresented in the popular media, regardless of their age group. In addition, findings from content analyses indicate that when gay and lesbian characters are featured in prime time, they are unlikely to engage in sexual behavior (Fisher et al. 2007) and that gay male characters are typically emasculated and feminized (Linneman 2008). It remains unclear how LGBT aging adults interpret or internalize the absence of positively portrayed sexuality among their peers in the mass media.

The Impact of Mixed Messages

Societal attitudes toward sexuality and aging have changed significantly within the last decade. Previous portrayals of sexuality and aging were nearly universally negative, in which older adults' sexuality was the subject of humor and derision. In contrast, contemporary characterizations in popular culture (e.g., movies, TV shows, and advertisements) can be categorized loosely as either absent or highly sexualized; the stereotypical portrayals of sexuality among middle-aged and older adults appear diametrically opposed. Interestingly enough, such black-and-white and either-or thinking are typically regarded as pathological and unproductive in psychotherapy (Beck 1976). Such dichotomous views can only foster limited, negative outcomes for aging adults, as well. In popular culture, there appears to be no middle ground or, forgive the pun, no gray area for sexual expression among older adults. Although a variety of media outlets provide information that help is available for aging adults who are dissatisfied with their sex lives or sexual function, the apparent solutions posed (e.g., simply take a blue pill) appear limited and overly simplistic. Because no empirical research is available, it also remains unknown to what extent middle-aged and older adults consume pornographic print or Internet-based media, and to what extent and under what conditions middle-aged and older adults themselves appear in pornographic content.

Among aging adults themselves, overly "positive" or sexualized images may bolster already unrealistic expectations that they are to remain beautiful, youthful, and robust even if they encounter debilitating illness or experience a tragic personal or financial loss. Because older adults are not represented in their full diversity in the media, aging adults who are LGBT, members of a minority group, disabled, depressed, chronically mentally or physically ill, impoverished, or institutionalized will have no available role models, and may be more likely to look upon their own personal situation with sadness or contempt. Thus, the use of vivacious, healthy, youthful, white elderly models and spokespeople may cause more problems for older adults than members of popular culture would like to admit. As clinicians, we can expect to see the impact of these and other socially mediated stereotypes in our offices.

It bears repeating that the only form of sexual expression valued in popular culture for middle-aged adults is limited primarily to that of penetrative intercourse among attractive, white, middle-class, heterosexual couples, including women who typically appear younger than their chronological age. These unrealistic and circumscribed portrayals, if internalized, pose significant challenges for middle-aged adults who do not, cannot, or choose not to model the consistently narrow range of behaviors modeled by individuals in the media. Because both middle-aged and older adults represent a demographic that spends a significant amount of time in contact with mass media, it is important for clinicians to review and potentially refute the cognitive schemas that middle-aged and older adult patients have consciously or unconsciously adopted from popular culture.

Understanding Research Methods

In any introduction to sexuality and aging, it is also important to address differences in clinical research methods. The use of a scientist–practitioner model has earned the field increased respect as well as a relatively large knowledge base. Despite this increased knowledge base, however, inherent problems exist in conducting and interpreting sexuality research, much less in conducting and interpreting elderly sexuality research. Clinicians who are informed about the differences between empirical research (including both quantitative and qualitative approaches) and case studies are better able to evaluate these methods' findings and to apply them to practice. Although these different research methods are all valuable in their own right, the findings from each must be judged according to their own strengths and weaknesses.

Empirical research often is regarded as the least clinical of all types of research. However, acknowledging the ability of this approach to track general norms or societal trends can provide essential information to guide clinical practice. For example, quantitative research findings regarding epidemiological estimates of older adults who have contracted :mv (including the finding that more than half of all new AIDS cases in Palm Beach County, Florida, are among adults over the age of 50) can be used to dispel established, agist stereotypes. These empirical findings suggest that older adults are sexually active people who may engage in high-risk behaviors such as intravenous drug use, sex with multiple partners or prostitutes, and homosexual and heterosexual anal sex. Thus, seemingly "sterile" quantitative research findings can be used to enable clinicians to feel more confident and justified in their willingness to conduct a thorough clinical interview.

Unfortunately, such empirical research findings can also be highly influenced or biased by differences in subject sampling, instrument selection, and self-report biases. One major problem with empirical research is that one finding cannot necessarily be generalized to all members of this heterogeneous population. Regarding a likely increase in the occurrence of eating disorders among elderly women (Hsu and Zimmer 1988), it becomes important not to generalize these findings to minority group women. Because the vast majority of these studies have employed white subjects, and because African-American women have been shown to have greater satisfaction with their body size and image (Hebl and Heatherton 1998), it remains unclear whether elderly African-American women, as compared to their white counterparts, would be as likely to manifest eating disorders in later life. These limitations in subject sampling often lead to problems in overgeneralizing research findings. In other words, if not interpreted properly, empirical research findings can be misconstrued and promote inappropriate stereotypes themselves.

The use of instrument or measurement selection in empirical research also can be problematic. If a test for knowledge of elderly sexuality has only true and false items, and the older adults in the study demonstrate little knowledge of elderly sexuality as revealed by a low average score, it remains unclear whether the older adults who took this test were generally unaware of the physiology of elderly sexuality. It also is possible that the participants were unable to understand the medical terms selected by the researcher. They also may have been irritated that the test was so long, which led them to stop answering items carefully after approximately half an hour. Other potential problems could be that the participants were unable to read the test items because the experimenter failed to remind them to bring reading glasses or even because they were offended by questions regarding masturbation and homosexuality because of their religious beliefs. Some of these influences may have led other participants to answer some questions randomly. When asked to respond to self-report items, individuals also are more likely to under- and/or overestimate their participation in sexual behaviors (Bradburn and Sudman 1979), perhaps in response to religious, societal, and cultural demands including ideals of purity and machismo (Catania et al. 1989).

In contrast to such quantitative research, which typically relies on printed, forced-choice measures, the use of qualitative research is more likely to employ open-ended questions and one-on-one subject interviews. This qualitative approach has become more common in elderly sexuality research because it generally allows for greater exploration of individual subjects' thoughts, feelings, and motivation. However, the use of such open-ended questions also relies on experimenter skill, consistency and training among interviewers, increased time requirements for both experimenters and participants, and the need for complicated statistical procedures to code and analyze the data. The use of qualitative approaches also presents its own unique problems and challenges. The age and sex of the interviewer can influence the results (i.e., gay men appear more likely to discuss their sexual activities with male than female interviewers), as can the actual mode of data collection. For example, study participants are more likely to admit that they engage in specific sexual behaviors if they are interviewed over the telephone than in person (Catania et al. 1989). Issues of privacy, concerns about self-presentation, and subject motivation (e.g., what can we infer about someone who is willing to volunteer a couple of hours to participate in a study about sexuality?) also appear heightened in these qualitative study formats.

Another important means of gathering information about elderly sexuality is through the use of case studies. The case also appears most closely aligned with traditional clinical practice, and allows for an in-depth understanding and appreciation for one patient's experience. All aspects of the individual involved are reviewed and discussed, including personal history, interpersonal relationships, cultural background, religious views, family dynamics, socioeconomic status, internalized beliefs and values, ethnic background, treatment progress, object relations, cognitive schemas, and any other relevant information. Although averages derived from quantitative research do provide critical information and can dispel unrealistic stereotypes, they inherently ignore individuals' unique histories and personal situations. The use of a case study also allows clinicians to recognize similarities and differences in their own approach to a patient's problem and to gain exposure to sometimes obscure or previously unencountered clinical issues. In contrast, clinicians also must differentiate between case studies and clinical anecdotes. While interesting and often illuminating, clinical anecdotes represent only a fragment of a patient's (and practitioner's) experience and must be viewed with the appropriate critical stance.

A Brief Review of the Literature

A brief overview of the literature on sexuality and aging suggests that we should not ascribe to societal stereotypes that older adults are helpless, passive, asexual beings not entitled to companionship, love, and sex. A number of researchers are quick to point out that many men engage in sexual intercourse well into their 80s and 90s, that older women tend to enjoy satisfying sexual relations in later life if they enjoyed satisfying sexual relations in their younger years and they have a healthy partner (Lindau et al. 2007; Waite et al. 2009). It also is important to note that the majority of older adults are healthy, community-living members of society. For those who face chronic illness, disability institutional settings, and even end-of-life issues, sexuality continues to appear important. Empirical evidence suggests that many older adults in nursing homes (Ginsberg et al. 2005) and hospice care (Cort et al. 2004) place significant value and importance upon sexual thoughts, fantasies, and behaviors.

In contrast to these positive reports, other epidemiological and research findings suggest that ED is the most common type of sexual dysfunction and dissatisfaction among both middle-aged and older men. Findings also indicate that up to one in three older women experience vaginal dryness and pain during intercourse, but feel uncomfortable discussing sexual issues with their health care provider (AARP 2010; Lindau et al. 2007). Older adults also are more likely to experience sexual dysfunction as a result of adverse drug reactions, as compared to young and middle-aged adults. Estimates suggest that more than 80% of Americans take at least one prescription drug, over-the-counter medication, or dietary supplement each week (Kaufman et al. 2002), with older adults taking an average of four different prescription drugs per day (Barrett 2005).

In some cases, the absence of information in the literature is just as notable as the presence of other information. Most strikingly, virtually no research exists regarding the sexuality of LGBT and minority group elders, older adults living with disabilities, older adults in hospice care, and the oldest-old members of our society. This lack of attention to these populations in the literature may reflect a general societal disavowal of these groups that has been unconsciously internalized by the research community. Or, perhaps less insidiously, this absence of coverage in the literature may reflect the logistic difficulties sometimes involved in recruiting such minority group members for research.

Summary

The field of sexuality and aging is heterogeneous and diverse. Clinicians can expect only that their clients may come from a variety of backgrounds, with a myriad of sexual histories, current involvement in any number of sexual activities, and a potential range of physical and mental health issues as well as a range of knowledge about these challenges and their treatment. With greater knowledge of sexuality and aging, we as practitioners will be better equipped to assist our clients as they face the physiological, demographic, cultural, and interpersonal changes associated with increasing age.

Chapter 2
Knowledge of Sexuality and Aging

Many men and women engage in vaginal sex, oral sex, and masturbation well into their 80s and 90s. Significant gender differences appear as older women are more likely to be limited by the availability of a partner and older men are more likely to be limited by physical health and erectile dysfunction.

Gathering information about an older client's sex life is challenging, rewarding, and a requisite part of an intake interview. In order to make this practice more effective and efficient, clinicians themselves can become more informed about basic aspects of elderly sexuality. Various large-scale national and international research studies indicate that older adults report being sexually active well into their 80s and beyond. Predictors of middle-aged and older adults' sexual activity include their prior level of sexual activity and satisfaction, the availability of a partner, and physical health. Knowledge of sexuality has also expanded, in which new empirical findings from large-scale studies suggest that aging adults engage in a wider variety of sexual behaviors, including masturbation, anal sex, and being sexually active with multiple partners, than previously believed. However, significant gaps in the literature regarding minority group and LGBT members' sexuality and aging remain.

Many adults appear uninformed about the physiological changes typically associated with aging and sexuality; many do not know that erectile dysfunction may be treatable or that masturbation can promote improved vaginal lubrication in women. Many aging adults, as well as their health care providers, are unaware that commonly prescribed medications and over-the-counter drugs can significantly impair sexual function. Gathering information about a patient's past and current sex life can reveal important information about close relationships, body image, knowledge of sexuality and aging, trauma, medical history, gender roles, and self-esteem. As informed clinicians, the chance to discuss issues of sexuality with both middle-aged and older patients presents unique opportunities and benefits.

J. Hillman, *Sexuality and Aging: Clinical Perspectives*,
DOI 10.1007/978-1-4614-3399-6_2, © Springer Science+Business Media New York 2012

Frequency and Types of Sexual Behaviors

Early Research Findings

The first empirical studies to provide information about the sexual practices of older adults were conducted in the 1950s by Kinsey and his colleagues (Kinsey et al. 1948, 1953). It is notable that before this time, the sexual life of older adults was virtually unexplored in any scientific way. Societal prohibitions against the discussion of such topics certainly influenced the overall lack of research prior to this time, and led to an uproar among some segments of the public when these landmark studies' results were released.

What Kinsey and his colleagues revealed was that among their sample of community-living older adults, men over the age of 60 engaged in intercourse slightly more than once a week on average, with no sudden decline in sexual activity related to aging. Women over the age of 60 were found to engage in intercourse less frequently than their male counterparts, but with similar patterns of sexual behavior to those reported in their late teenage years. In the mid-1960s, Masters and Johnson (1966) also pronounced that men and women were biologically capable of engaging in sexual intercourse at any age.

Another series of researchers presented more detailed information regarding older adults' sexual behaviors during the mid-1970s and 1980s. George and Weiler (1981) recruited more than 340 elderly husbands and wives in a longitudinal study of sexual behavior. Excluding the effect of losing a partner by widowhood, the frequency of sexual activity among these older men and women remained remarkably stable over the 6-year study period. These spouses reported that they engaged in sexual intercourse between one and two times a week on average. Pfeiffer and Davis (1972) found that elderly women were more likely to engage in sexual relations if they were married.

Specifically, an older woman's marital status, particularly whether she was divorced or single, was a better predictor of a decrease in her reported frequency of sexual intercourse than was her age itself. Although this finding appears remarkably obvious, it marked a drastic and important change in the way in which sex researchers began to interpret their basic frequency reports of sexual intercourse among older adults. Researchers, like clinicians, began to focus on some of the individual aspects of older adults that made them more or less likely to engage in specific types of sexual behavior. In other words, understanding the context in which an older person's sexual expression takes place began to take precedence over the acquisition of absolute numbers or base rates.

Other investigators also began to explore sexuality and aging beyond the singular act of intercourse. Butler and Lewis (1976) indicated that the sexual behaviors engaged in by elderly persons often encompassed more than intercourse, and that older adults often placed greater emphasis on cuddling, fondling, and mutual manual stimulation. Botwinick (1984) also expressed the importance of examining sexual gratification in regard to a variety of sexual behaviors, including masturbation.

However, empirical findings regarding the percentage of older adults who engage in masturbation have varied greatly. For example, Starr and Weiner (1981) revealed that approximately one-half of the older participants in their study engaged in masturbation on a regular basis. In contrast, the majority of older participants in a study conducted by Waslow and Loeb (1979) reported that they refrained from self-stimulation, primarily out of religious concerns and an overall lack of privacy in an institutional setting. In sum, the few studies that did examine the extent to which older adults engaged in masturbation varied widely in their selection of participants and the extent to which they addressed contextual issues (e.g., availability of a partner, religious beliefs, and prohibitions), and their subsequent findings appear just as varied.

US Population Data

Despite the necessity of evaluating older adults' sexual behavior on an individual basis, a number of recent research studies do provide valuable information about general trends in sexual relations across the life span, and about some of the factors that are predictive of sexual activity with age. Again, emphasis should be placed on the fact that these research studies only highlight general trends, and that they do not indicate absolute ideals or expectations for midlife and older adults. What the findings of these studies do offer, however, is a wealth of information that contradicts the commonly held stereotype that older adults are asexual beings who do not participate in or desire sexual relations.

Within the last decade, findings from a number of large, nationally representative surveys now allow for an examination of middle-aged and older adults' reported participation in a variety of sexual behaviors. (Previous studies provided valuable insight into older adults' frequency of sexual intercourse and other behaviors, but those studies typically were smaller, non-representative community-based or convenience samples). Two of the largest studies that provide a representative cross-section of the US population are the National Social Life, Health, and Aging Project (NSHAP; Lindau et al. 2007; Waite et al. 2009), including responses from 3,005 men and women aged 57–85, and the 2010 American Association of Retired Persons (AARP) Survey of Midlife and Older Adults (Fisher 2010), derived from responses from 1,670 men and women between the ages of 45 and 90.

The NSHAP included in-home interviews and oversampled Black, Hispanic, male, and 75–85 year-old participants; whereas the AARP study represents the first online research study of its kind to provide a representative US sample, including an oversampling of Hispanic participants. This oversampling allowed for greater numbers of underrepresented groups to be included in the surveys, and provided enough responses to provide statistically meaningful analyses. (Individuals recruited for the AARP survey also were provided with Internet access if they did not already have it). Although the two studies differ in the ways in which the age ranges of participants were broken down for analysis, with the NSHAP reporting data for men and

Table 2.1 Participation in sexual behaviors by age and gender

National Social Life, Health, and Aging Project (NSHAP) 2007 US survey

| | | Age range | | |
| | | Percentage agreement within previous year | | |
	Behavior	57–64	65–74	75–85
Men	Any kind of sex	84	67	38
	Vaginal sex[a]	40	31	23
	Oral sex[a]	62	48	23
	Anal sex[b]	–	–	–
	Foreplay[a,c]	94	90	92
	Masturbation	63	53	28
Women	Any kind of sex	61	40	17
	Vaginal sex[a]	34	31	24
	Oral sex[a]	53	46	36
	Anal sex[b]	–	–	–
	Foreplay[a,c]	89	88	89
	Masturbation	32	22	16

American Association of Retired Persons (AARP) 2010 national survey

| | | Percentage agreement within the last 6 months | | | |
		45–49	50–59	60–69	70+
Men	Vaginal sex	69	72	60	40
	Oral sex	63	56	30	17
	Anal sex	19	23	3	1
	Foreplay[c]	82	85	84	73
	Masturbation	66	72	50	47
Women	Vaginal sex	58	51	45	13
	Oral sex	49	38	25	9
	Anal sex	8	4	5	1
	Foreplay[c]	74	69	61	41
	Masturbation	50	48	37	18

European Male Ageing Study (EMAS) European, cross-national 2010 survey

| | | Percentage agreement within the last month | | | |
		40–49	50–59	60–69	70+
Men	Vaginal sex[a]	96	93	81	60
	Foreplay[c]	93	92	81	73
	Masturbation	57	46	36	26

[a]Sample limited to those participants with an available partner
[b]Not assessed
[c]Included kissing, hugging, sexual touching, petting, and caressing

women aged 57–64, 65–74, and 75–85, and the AARP study reporting data for men and women aged 45–49, 50–59, 60–69, and 70 and older, a number of age and gender differences emerge along a variety of dimensions. Please refer to Table 2.1 for selected findings.

To provide an overview of relevant findings from these US population-based studies:

- Both older women and older men report participation in vaginal sex, oral sex, and masturbation well into their 80s and 90s.
- Older men and women are significantly less likely to participate in vaginal, oral, and anal sex and masturbation than middle-aged men and women.
- Men, whether they are middle-aged or older, are more likely to have vaginal, oral, and anal sex, and to masturbate than women.
- Older adults appear unlikely to "substitute" oral sex or masturbation for vaginal sex, when it is unavailable, although rates of participation in foreplay remain consistent across age.
- Approximately 10% of aging men and women report a sexual orientation other than heterosexual.
- Six percent of men and one percent of women indicate that they have multiple sex partners, while less than 5% report usually or always using condoms.
- Significantly, more Hispanic men report participation in weekly vaginal sex than non-Hispanic men.

More specific findings reveal that older adult men and women in America, including those with and without an available partner, are less likely to report that they have any kind of sex than their middle-aged counterparts. To provide some base rates, findings suggest that at least 85% of middle-aged men in the USA have sex at least once annually, whereas 67% of the young-old men (aged 65–74) and only less than half or 38% of the middle-old men (aged 75–85) have sex at least once each year. Women in the USA report having sex significantly less than men, with at least 61% of middle-aged women, nearly half (40%) of the young-old women (aged 65–74), and only 17% of the middle-old women (aged 75–85) having sex at least once a year. It also is important to note that these reported "base rates" for participation in sexual activity may be skewed due to differences across gender and age regarding the availability of a partner. Among those in the AARP (Fisher 2010) survey, men aged 70 and older were twice as likely to report having a recent or currently available sexual partner (63%) than women aged 70 and older (34%). A number of factors may contribute to this significant gender difference, including the longevity of women compared to men and the propensity of men to select younger female partners. Even into the eighth and ninth decade of life, however, older men and women report participation in vaginal and oral sex, as well as masturbation.

Fortunately, both the NSHAP (Lindau et al. 2007; Waite et al. 2009) and AARP (Fisher 2010) surveys provide information about the sexual behavior of aging men and women with available partners. Partnered middle-aged and older women and men's reported participation in vaginal, oral, and anal sex also can be compared to their reported use of foreplay and masturbation, allowing for some important conclusions. For example, as women age, their participation in vaginal sex does not decrease significantly as long as they have an available sexual partner. In contrast, among men with an available sexual partner, their participation in vaginal sex declines significantly with increasing age, suggesting that some aspect of male sexuality, likely impaired health or ED, is responsible for a decline in participation. It also is important to note that aging adults do not appear to substitute oral sex for vaginal sex; no increases in the use of oral sex coincide with decreases in vaginal sex. Similarly, rates of participation in masturbation do not increase (as a substitute) as declines in vaginal sex occur. It appears that masturbation is an activity separate from partnered sexual activity. It also is important to note, however, that this finding may be due to cohort effects; the average, aging baby boomer may consider the use of other forms of sexual behavior, such as oral sex, or self- or mutual masturbation, when vaginal sex is not part of their repertoire.

Additional findings from the NSHAP (Lindeau et al. 2007; Waite et al. 2009) and AARP (Fisher 2010) surveys are notable. While 92% of the AARP participants identified their sexual orientation as heterosexual, 3% reported that they were gay, 1% reported lesbian, 1% indicated bisexual, and the remaining reported "other" which included "bicurious" and gay individuals who were married. Among those respondents who were Hispanic, they were significantly more likely to report having vaginal sex at least once a week (39%) when compared to all other respondents (28%). This difference remained consistent when examining participants who had an available partner. In this case, 54% of Hispanic respondents with a partner reported having vaginal sex at least once a week when compared to 41% of all other respondents with an available partner. In addition, of the respondents aged 70 and older, 53% of male Hispanics reported having sex at least once in the past six months compared to 40% of all other respondents. Among respondents aged 70 and older, rates of participation in sex in the last six months were similar for women (approximately 13%) whether they were Hispanic or not. Although the relatively small number of respondents did not allow for a meaningful statistical analysis of these apparent differences by age and gender, it appears that both Hispanic and non-Hispanic men are more likely to engage in vaginal sex in older adulthood than their female counterparts.

General information also could be determined about middle-aged and older adults regarding the presence of multiple partners and condom use. Among all AARP participants, 4% report that they have been sexually active with more than one partner at a time. By gender, 6% of men and 1% of the women respondents indicated that they had simultaneous, multiple partners. Findings from the NSHAP (Lindau et al. 2007; Waite et al. 2009) indicate that less than 5% of older men and 4% of older women "usually or always" use a condom when having sex. Unfortunately, it is not possible to determine from the published survey data what proportion of adults with multiple sex partners use condoms, nor is it possible to identify specific characteristics of those respondents (beyond gender) who were more likely to report having multiple sex partners.

Although these two studies provide vital information about US citizens' participation in various sexual activities across the life span, a number of significant limitations exist. For example, a variety of ethnic minorities (e.g., Asian, Black, and Native American), LGBT individuals, and the oldest-old (individuals aged 85 and older) were not oversampled in order to provide meaningful frequency data for sexual behaviors. Certainly, these two studies represent cutting edge work in the field, and require significant amounts of funding, but the lack of attention to these underrepresented groups is suggestive of Western culture's predominant acceptance of sexuality among younger, Caucasian, heterosexuals. Future work is needed to establish base rates for sexual activity among other US subpopulations.

Cross-national Findings

Although caution must be taken when comparing the base rates observed in the aforementioned US surveys, a variety of cross-national European studies provide additional information regarding the frequency with which older adults engage in sexual activity. The European Male Ageing Study (EMAS; Corona et al. 2010) obtained information from more than 3,300 community-living middle-aged and older men (aged 40–79 years) from eight different European nations including Italy, Belgium, Sweden, the U.K., Poland, and Hungary. (A number of findings from the EMAS are presented in Table 2.1.)

What is notable from the EMAS study is that European middle-aged and older men with available partners report that they engage in sexual intercourse more frequently than their American counterparts. However, the age-related declines in reports of vaginal sex into older adulthood revealed among American males also appeared among European males. Both American and European males report similar, high levels of participation in foreplay appear to hold continued value above and beyond that associated with a simple precursor to vaginal intercourse.

In relation to masturbation, the European middle-aged and older men's reported participation in masturbation was lower than American men's reports. However, this lower likelihood of participation in masturbation among European men was reported for the previous month versus the past six months in the AARP (Fisher 2010) study and in the past year for the NSHAP. Because European men showed a similar, slight decline in participation in masturbation with increased age, it appears unlikely that they compensated for a lack of participation in partnered or vaginal sex with masturbation. Overall, European and American middle-aged and older men report some differences in their likelihood of participation in various sexual behaviors, but general declines in sexuality activity into older adulthood appear consistent between both groups. Also similar to the American population-based studies, virtually no information is available regarding European middle-aged and older adults who identify themselves as non-heterosexual, as a member of a minority group, or who are among the oldest-old (over 85 years of age.)

In terms of additional cross-national differences, findings from Laumann and colleagues' (Laumann et al. 2006) global study reveal significant discrepancies in sexual behavior and satisfaction between Western and Asian participants. For example, although women's ratings were lower than men's (also see Youn 2009), both women and men from China, Japan, and Thailand reported significantly lower rates of sexual satisfaction and placed significantly less importance upon sex than their Western counterparts. Additional analyses of the survey data indicate, consistent with smaller cohort studies (Cain et al. 2003), that older Asian men and women engage in sex with significantly less frequency than their Western counterparts. In response to the integrative biopsychosocial model, it is essential to go beyond the simple identification of cultural differences to explore potential, underlying social and psychological mechanisms that may account for those differences (c.f., Lewis 2004). This difference in frequency of sexual intercourse between Asian and Western elders may be attributed, in part, to traditional Asian belief systems (e.g., Confucianism, Taoism, and Buddhism) that place value upon sexual behavior primarily for procreation (Guan 2004; Moore 2010.)

What Is Average and What Is Normal?

In discussing the relative frequency with which middle-aged and older adults engage in sexual activity, it becomes vital to make the distinction between "average" frequencies and participation in types of behavior and "normal" frequencies and types of behavior. As among other age groups and subject populations, it is vital that both patients and practitioners understand the difference between average and normal participation rates. Average frequencies simply represent our best guess at a numerical mean for a specific sexual behavior, based on a specific sample of older adults using a specific research methodology. For example, in one of the earliest reports of sexuality and aging, Starr and Weiner (1981) reported that the participants in their study aged 80 years and older engaged in sexual intercourse 1.2 times per week, on average. However, these figures told us nothing about whether these particular older adults were satisfied with their sexual relationships, whether they desired more or less sexual contact, whether or not they had consistent access to a consenting sexual partner, and if they were healthy enough to engage in certain sexual activities. In other words, what is "normal" or personally acceptable to one older adult may or may not be personally acceptable to another.

Unfortunately, many patients regard clinical averages as a benchmark that they must match or exceed in order to demonstrate that they are aging well. What is normal about sexuality must be assessed on a case-by-case basis, with substantial emphasis placed on the perceptions, feelings, and expectations of the middle-aged or older adult in question. Patients often express incredible relief when they learn from their health care provider that the number that they read in a magazine article (e.g., the AARP magazine) indicating how often they "should be having sex each week" does not indicate what is normal for them. Productive collaborative work between patient and practitioner begins when patients are able to focus on the nature and quality of their own sexual relationships instead of the relations that they believe they should be having in order to meet some primarily arbitrary, albeit "normal, national average."

Alfredo

Consider the following case example. Alfredo was a 76-year-old, decorated Air Force veteran of World War II. He attended group therapy at a day hospital program for older adults in order to help him work through residual issues from posttraumatic stress disorder. During the war, Alfredo was one of the few men to survive an unexpected enemy attack on his squadron. He interpreted this event as a divine sign that he was specially chosen to survive. This narcissistic interpretation of events helped him to assuage his guilt and to proceed through life with confidence, but at a costly emotional price. Alfredo was particularly fearful of aging because "such a special person as myself should be allowed to remain on this earth for as long as possible." Regardless of the weather, Alfredo always seemed to wear his leather bomber jacket. He spoke daily about his vigorous exercise routine, feeling "as healthy as a horse," and still being lucky enough to have all of his hair. Although many of Alfredo's peers from the group therapy hospital program were irritated by his need to flaunt his wealth, status, and health, they tolerated his displays as if they knew that he could not easily tolerate this narcissistic insult.

Unlike his peers in group therapy, his 72-year-old wife, Aleni, was less tolerant of Alfredo's narcissistic displays. Aleni sometimes traveled to the hospital with her husband for treatment team meetings and occasional couple's sessions. In contrast to her husband, Aleni appeared comfortable with her own aging. She appeared happy and confident with her healthy and slightly plump figure. However, she sometimes appeared overly tired, particularly on the days after Alfredo insisted that they take a strenuous sightseeing trip or bike ride. Aleni also admitted to the staff psychologist that her husband was equally demanding in their bedroom. Aleni admitted that although she was pleased that her husband found her attractive and sexy, she sometimes wondered whether he really wanted to make love to her, or if he just wanted to prove something to himself.

When asked to broach the subject as a couple, Alfredo remarked that he just wanted to engage in sex with his wife as much as everybody else. When asked how he knew what "everybody else" did, he spoke about an article that he read in a popular men's health magazine. The article presented the results from a reader's poll in which the few elderly men who responded reported that they had sexual intercourse three times a week on average. Alfredo was highly motivated to remain as healthy, sexy, and competitive as the other elderly men. Although it certainly was not possible to address Alfredo's core narcissistic issues in a few couples session, Alfredo was able to understand that this average number of three essentially was arbitrary. When asked directly, Alfredo also noted that the article did not provide any information about whether these men who engaged in sex many times a week were satisfied with their physical or emotional relationships. The social worker further explained that the elderly men who chose to answer a survey for this young men's health magazine probably felt pressured to artificially inflate the number of times that they had sex each week.

More importantly, Aleni was able to tell her husband that she only wanted to have sex with him when they *both* wanted to feel close. With support from the psychologist, she was able to assert her rights to her own body. She told Alfredo that she would think more of him as a man if they had sex less often, but with more emotion and intensity instead of "just going through the motions so you can say you had it with me." She told him that she wanted to be a special part of his life, and not just some number that he had to live up to. The psychologist was able to draw a parallel to Alfredo's combat experience; she remarked that while Alfredo was very proud of how many enemy kills he had painted on the side of his aircraft, his wife wanted to be more than just a number to him in his own private war against aging. Alfredo initially balked at the idea, but soon recognized that his wife was being honest and forthright. He learned that his wife based his masculinity and youthfulness on the quality, and not the quantity, of their sexual relations. He also reported that he might feel more relaxed knowing that there was one less area in his life in which he had to measure up to some youthful standard in order to prove that he was special.

Predictors of Sexual Behavior and Satisfaction

In order to aid clinicians in their work with middle-aged and older adults, it becomes important to gather additional information beyond that of simple frequency data or base rates for participation in sexual activity. Equally important underlying factors are related to an aging individual's physical health, mental health, and level of sexual satisfaction, including the likelihood of sexual dysfunction. In addition to findings from the EMAS, the AARP survey, and the NSHAP, additional information can be gleaned from a Finnish representative national sex survey (Kontula and Haavio-Mannila 2009) and the Australian Longitudinal Assessment of Ageing in Women (LAW) Study (Howard et al. 2006). Relevant participant reports are provided in Table 2.1 by age and gender.

Availability of a Partner

Regarding predictors of middle-aged and older adult's sexual activity, Matthias et al. (1997) were some of the first researchers to identify aspects of older adults' lives that may limit or curtail participation in sex. They discovered that older men and women responded differently to life stressors in terms of how they impacted upon their sex lives. For example, marriage appears to be a significant predictor for sexual activity (with the exception of masturbation) among older women, but not for older men. Among women, being single, widowed, or divorced meant that they were less likely to have an active sex life than their married counterparts. As noted previously, as women move from middle-adulthood to older adulthood their likelihood of having an available sexual partner declines significantly.

A variety of factors account for this difference in partner status as a predictor of sexual activity. First, the US culture maintains a double standard in which men are viewed as hypersexual beings who require a variety of sexual outlets. Although it is more acceptable for men to satisfy those sexual needs within the context of marriage, seeking sexual gratification outside of marriage is tolerated and sometimes viewed as exciting or accepted. The vast majority of prostitutes are women, who often work for married, male clients. Women, alternatively and traditionally, are expected to honor their marriage vows for their entire lives. A female adulterer typically is viewed with contempt, as an ungrateful and selfish spouse. Although recent statistics suggest that women, particularly women who work outside the home, are becoming increasingly likely to engage in affairs in similar numbers as their male counterparts (Fisher 2010), this represents a likely cohort effect in which young adult women appear significantly more likely to engage in sexual affairs than older women.

Perhaps more importantly, the numbers of single older women in the USA greatly exceed the numbers of single older men. Women have a longer life span than men and tend to marry men who are older than them. Due to the basic principles of supply and demand, it is much easier for an older man to seek out a romantic or sexual relationship with a single middle-aged or older woman than it is for her to establish such a relationship with a single middle-aged or older man. Many older men in nursing homes, for example, find that they are not used to all of the attention that they receive from the multitude of single women in their midst. One elderly gentleman in a Florida retirement community took five elderly women at a time with him for breakfast at a local restaurant. Another five women had to "wait their turn" to be one of his guests at lunch. One of these women was noted wryly as saying, "Well, who am I to hurt his feelings and turn down a free lunch?" Another older female resident regarded the situation plaintively and replied, "Sometimes you've got to take what you can get." Clinicians certainly must be aware of the self-esteem issues that come into play in such highly competitive environments.

Physical Health

Another factor that is related to sexual activity among middle-aged and older adults is physical health. Older adults suffering from arthritis, high blood pressure, heart disease, stroke, diabetes, and kidney problems were found to engage in less sexual activity than those older adults who suffered from fewer of these chronic ailments (Matthias et al. 1997). Physical disability, in relation to functional status, also was related to engagement in sexual activity. Both community-living and institutionalized older adults who reported that they had difficulty getting in and out of bed, bathing themselves, and getting dressed and undressed had fewer sexual interludes than those who were independently mobile. Although this finding certainly is not surprising, it can provide relief to older adults who are suffering from medical problems; they learn that they are not alone or unique in their plight. In other words, it is normal for an older person who suffers from physical disabilities to engage in sexual behavior with lesser frequency than an older person in the best of health. However, it also is vital that older adults who suffer from physical illness and disabilities know that such difficulties do not automatically preclude them from satisfying, enjoyable, and frequent sexual relations if they so choose.

For middle-aged and older adults who are community-living and generally physically independent, physical health still plays a significant role in sexual activity, as reported in a variety of population-based studies including the NSHAP (Lindau et al. 2007), the EMAS (Corona et al. 2010), the AARP 2010 survey (Fisher 2010), and the LAW (Howard et al. 2006). In the AARP survey, the vast majority or 78% of the total sample report that their health is at least "good," with nearly half or 40% of the total sample reporting that their health is "very good" or "excellent." Only 21% report their health as "fair" or "poor," and men, regardless of their age, report being in better health than women. (Of course, it is unclear whether this is a self-report bias in which men choose to view their health more positively than women, if their health status is better than women's, or if women have higher expectations for their health status than men.) More importantly, 40% of those women and men who rated their health as "excellent" were likely to have sex weekly, versus only 14% of those who rated their health as "poor." Primary health problems in this community-based sample included high blood pressure (44% of the respondents), back problems (36%), arthritis (32%), diabetes (16%), and depression (16%). It also is notable that, by definition, one of depression's primary symptoms is a lack of interest in sex. The links between high blood pressure, diabetes, and erectile dysfunction will be discussed in depth in Chaps. 5 and 8.

Sexual Dysfunction

Although sexual dysfunction includes a myriad of disorders and symptoms, most population-based surveys do include some assessment of erectile dysfunction for men and difficulty with lubrication (e.g., vaginal dryness) for women. (Please refer to Table 2.2.) Among middle-aged respondents, the percentage of men who reported some degree of ED ranged from a low of 6% in the European EMAS (Corona et al. 2010) to a high of 31% in the US NSHAP (Lindau et al. 2007). Among the older male respondents, rates of ED ranged from 30% in the Finnish National Sex Survey (Kontula and Haavio-Mannila 2009) to a high of 56% in the US AARP survey (2010) and 64% in the European EMAS (Corona et al. 2010). One trend is clear; self-reported rates of ED increase significantly as men age. Conservatively, approximately half of the men over the age of 70 report some degree of ED. Of course, it also is important to note that among middle-aged and older men, rates of reported premature ejaculation, a different type of sexual dysfunction, decline significantly with age. In other words, increased age appears to provide some protection from premature ejaculation as well as increased likelihood of ED.

In terms of self-reported difficulties with vaginal lubrication among middle-aged women, rates of vaginal dryness range from 13 to 43% (Finnish National Sex Survey; Kontula and Haavio-Mannila 2009 & LAW; Howard et al. 2006, respectfully). Among older women, rates of vaginal dryness range from a low of 31% in the

Table 2.2 Sexual dysfunction and satisfaction by age and gender

US National Social Life, Health, and Aging Project (NSHAP) 2007 survey

		Percentage agreement		
		Age range		
		57–64	65–74	75–85
Men	Erectile dysfunction			
	Self-reported	31	45	44
	Sexual satisfaction[a]	96	93	95
Women	Vaginal dryness	36	43	44
	Sexual satisfaction[a]	76	78	75

Finnish National Sex Survey, 2009

		Age range		
		45–54	55–64	65–74
Men	Erectile dysfunction			
	Self-reported	8	16	30
	Sexual satisfaction	82	79	80
Women	Vaginal dryness	13	36	31
	Sexual satisfaction	80	70	75

American Association of Retired Persons (AARP) 2010 national survey

		Age range			
		45–49	50–59	60–69	70+
Men	Erectile dysfunction				
	Self-reported	13	18	38	56
	Diagnosed	6	16	29	48
	Sexual satisfaction	60	50	52	26
Women	Vaginal dryness[b]	1	5	5	3
	Sexual satisfaction	48	40	41	27

European Male Ageing Study (EMAS) cross-national 2010 survey

		Age range			
		40–49	50–59	60–69	70+
Men	Erectile dysfunction				
	Self-reported	6	19	38	64
	Sexual satisfaction	60	57	51	43

Australian Longitudinal Assessment of Ageing in Women (LAW) 2006 study

		Age range			
		40–49	50–59	60–69	70–79
Women	Vaginal dryness	43	43	43	43
	Sexual satisfaction	55	41	38	23

Note. Regarding sexual satisfaction, participants indicated that they were "satisfied, somewhat satisfied, highly satisfied, extremely satisfied, or had pleasure" in their sex life

[a]Limited to participants who reported having sex in the past year

[b]Participants who reported taking a prescription medication for vaginal dryness

Finnish National Survey (Kontula and Haavio-Mannila 2009) to a high self-reported rating of 44% in the US NSHAP (Lindau et al. 2007) survey. Although approximately one in three women reported having difficulty with vaginal lubrication, unlike the significant increase in incidence of ED in men with age, vaginal dryness does not appear to become an increasing problem among women with age. (Significant additional information is devoted to both women's and men's issues in sexuality and aging in Chaps. 7 and 8, respectively.)

Among women who actually reported receiving treatment for vaginal dryness, the rates dropped precipitously to 1% among middle-aged women and 5% of older women (AARP; Fisher 2010). Similarly, fewer men, in all age ranges, reported being officially diagnosed with ED than those who made a self-report or "self-diagnosis" of ED. Additional data from the AARP (2010) survey indicate that approximately 69% of the respondents who reported having some type of sexual dysfunction sought help from a physician, mental health provider, or sex therapist, and approximately half of those individuals who sought help (53%) showed improvement. For those 30% who did not seek treatment, virtually half indicated that they felt uncomfortable discussing sexual issues and nearly 20% cited high financial cost as a limitation. Because these are self-report measures, it is unclear exactly what led these men and women to seek or not to seek treatment. What is clear, however, is that clinicians can assume that many of the middle-aged and older patients they see may have an undisclosed sexual problem that the potential to be improved significantly with treatment.

Sexual Satisfaction

Across both the USA and European population-based studies, overall rates of sexual satisfaction for women were varied, with a range of 40–80% of middle-aged women indicating that they were satisfied, and a range of 23–78% of older women indicating that they were satisfied. It is likely that the lack of availability of a partner influenced lower numbers of older women to respond negatively about their sex lives. For example, in the US NSHAP (Lindau et al. 2007) survey, sexual satisfaction among older adults was assessed only among those who had sex within the past year, and at least 75% of older women responded affirmatively that they were satisfied. (Among a convenience sample of approximately 100 older lesbians, 33% of the women sampled that they were celibate, and the vast majority indicated that this was not by choice. Only one-third of the total sample indicated that they were satisfied with their sex lives, but it was unclear if these were the women who had available partners. Virtually no other information is available regarding sexual satisfaction among LGBT versus heterosexual aging adults; Goldberg et al. 2005.)

Among the large-scale American and European studies, male rates of sexual satisfaction also were varied, with 50–96% of middle-aged men reporting positive appraisals of their sex life, and 43–95% of older men being pleased. Among older men, aged 65 and older, who had sex within the past year (NSHAP, Lindau et al. 2007), at least 93% reported sexual satisfaction, whereas rates of satisfaction among all older men with and without a partner ranged widely between 26 and 80%. It also is important to note that rates of ED rose from those in the middle-aged to older male cohort.

The AARP study (Fisher 2010) provides additional information regarding predictors of aging adults' sex lives and found that for men, frequency of sexual intercourse, sexual dysfunction, and physical health emerged as additional predictors of sexual satisfaction. Although more of the sexually active participants were younger, healthy men, it appears that current sexual involvement is not necessarily a requirement for satisfaction with one's level of sexual activity. Results from these studies also suggest that engaging in sexual activity is no guarantee of personal satisfaction. In some cases, sexual satisfaction and participation in sexual activity may be completely unrelated. This is an important concept for both patients and practitioners alike. One can expect that even among the oldest-old, a relatively high level of sexual activity (or virtual inactivity) can be a normal part of an older person's life. Greater variability in sexual expression, not an inevitable cessation in sexual activity, appears to be the norm with aging. It also remains unchallenged that despite societal myths, older adults often possess sexual urges and interests late into the final decades of life.

To highlight some of the clinically relevant findings from population-based studies:

- Physical health appears related to participation in sexual intercourse for both men and women.
- Approximately one-third of middle-aged and older women report experiencing vaginal dryness.
- Approximately one out of three middle-aged men and one out of two older men report some degree of ED; older men are significantly more likely to experience ED than middle-aged men.
- Nearly half of middle-aged and older women and men fail to seek help for sexual problems, citing embarrassment and financial expense as reasons.
- More than half of those men and women who do seek help from a professional report improvement in sexual function.
- The presence of an available partner is a greater predictor of sexual activity for aging women, whereas physical health and ED are greater predictors of sexual activity for aging men.
- Although older women and men with partners are more likely to report sexual satisfaction, significant numbers of middle-aged and older adults without partners do report being satisfied with their sex lives.
- Virtually no information from large-scale studies is available regarding the reported sexual satisfaction or dysfunction among older LGBT individuals.

When conducting an intake interview, particularly with an older adult, collecting information about that person's sexual history and current level of sexual activity is often one of the last things on a health care provider's mind. Practitioners usually have a brief, circumscribed period of time for this initial interview in which they are consumed with assessing their patient's mental status, gathering pertinent medical information, building rapport, developing a treatment plan, evaluating the underlying family or group dynamic, mobilizing their patient's support system, and discussing the precipitating incident that brought the patient for help. Because older adults often have a wealth of life experience compared with many of their younger counterparts, gathering an elderly patient's social history can be very time consuming. Difficulties in gathering such information can be compounded further, particularly if the older adult displays a sensory deficit or presents with an impaired mental status. Despite the strenuous demands placed on a clinician when conducting such an interview, it remains vital that the discussion of an older patient's sexual activity, both current and historical, be granted high priority. The same premise holds true for middle-aged as well as older adults. One way that a clinician can become better prepared to elicit such information is to become familiar with some basic knowledge of sexuality and aging.

Sexuality Across the Life Span

Sexual behavior and its related goals, frequencies, types, and expectations all can be expected to change throughout the life span. Among adolescents, sexual behavior can provide a formal sense of identity, an opportunity to test limits, an experience of emotional intimacy, and the chance to explore and become comfortable with one's own body. For young adults in their 20s and 30s, sexual behavior can provide an outlet for tension, opportunities for recreation and pleasure, the expression of love and intimacy, the consummation of a marriage, and last, but certainly not least, the ability to become parents. Psychobiology suggests that sexual behavior among young adults serves the primary purpose of procreation and the solidification of pair bonds. Sex was biologically designed to "feel good" in order to promote childbearing and child rearing among healthy adults. In midlife, with the advent of menopause for women and some parallel hormonal changes in men, the biological goals of sexual behavior appear to change. Sexuality generally does not lead to parenthood; the pleasure, emotional intimacy, individual expression, and the desire to satisfy individual, familial, and societal expectations typically take precedence as the primary motivating factors for sexual behavior.

In later life, as compared with early adulthood and midlife, sexual activity enters a new realm in which its expression is related more directly to the personal motivation, needs, and satisfaction of the participants. Sexual activity in later life may be associated with desires to:

- Foster emotional intimacy.
- Experience and enjoy physical pleasure.
- Satisfy continuing biological urges.
- Assert independence and to experiment with new things.
- Feel youthful.
- Challenge societal myths and stereotypes.

- Reestablish a sexual identity.
- Heighten bodily awareness.
- Engender comfort and familiarity with a changing body.

Issues of privacy, physical health, and the availability of a partner become a significantly greater factor in the expression of sexual behavior for older adults. Even though clinicians and researchers often focus on the actual rates of sexual intercourse among older adults, it remains even more important to remember that the motivating (and limiting) factors for sexual relations among both older and middle-aged adults may be *similar* to or *very different from* those of their younger counterparts.

What Is Abnormal?

Although anecdotal reports exist to substantiate that older adults engage in fetishes, bondage, bisexuality, cross-dressing, ménage a trios (threesomes), swinging, and other deviations from traditional sexual norms, virtually no research findings exist regarding the frequencies or predictors of these behaviors.

What becomes more important when dealing with a middle-aged or older client who presents with an atypical or deviant sexual behavior is to determine to what extent this behavior is abnormal, and whether attempts should be made to help the individual change the type or frequency of this behavior. As in work with younger clients, it is important that clinicians not project their own values and expectations into their evaluation of a specific sexual activity or relationship. Important aspects to consider include:

- Is this behavior hurtful or harmful to the older person?
- Is this behavior hurtful or harmful to other people?
- Does this behavior disrupt the patient's daily functioning?
- Does this behavior disrupt the older person's interpersonal relationships?
- Is this behavior illegal?
- Are there violations of privacy or consent for other people?
- Does this behavior run counter to the patient's religious beliefs?
- Is involvement in this behavior ego syntonic or ego dystonic?
- Does performing this behavior put the patient at high risk for acquiring sexually transmitted illnesses such as HIV?

Discussing involvement in such intensely private behaviors requires tact and sensitivity, regardless of the age of the patient. Many older adults are from a cohort in which sexual deviance, much less traditional sexual behavior, is considered an embarrassing topic to discuss even with health care professionals. Clinicians can expect that if an older patient does broach the subject of potentially deviant sexual behavior (e.g., voyeurism, bondage), it probably indicates a high level of trust in the professional relationship and the high level of stress and discomfort that the patient probably wants to alleviate. Among baby boomers, in contrast, who have lived through the sexual revolution, lesser hesitation in discussing sexual matters with a professional may be present. It remains vital not to make generalizations or assumptions, however, simply based upon a patient's age and generational cohort. As noted previously, both middle-aged boomers and older adults were less likely to report a formal diagnosis for ED or vaginal dryness than a self-report, indicating hesitance to discuss sexual issues with health care providers, regardless of their age cohort.

Anna

Anna, a 68-year-old married woman, was in weekly psychotherapy for treatment of moderately severe clinical depression. About 6 months into treatment, Anna had begun to gain the weight that she had lost, to develop some same-sex friendships outside of her marriage, to exercise on a weekly basis, and to become more assertive with her adult children about not wanting to babysit her grandson on a full-time basis. The transference in the therapeutic relationship was positive and Anna had begun to engage in more, appropriate self disclosure with each subsequent session. Once Anna's mood began to stabilize, she and her therapist began to explore other issues.

In her next session, with only a few minutes remaining in the therapy hour, Anna announced quietly that she had a secret that she wanted to share. She said that she had never told anyone else about it, either inside or outside of her family, and that she was very hesitant to talk about it. Her therapist suggested, "Well, because you feel so hesitant about telling this secret, maybe it would be easier to start by telling me about how you think you would feel if you told me about it, or what you think my reaction would be." Anna sighed and responded, "Well…I guess I am afraid that you would think I am a terrible wife…and that there is something really wrong with my husband." Her therapist retorted, "Even if I were to think that way, which I'm not sure I would, would that really be so terrible?" Anna smiled sheepishly and then frowned. She looked down at the floor, and said that a few months ago she walked into the master bedroom without knocking, and found her husband standing in front of the bureau, looking in the mirror, wearing a pair of her panties and one of her bras.

Anna kept looking at the ground and said that she knew then that her husband must "be a queer … a homo." She balled her hand into a fist and pounded it against her leg. Her gaze remained fixed on the ground. Anna began to cry and said, "It makes me feel so sick. … After it happened, Curt got dressed and ran out after me. He tried to tell me that he never wanted me to see him that way, that he was still a man, that he still loved me. I haven't really talked about it with him. We just pretend like things are going along as usual … I mean, we haven't made love much in the past few years anyway; I guess this is part of the reason why. I'm surprised he could force himself on me all of those years … I mean, if he really didn't like women after all … maybe he just wanted to 'do the right thing' and have children. I thought …." Anna's voice grew thick and she began to sob.

Anna's therapist asked her if she would like to know what her husband's behavior was called. After nodding yes, Anna was told that her husband was engaging in *cross-dressing*. Anna looked up and said, "So there is a name for this … thing he does, or they do?" Her therapist decided that additional information regarding her husband's behavior was warranted before delving into the emotional baggage associated with it, in order to dispel any additional myths. Anna's therapist continued, "Cross-dressing is when a man wears women's clothing in order to help feel aroused. Sometimes these men wear only women's underwear, and sometimes under their own clothes. The other important thing to know about this behavior is that most of the men who engage in cross-dressing are not homosexual or 'queer,' but heterosexual; they love women and find women sexually attractive, not men." Anna sat up a little straighter in her chair and asked, "So, you mean that Curt isn't gay or queer or whatever, like he was saying?" Her therapist responded, "Yes, based upon most research and my clinical experience, it is more likely that he is not [gay].... Now, you just told me something very important that you have kept a secret for a very long time, and we have only a few minutes left in our session. We unfortunately will have to stop very soon. I think it's important that we wrap up and prepare to talk more about it next time…How did you think I would respond when you told me about your husband's cross-dressing?"

It is notable that Anna missed her next two appointments. When asked matter of factly about her absence, Anna was able to articulate, "I guess I wasn't ready to talk about it just yet." Her therapy changed dramatically as the original precipitant to her depression was revealed, and Anna began to express her anger, fear, humiliation, and concerns about her husband's behavior. She also began to discuss adjunctive couples therapy in order to confront Curt's behavior and its effect on her. (Anna said, "Why does he want to do that? I mean, that's sickening to me…aren't I good enough for him to want me the way I am?") Anna's therapist also was able to provide important factual information about fetishes and the ways in which they are formed. For example, the initiation of Curt's cross-dressing probably had little or nothing to do with Anna, and it may have begun even before he knew her.

In this case, Anna was able to discuss her feelings openly and to become angry about Curt's behavior instead of remaining paralyzed by it. It also was crucial that Anna's therapist did not become "paralyzed" in session over this revelation with her own countertransference; Anna's therapist herself was shocked that her first encounter with cross-dressing occurred in the context of therapy with an elderly couple.

A Self-Test for Knowledge

Based on the work of Charles White (1982), the Aging Sexual Knowledge and Attitudes Scale (ASKAS) has been used extensively among adult children of older adults, college students (Allen et al. 2009), doctoral-level psychology students, medical students (Snyder and Zweig 2010), health care providers including gynecologists (Langer-Most and Langer 2010), nursing home and long-term care staff members (Bouman et al. 2006; Hinrichs and Vacha-Haase 2010), health care educators, and older people themselves. Helping mental health and primary care professionals become versed in sexuality and aging appears vital. As noted in the (NSHAP, Lindau et al. 2007), approximately 1 in 3 middle-aged and older men, and only 1 in 5 middle-aged and older women reported that they discussed sexual issues with a primary care physician since their 50th birthday. Findings from various research studies incorporating the ASKAS (White 1982) reveal that both medical school students (Snyder and Zweig 2010) and even gynecologists (Langer-Most and Langer 2010) maintain only moderate levels of knowledge of sexuality and aging.

Although the ASKAS has been used primarily in research settings, it also can be employed successfully as a valuable educational tool for older adult patients. Asking a couple to complete the knowledge section of the scale separately and to then score their answers jointly often allows for a discussion of previously taboo topics. It also allows the practitioner to gain insight into the knowledge base of her clients. Many clients admit relief when they learn that some of the very stereotypes that they had ascribed to were false. Others realize for the first time that it is acceptable to discuss intimate sexual issues and concerns with their health care provider. This 35-item knowledge subtest of the ASKAS can be administered as a self-report paper-and-pencil test, or as a clinician-administered interview. This assessment instrument employs nonscientific language and is suitable for individuals who have acquired less than a high school education. Presented in the appendix at the end of this chapter, the knowledge section of the ASKAS can be used as a self-test for clinicians as well. Some selected items from this knowledge section will be reviewed here.

Although entire chapters are devoted to both men's and women's issues in sexuality and aging, a basic review of the physiological sex-related changes in aging adults will be presented here briefly. As men age, they can expect to experience a decline in testosterone production, a decline in sperm production, changes in the amount and consistency of their seminal ejaculate, diminished force of ejaculation, a likely increase in the size of the prostate gland, longer periods of time to stimulate sexual excitement and erections, longer-lasting erections (with decreased likelihood of premature ejaculation), less-frequent ejaculations, and longer refractory periods. The need for increased stimulation, typically tactile stimulation, to achieve an erection also is likely to occur with increased age. It also is important to note that all men, regardless of age, have an orgasm that is independent from the act of ejaculation; most individuals assume that ejaculation and male orgasm are inseparable. In other words men, including those who have had surgeries or injuries that impair or prevent their ability to ejaculate, are still able to experience orgasm (e.g., Martinez 2005). In contrast, as women age the most common physiological changes include, particularly after menopause, a decline in production of estrogen and progesterone, shrinking of the external genitalia, a decrease in growth of pubic hair, shrinkage and thinning of the vaginal wall, a decline in natural vaginal secretions, and longer periods of time to achieve sexual excitement and related vaginal lubrication (see Croft 1982; Willert and Semans 2000).

Sample Items from the Knowledge Section of the ASKAS

2. Males over the age of 65 typically take longer to attain an erection of their penis than do younger males.

TRUE:

Because of changes in the internal structure of the penis over the age of 60 (i.e., most men develop more venous blood vessels that are larger in diameter), most middle-aged and older men require increased blood flow to the penis in order to attain an erection. It takes more time for an older man to have an erection, related in large part to the increased time it takes to provide increased blood flow to the penis. Many middle-aged and older men also find that they need more tactile stimulation in order to attain a full erection. The need for additional physical stimulation of the penis in order to obtain an erection may sometimes require significant negotiation between a man and his partner as the "script" for foreplay, which often becomes ritualized or rigid (e.g., Willert and Semans 2000), must change and evolve.

4. The firmness of erection in aged males is often less than that of younger persons.

TRUE:

Because of changes in the blood flow to the penis, older adult males often have a less firm and erect erection than younger males, sometimes impeding their ability to penetrate a partner during intercourse. As noted, conservative estimates suggest that up to one-half of US men over the age of 65 have some degree of erectile dysfunction.

7. The older female may experience painful vaginal intercourse due to reduced elasticity of the vagina and reduced vaginal lubrication.

TRUE:

Many older adult women benefit from the use of lubricants (e.g., KY Jelly) in order to alleviate any discomfort during intercourse associated with vaginal dryness. Some researchers suggest that up to two-thirds of elderly women experience discomfort during intercourse, related primarily to a lack of lubrication. The more positive aspect of this finding is that the problem is usually readily treatable with over-the-counter lubricants, prescription lubricants, and hormone therapies.

8. Sexuality is typically a lifelong need.

TRUE:

Masters and Johnson (1966) were among the first researchers to point out that older adults have as much interest in and need for sexual contact as their younger counterparts. Even though an older person can no longer produce offspring, the underlying biological urge to engage in sexual (and sensual) activity does not appear to diminish significantly with age. In the recent survey by the AARP (Fisher 2010), a number of healthy older men and women were found to engage in sexual activity well into their 80s and 90s.

9. Sexual behavior in older people increases the risk of heart attack.

FALSE:

Unless an older adult is under a physician's orders to limit his or her physical activity, sexual activity can be actively pursued without fear of life-threatening exertion.

Among healthy older adults, sexual activity can provide some of the benefits of cardiovascular exercise. There also is evidence that sexual activity in older persons has beneficial physical effects on the participants. Increased blood flow to the genital area can provide an older woman with increased vaginal lubrication and can provide an older man with more sustained erections in future sexual activity.

11. The relatively more sexually active younger people tend to become the relatively more sexually active older people.

TRUE:
As discovered in earlier research, one predictor of an older adult's sexual activity is his or her prior level of sexual activity. Men and women who did not engage in sexual relations in their younger years are less likely to engage in sexual relations during their later years (Call et al. 1995). Of course, availability of a partner and physical health, among other factors, can limit this trend. It also is interesting to note that newer research decade generally ignores this factor of prior sexual activity and focuses instead upon aspects of physical health (Bancroft 2007).

13. Sexual activity may be psychologically beneficial to older person participants.

TRUE:
Many older adults cite that they enjoy their sexual relationships even more than they did when they were younger. Others point out that even though they may engage in sexual activities less frequently than they did when they were younger, they cherish and enjoy them more. It also is common to hear older adults mention that they typically feel more comfortable with their partner and no longer have "unrealistic expectations" about the sex act. Participation in foreplay, including kissing, hugging, and caressing appears to remain consistent throughout later life (Fisher 2010).

16. Prescription drugs may alter a person's sex drive.

TRUE:
A variety of prescription medications for depression, blood pressure, and diabetes can negatively impact on an older person's level of sexual interest and sexual functioning. Clinicians should be aware of all medications that middle-aged and older adult patients are taking.

21. The most common determinant of the frequency of sexual activity in older couples is the interest or lack of interest of the husband in a sexual relationship with his wife.

TRUE:
For married older adults, studies do suggest that the rates of sexual intercourse are determined primarily by the interest level of the husband. It remains unclear, however, to what extent this can be accounted for by a cohort effect. In this current elderly cohort, the husband is primarily responsible for making sexual advances; "good wives" were unlikely to initiate sexual behavior. In this cohort, the elderly wife also has been inculcated with the expectation that "a good wife does not say 'no' to her husband." Both of these beliefs further increase the likelihood that the

frequency of sexual activity among current older couples is influenced primarily by the husband's level of interest and desire.

22. Barbiturates, tranquilizers, and alcohol may lower the sexual arousal levels of aged persons and interfere with sexual responsiveness.

TRUE:
Many over-the-counter medications and substances can alter sexual function. Because older adults metabolize alcohol more slowly than younger adults due to changes in body composition (e.g., increased fat-to-muscle ratios) and a general decline in liver function (Schmucker 2005), even one or two alcoholic drinks can negatively affect sexual performance. Heavy consumption of cigarettes also may diminish sexual desire, and nicotine also has been related to ED among older men.

23. Sexual disinterest in aged persons may be a reflection of a psychological state of depression.

TRUE:
One of the most common symptoms of depression is a loss of interest in sexual activity. Among older adults without available sexual partners, it is appropriate (and recommended) that clinicians ask that older adult if her or his own interest in sexuality has changed. One might ask, "Even if you haven't been engaged in sexual relations for a few years, do you find yourself thinking about sex less frequently than you used to in the past few months?"

28. Fear of the inability to perform sexually may bring about an inability to perform sexually in older males.

TRUE:
Erectile dysfunction certainly can be initiated or compounded by psychological causes. Fear of intimacy and fear of attaining or not being able to attain an erection have been identified as some of these common concerns. "Widower's syndrome" also been identified, in which a recent widower who is not yet emotionally prepared for intimacy with another woman (particularly if he had been married in a relationship spanning many decades) experiences ED with a new partner. Clinicians also must consider and rule out organic causes for erectile dysfunction among their middle-aged and older adult patients, even if psychological factors appear present.

29. The ending of sexual activity in old age is most likely and primarily due to social and psychological causes rather than biological and physical causes.

TRUE:
Many of the physical problems associated with aging can be addressed. Even older adults with chronic illness can engage in a variety of sexual behaviors if they so choose. Many older adults cease to become sexual beings simply from perceived societal pressures and stereotypes. Societal myths suggest that older adults are asexual beings who are not entitled to the benefits and joys of sexuality (and sensuality) that seem to belong exclusively to the young and physically fit.

35. Masturbation in older males and females has beneficial effects on the mainte-
    nance of sexual responsiveness.

TRUE:

For men, masturbation can improve blood flow to the penis and help prevent future
incidents of ED. For women, masturbation has been shown to increase blood flow
to the vaginal area and promote premenopausal levels of lubrication. Some clinicians have successfully "prescribed" masturbation to middle-aged and older
women, who later reported that they were able to experiment with and enjoy self-
stimulation for the first time in their lives after receiving approval from their health
care provider.

Reviewing aspects of sexuality and aging is vital. As noted, studies show that
doctoral-level psychology and medical students (Snyder and Zweig 2010), health
care providers (Langer-Most and Langer 2010), and health care educators (Glass
and Webb 1995), as well as older persons themselves (Adams et al. 1990; Smith and
Schmall 1983) often have limited knowledge of sexuality and aging. As clinicians,
we owe it to our patients and ourselves to be informed.

The Importance of Sexual Histories at Intake

Although many older patients, like their younger counterparts, seem to reveal important aspects of their sex life only after they have been in treatment for an extended
period of time, most information about an older person's sex life can be gathered
effectively during the initial intake interview. (Unfortunately, various research findings suggest that both general practitioners and psychiatrists are significantly less
likely to take sexual histories and discuss sexual issues in general with their older, as
compared to younger, patients; Bouman and Arcelus 2001; Gott et al. 2004.)

It is important that all clinicians place value on this information and recognize
that obtaining such information represents much more than just finding about how
many times a week their patient or client has sex with a significant other. Additional
studies suggest that aging adults with sexual dysfunction would welcome and even
prefer the initiation of such questions by their health care provider (e.g., Gott and
Hinchliff 2003). Asking aging patients about their sex lives provides clinicians with
valuable opportunities early in treatment to discuss and assess:

• Emotional intimacy
• Body image
• Physical health
• Medical history
• Self-esteem
• Quality of interpersonal relationships
• Ascribed sex roles
• Sexual identity and orientation
• Attitudes toward aging

- Knowledge of sexuality and aging
- Religion
- Potential sexual dysfunction
- Potential partner abuse
- Sense of humor

How to Gather Information

Interviewing an older client about sexual matters can pose unique problems as well opportunities. Many older adults are reticent to discuss such personal matters without proper groundwork by the clinician. Before addressing issues of sexuality and sexual behavior, it often can be helpful to simply ask the patient permission to inquire about such matters (e.g., "May I ask you some questions about your love and sex life?") Asking the patient for such permission can instill a sense of respect and concern; the patient is given control in a potentially anxiety-provoking situation.

Another tactic is to present the questions as part of the standard interview that is used with all patients who come to the clinic. Likening sexual information to medical information also can allow older patients to feel more comfortable discussing such personal issues. In some cases, admitting that "it is not always an easy thing for health care providers or patients to talk about, but it is important that we gather some information about your love and sex life" normalizes the stress and anxiety that an older person may feel about discussing sexual issues with a clinician.

When asking questions about sexual behaviors, it is important to word questions in "the affirmative." In other words, it is helpful to provide older adults with an opportunity to answer the question without appearing as though they are admitting something wrong, immoral, or embarrassing. For example, instead of asking, "How many affairs have you had?" it would be preferable to say, "Some people become unhappy in their relationships, for any number of reasons, and engage in extramarital affairs or seek out other lovers. Can you tell me about any affairs you may have had?" It also is beneficial to allow for humor in the process. Sometimes humor (on the part of the clinician or the patient) is appropriate and helps to dispel tension.

Requisite Information

It is important to gather a variety of information regarding both middle-aged and older adults' sexuality. Sometimes it helps to have a checklist or other guideline available during the course of the clinical interview. Some relevant questions for both older men and women include (see also Galindo and Kaiser 1995):

- What would you consider a satisfactory sex life? Some people are satisfied while other people are dissatisfied with their sex lives. How do you feel about your sex life?

- How long has it been since you have engaged in any sexual activity?
- Do you have any current sexual partners? Do you have an exclusive relationship or do either of you have other partners as well? (Gather activity about the sex, numbers, and potential high-risk behavior of the partners such as prostitution or drug use.)
- Does your religion influence how you feel about sex or influence your current sexual activities in any way? Can you tell me about your religious views?
- Having sex means different things to different people. For some people it means having sexual intercourse and to others it means holding hands. What different types of intimate sexual activity [vaginal intercourse, oral sex, anal sex, petting, cuddling, holding hands] do you engage in?
- How often do you masturbate or touch yourself to feel good? How do you feel about it? (This is often a good opportunity to tell an older adult about the potential emotional and physical benefits of masturbation.)
- As people age, they sometimes experience pain or discomfort during sexual intercourse. Do you ever experience such pain or discomfort? Have you discussed this with anyone [like a partner or health care provider], or have you tried to do anything about it?
- The average person experiences some kind of sexual difficulty at some point in their life. Could you tell me about any trouble or problems that you, or a partner of yours, may have had in the past? Do you have any concerns about your sexual behavior or functioning right now that you could tell me about?
- Some people experience changes in their body as they age, either slowly over time or more suddenly through illness or surgery. How do you feel about your body? How does your partner feel about your body?
- Do you use any lubricating gels or liquids when you engage in sex? What kind do you use? (Does the client use something inappropriate like an oil-based lubricant such as Vaseline with a condom, or does she use something water soluble like KY jelly?)
- Have you ever used condoms when you have sex? Do you use condoms now? Why or why not? How does your partner(s) feel about condoms?
- Do you have any concerns or worries about your sexual performance or your sex life in general? Do you have any concerns or worries about your [a] partner?
- [Whether or not you have had sexual relations lately] have you been thinking more or less about sex than you typically have in the last few months?
- Do you have enough privacy for the sexual activities that you want to engage in?
- Is there anything about your sex life that you wish were different?
- Do you ever feel hurt or threatened by your [a] partner? (Unfortunately, sexual and physical abuse are not limited to younger adults.)
- How easy or difficult is it for you to talk about your sexual behavior with your partner? Your physician? With me right now? (You can inform patients that they are doing a wonderful job talking about such a personal topic, if in fact they are. This also can provide an opportunity to talk about how many people

feel uneasy talking about sexual matters and to empathize with their fears and concerns.)
- Many people have fantasies about sex. What kind of sexual fantasies can you tell me about?
- People often have questions they would like to ask about sex. Do you have some questions about sexual activity that you would like to ask me?

Some gender-specific questions also should be addressed. For example, older women can be asked about potential discomfort during intercourse and age of menopause. Similarly, middle-aged women can be asked about potential discomfort or pain during sex (i.e., dyspareunia) and about menopausal symptoms. Both middle-aged and older men can be asked about their erections and their frequency of urination. Depending on the numbers of types of partners that both aging men and women have (especially if high-risk behaviors are involved), pointed questions should be asked about condom use. Older people also should be asked if they know how to properly put on a condom, and whether they know that a new condom should be used for each subsequent act of intercourse. Many middle-aged and older adults underestimate their risk of contracting a sexually transmitted disease outside of a long-term monogamous relationship, or even within the context of infidelity in a long-term relationship, and focus mainly on their freedom from fear of unwanted conception.

Because so many prescription and over-the-counter medications can be associated with a decrease in sexual desire and function (see Galindo and Kaiser 1995), clinicians should make a concerted effort to gather detailed information about a patient's medication history. As gathering such medication information is already a prerequisite for a thorough geriatric intake, no additional time will be lost during an interview with an older patient who may (or may not) initially discuss concerns about changes in their sexual functioning. Many middle-aged and older adults are surprised to learn that a variety of drugs, including those listed in Table 2.3, can interfere with sexual functioning. Older adults are at increased risk for such side effects when compared to younger adults because they take, on average, more than four prescription drugs each day, metabolize drugs less effectively, and less than one-third inquire consistently about potential side effects or drug interactions (Barrett 2005).

With older adults often taking a large number of prescription and over-the-counter medications, it may be difficult to get a proper assessment of daily intake. One way to circumvent these problems is to ask all patients to bring a "brown bag" that contains all of their current medications with them to the intake interview. Names of medications, as well as the names of the prescribing physicians, can be taken directly from pill bottles. Important information also can be gleaned as to whether older clients are taking their medication as instructed on the label, whether they are non-compliant based upon the number of pills remaining in the bottle, or whether they misunderstood the directions for administration in the first place.

Table 2.3 Drugs with reported sexual side effects

Medication class	Name	Trade name	Common side effects
Prescription medications			
Antidepressants	Clomipramine	Anafranil	Anorgasmia, dysorgasmia
	Fluoxetine	Prozac	Dysorgasmia
	Paroxetine[a]	Paxil	Anorgasmia, dysorgasmia
	Sertraline[a]	Zoloft	Anorgasmia, dysorgasmia
	Venlafaxine	Effexor	Anorgasmia, dysorgasmia
Antianxiety	Alpraolam[a]	Xanax	Decreased libido
	Lorazepam[a]	Ativan	Decreased libido
	Temazepam[a]	(Generic)	Decreased libido
Antihypertensives	Atenolol[a]	(Generic)	ED
	Clonidine	Catapress	Decreased libido, ED
	Digoxin[a]	Lanoxin, Digibind	ED
	Hydrochlorothiazide[a]	Lopressor	Decreased libido, ED
Antiparkinsonian	Levodopa	Sinemet	ED
Antipsychotics	Haloperidol	Haldol	Dysorgasmia, ED
	Mesoridazine	Serentil	Decreased libido, dysorgasmia
	Trifluoperazine	Stelazine, Suprazine	Decreased libido, ED
Antiseizure	Phenytoin[a]	Dilantin	ED
Chemotherapy	Tamoxifin	Nolvadex	Vaginal dryness, ED
Mood stabilizers	Lithium	Eskalith, Lithonate	ED
	Topamarite	Topomax	ED
Sleeping aids	Zolpidem	Ambien	ED
Miscellaneous	Celebrex	Celecoxib	ED
	Donepezil[a]	Aricept	ED
Over-the-counter medications and herbs			
Antihistamines	Diphenhydramine	Benadryl	ED
	Drixoril	Tavist-D	ED
	Cimetidine	Tagamet	Decreased libido, ED
	Ranitidine	Zantac	ED
Herbs	Alkaloids	Rauwolfia	ED
	Hypericum	St. John's Wort	Dysorgasmia
Recreational drugs			
	Ethyl alcohol	Beer, wine, liquor	Decreased libido, ED
	Nicotine	Cigarettes, chew	ED
	Narcotics	Cocaine, heroin	Decreased libido, ED

Notes: This list of drugs is meant to be illustrative rather than comprehensive. Attention to older patients' sometimes atypical side effects remains essential

ED erectile dysfunction

[a]Identified as one of the 50 most frequently prescribed oral drugs among older adults (Steinmetz et al. 2005)

Summary

Gathering information about an older client's sex life is challenging, rewarding, and a requisite part of a geriatric intake interview. In order to make this practice more effective and efficient, clinicians themselves can become more informed about basic aspects of sexuality and aging. For example, older adults can be sexually active well into the last decades of life. The best predictors of older adults' sexual activity are prior levels of sexual activity and satisfaction; the availability of a partner, particularly for women; and their physical health. Many middle-aged and older adults are uninformed about the physiological changes associated with aging and sexuality; many do not know that ED may be treatable or that masturbation can promote improved sexual lubrication in women. Many middle-aged and older adults, as well as health care providers, are unaware that commonly prescribed medications and over-the-counter drugs can significantly impair sexual function. As noted, gathering information about an older client's past and current sex life can reveal important information about close relationships, body image, knowledge of sexuality and aging, medical history, and self-esteem. As informed clinicians, the chance to discuss issues of sexuality with our middle-aged and older clients presents us with unique opportunities and benefits.

APPENDIX: Knowledge Section of the ASKAS[1]

Answer Key: T = True; F = False; DK = Don't know
Correct answers are in **BOLD**.

1. T/**F**/DK Sexual activity in aged persons is often dangerous to their health.
2. **T**/F/DK Males over the age of 65 typically take longer to attain an erection of their penis than do younger males.
3. **T**/F/DK Males over the age of 65 usually experience a reduction in intensity of orgasm relative to younger males.
4. **T**/F/DK The firmness of erection in aged males is often less than that of younger persons.
5. **T**/F/DK The older female (65+ years of age) has reduced vaginal lubrication secretion relative to younger females.
6. **T**/F/DK The aged female takes longer to achieve adequate vaginal lubrication relative to younger females.
7. **T**/F/DK The older female may experience painful vaginal intercourse due to reduced elasticity of the vagina and reduced vaginal lubrication.
8. **T**/F/DK Sexuality is typically a lifelong need.
9. T/**F**/DK Sexual behavior in older people increases the risk of heart attack.
10. T/**F**/DK Most males over the age of 65 are unable to engage in sexual intercourse.
11. **T**/F/DK The relatively more sexually active younger people tend to become the relatively more sexually active older people.

---

[1] These items from the ASKAS appear from White (1982) with permission.

12. **T**/F/DK There is evidence that sexual activity in older persons has beneficial physical effects on the participants.
13. **T**/F/DK Sexual activity may be psychologically beneficial to older person participants.
14. T/**F**/DK Most older females are sexually unresponsive.
15. T/**F**/DK The sex urge typically increases with age in males over 65.
16. **T**/F/DK Prescription drugs may alter a person's sex drive.
17. T/**F**/DK Females, after menopause, have a physiologically induced need for sexual activity.
18. **T**/F/DK Basically, changes with advanced age (65+) in sexuality involve a slowing of response time rather than a reduction of interest in sex.
19. **T**/F/DK Older males typically experience a reduced need to ejaculate and hence may maintain an erection of the penis for a longer time than younger males.
20. T/**F**/DK Older males and females cannot act as sex partners as both need younger partners for stimulation.
21. **T**/F/DK The most common determinant of the frequency of sexual activity in older couples is the interest or lack of interest of the husband in a sexual relationship with his wife.
22. **T**/F/DK Barbiturates, tranquilizers, and alcohol may lower the sexual arousal levels of aged persons and interfere with sexual responsiveness.
23. **T**/F/DK Sexual disinterest in aged persons may be a reflection of a psychological state of depression.
24. **T**/F/DK There is a decrease in frequency of sexual activity with older age in males.
25. T/**F**/DK There is a greater decrease in male sexuality with age than there is in female sexuality.
26. **T**/F/DK Heavy consumption of cigarettes may diminish sexual desire.
27. **T**/F/DK An important factor in the maintenance of sexual responsiveness in the aging male is the consistency of sexual activity throughout his life.
28. **T**/F/DK Fear of the inability to perform sexually may bring about an inability to perform sexually in older males.
29. **T**/F/DK The ending of sexual activity in old age is most likely and primarily due to social and psychological causes rather than biological and physical causes.
30. T/**F**/DK Excessive masturbation may bring about an early onset of mental confusion and dementia in the aged.
31. T/**F**/DK There is an inevitable loss of sexual satisfaction in postmenopausal women.
32. **T**/F/DK Secondary impotence (or non-physiologically caused) increases in males over the age of 60 relative to younger males.
33. **T**/F/DK Impotence in aged males may literally be effectively treated and cured in many instances.
34. **T**/F/DK In the absence of severe physical disability, males and females may maintain sexual interest and activity well into their 80s and 90s.
35. **T**/F/DK Masturbation in older males and females has beneficial effects on the maintenance of sexual responsiveness.

Chapter 3
Attitudes toward Sexuality and Aging

Our youth-oriented culture views the process of aging as a tragic, narcissistic injury instead of as an opportunity for personal growth and change. Unfortunately, many older adults internalize and generalize these negative attitudes toward their own sexuality and diminish or cease sexual expression out of fear, disgust, and shame. It becomes vital for practitioners to examine their own countertransferential reactions in relation to their work with aging adults.

After one considers the availability of a partner and general health status, one of the most important determinants of older adults' sexual activity is the positivity of their attitudes toward sexuality and aging. Attitudes toward sexuality and aging range from permissive to restrictive, curious to avoidant, and can be global or specific in relation to particular behaviors such as masturbation, oral sex, sexuality among LGBT individuals, or sexuality within an institutional context. Such attitudes also have both explicit (i.e., conscious) and implicit (i.e., automatic and unconscious) components. Emergent research suggests that different populations maintain a variety of both positive and negative attitudes that affect aging adults. Health care providers including psychologists, counselors, physicians (including those in primary care, gynecology, urology, primary care, and oncology), social workers, nurses, pharmacists, occupational therapists, health educators, nursing home directors, and nutritionists all play a role in influencing middle-aged and older adults' attitudes toward their own sexuality, both directly and indirectly.

As clinicians, it is not sufficient to be aware of both our clients' and general society's attitudes about sexuality and aging. We also must be cognizant of our own attitudes as individuals. To some extent, it does not matter as much whether our own attitudes are permissive or restrictive, as long as we become aware of them and override any personal biases (either positive or negative) to provide our clients with the most appropriate, objective information and care. As practitioners, our countertransferential reactions to aging clients can often serve as an important diagnostic barometer for what is happening within the therapeutic alliance. The clinical issue at stake is whether we can observe our own attitudes and reactions to

J. Hillman, *Sexuality and Aging: Clinical Perspectives*,
DOI 10.1007/978-1-4614-3399-6_3, © Springer Science+Business Media New York 2012

our middle-aged and older patients' sexuality, but maintain an objective and neutral stance in our work with them. A number of exercises and self-tests, presented here, can be used to help individuals examine and assess their own implicit and explicit attitudes.

Attitudes Toward Sexuality and Aging

The self-reported attitudes maintained about sexuality and aging among young adults, middle-aged adults, older adults, and health care providers appear quite variable. The vast majority of studies indicate that young adults typically hold negative or restrictive attitudes (e.g., Allen et al. 2009). Among middle-aged adults, attitudes toward sexuality and aging appear generally positive, but somewhat varied by gender. Findings from the US based AARP (2010) study indicate that among their middle-aged participants, only 3% of male and 7% of female respondents believed that "sex is only for younger people," and only 3% of the men and 6% of the women indicated that they would "be quite happy" if they never had sex again. However, nearly half or approximately 45% of both the male and female respondents believed that sex loses its importance as one ages. Additional qualitative studies of middle-aged women's attitudes toward sexuality indicate that the majority of these women regard sexuality as very important in their lives, but often regarded their own desire for sex within the context of their typically male partners' needs (Hinchliff and Gott 2008).

Among older adults, attitudes toward sexuality show increased variation and gender differences. In the AARP (2010) survey, both men and women over the age of 70 were equally likely (63%) to endorse the statement that "sex becomes less important to people as they age." However, only 5% of these older men and 8% of the older women believed that sex was "only for younger people." In terms of their own sex lives, attitudes remained mixed, in which 38% of the older women and only 15% of the older men claimed that they would be "quite happy" if they never had sex again. From these representative survey findings, it appears that the vast majority of both men and women, even into their 80s and 90s, have affirming attitudes toward sex and believe that it is meant for both older and younger people to enjoy. However, men appear less accepting than women of being content or happy with an asexual lifestyle in advanced age. It also is important to note that the presence of an available partner appears to be a confounding factor. Findings from the NSHAP (Lindau et al. 2007) survey indicate that among sexually active older men and women, only 5% rated sex as "not at all important," whereas 35% of women and 13% of the men without partners rated sex as unimportant. Perhaps older adults, and older women in particular, begin to assign less importance to sex once their access to partnered sex becomes limited.

One reason it is so vital to understand individuals' attitudes toward sexuality and aging is that research suggests that among individuals with more negative or restrictive attitudes, simply increasing their knowledge of elderly sexuality does not lead to more positive or even neutral attitudes. For individuals who already possess liberal attitudes, some studies suggest that no relationship exists between increased levels of knowledge and more positive or accepting attitudes. It is as though attitudes toward sexuality and aging are robust, and typically unaffected by one's knowledge of the subject, even among health care providers including nursing home staff, gynecologists, psychiatrists, medical students, and doctoral-level psychology students (Allen et al. 2009; Bouman and Arcelus 2001; Bouman et al. 2006; Langer-Most and Langer 2010; Snyder and Zweig 2010; Walker and Harrington 2002), as well as members of the general population (Allen et al. 2009). Health care providers also may be influenced by specific religious beliefs (Hillman and Stricker 1994), previous negative interactions with older people (Glass et al. 1986), and perceived social norms (Snyder and Zweig 2010) and profess more negative and restrictive attitudes toward elderly sexuality as a result. If conscious attitudes appear generally resistant to increases in factual knowledge via educational programming, it appears vital for practitioners to process and understand the basis of their own attitudes in order to have the best chance to engage with their older clients in an objective or positive manner.

Implicit Attitudes Toward Aging

Within the social psychological literature, it has been demonstrated consistently that both conscious and *unconscious* attitudes, including those about aging, are powerful constructs, typically linked with behavior. Findings from empirical studies present somewhat shocking findings in which both younger and older adult participants who were primed with stereotypes about elderly or older adults proceeded to engage in behavior consistent with those stereotypes. For example, when college students were primed with an elderly stereotype by engaging in a word-scramble task that included words like "bingo," "Florida," "wrinkles," and "gray", they were significantly more likely to walk slowly (Bargh et al. 1996) and even drive more slowly via a computer simulator (Branaghan and Gray 2010). What also is striking about these studies' findings is that the participants remained *unaware* that they were primed with the older adult stereotype. Participants in the Branaghan and Gray (2010) driving study actually guessed that the scrambled words had something to do with food or warm places, rather than old age. In other studies, in which agist stereotypes were consciously activated by showing pictures of elderly adults, both young and older adults taking place in the study displayed significantly longer reaction times in subsequent tasks (Dijksterhius et al. 2001; Kawakami et al. 2002). In essence, the activation of older adult stereotypes both in and out of awareness (i.e., explicit and implicit attitudes, respectfully) appears to generate significant behavioral changes consistent with those stereotypes.

Another way in which the strength of individuals' automatic, unconscious, or implicit attitudes toward aging has been assessed is via the Implicit Association Test or IAT (Greenwald et al. 1998; Hummert et al. 2002; Levy and Banaji 2002). Essentially, the IAT measures the extent to which an individual more readily or quickly makes connections between concepts that are already paired in one's mind, than with others. For example, are individuals more likely to equate aging with positively or negatively charged words? For people who have internalized agist societal beliefs, under a timed test they would make a link between "elderly" and "unattractive" more quickly than between "elderly" and "beautiful." The perceived discrepancy between "elderly" and "beautiful" would generate a slight but significant lag in response time, and the computerized IAT can assess differences in response times by milliseconds. (The IAT uses both words and images as target stimuli and can be used to assess implicit attitudes toward a variety of other target groups, including those related to race, religion, and sexual orientation.) Anyone who visits the researchers' Web site, www.implicit.harvard.edu, can assess their own unconscious attitudes or associations via the IAT at no cost.

To give the reader an approximation of the format (also see Gladwell 2005) and some immediate insight into one's own implicit attitudes toward aging, a simplified paper-and-pencil task is offered here in Exercise #1. Completion of this exercise, as well as the computerized version of the IAT, would be useful for anyone who works with older adults, aging adults, and clients themselves.

Attitudinal Exercise #1:
Approximation of the IAT

A. For each of the words listed below, make an "X" or tap your finger in the appropriate column in which that item belongs. You are asked to do this as quickly as you can, without making a mistake.

Elderly		Young
✗	Wrinkles	
	Florida	✗
	Infancy	✗
✗	Gray	
✗	Bingo	
	Cherub	✗
	Modern	✗

This task probably seemed easy and straightforward. Few individuals would hesitate about assigning related words to either the elderly or young category, or make a mistake.

B. Now assign each word to the appropriate category on the left or right, with an "X" or tap the area with your finger. Note that there are now two categories for each column. Continue to work as quickly as you can, without making a mistake.

Elderly or Bad		Young or Good
✗	Ancient	
✗	Joy	
✗	Oldie	
	New	✗
	Wonderful	✗
✗	Biddy	
✗	Agony	
✗	Hurt	
✗	Antiquated	
	Infancy	✗
	Laughter	✗
	Fresh	✗

This previous task probably took a bit more concentration, but still appeared generally straightforward.

C. Sort each of the following words into one of the two columns on the left or the right. Again, there are two categories in each column, but note that the terms "good" and "bad" are now switched, so that "elderly" and "good" now belong in the same group, and "young" and "bad" are now in the same group. Draw an "X" or tap the appropriate space with a finger.

Elderly or Good		Young or Bad
_____	Modern	X
_____	Hurt	X
_____	Fledgling	X
X	Senior	_____
_____	Nasty	X
X	Retiree	_____
X	Tender	_____
X	Pleasure	_____
X	Codger	_____
_____	Recent	X
_____	Failure	X
X	Ancient	_____
_____	Junior	X
X	Happy	_____
X	Fossil	_____
X	Frail	_____
_____	Spring Chicken	X
_____	Horrible	X

Although this simplified paper-and-pencil exercise cannot measure an individual's response time in milliseconds as the computerized IAT does, someone who completes this exercise can typically discern if this final word-sorting task (C), that asks one to group "elderly" and "good" words together, inconsistent with agist stereotypes, took more effort to complete, and if it fostered more initial errors, hesitation, or confusion than the second sorting task (B), that asked one to group "elderly" and "bad" words together, consistent with potentially internalized, agist stereotypes.

Findings from large-scale studies of the IAT reveal that young, middle-aged, and older adults themselves possess moderately to highly negative, implicit attitudes toward older adults (Hummert et al. 2002; Levy and Banaji 2002). Specifically, those over the age of 60 appear to automatically attribute more positive traits to younger than older adults at nearly the same rate as 20-year-olds do themselves. A critically important finding from this work is that older adults possess strong, negative attitudes toward their own in-group; they don't even appear to "like" themselves. One theory advanced from this research suggests that agist societal norms, which are highly negative and pervasive in most Western cultures, lead individuals of all age groups to internalize automatic, negative associations to aging. Although implicit attitudes may be deeply internalized and resistant to modification (e.g., Gregg et al. 2006), it is essential to note that one's conscious, purposeful responses to any negatively perceived group, including aging adults, can be monitored and controlled. Our outward behavior certainly does not have to correspond to negative, implicit attitudes (Dasgupta and Greenwald 2001).

Consistent with these empirical findings from cognitive and social psychology in which the activation of both conscious and unconscious, stereotypes of aging can significantly influence someone's subsequent attitudes and behavior, clinicians often find that clients' self-reported attitudes about aging can influence their behavior in a variety of domains. Although a number of factors certainly moderate the relationship between individuals' attitudes or feelings about something and their resulting behavior, some middle-aged and older clients believe, or have internalized, that the process of aging itself is suggestive of forced retirement, an unavoidable decline in physical health, an increase in sadness and depression, and a loss of perceived social value. For a number of older adults, such negative attitudes toward aging may become translated into a decline in sexual expression and a loss in an innate sense of sexuality and self-worth.

Geriatric Sexuality Breakdown Syndrome

Although her initial work took place nearly two decades ago, Kaas (1981) was the first researcher to identify and categorize the potentially negative impact of agism on an older adult's sexuality. She called this process, in which the social environment of older adults becomes an overwhelming force for behavioral change and even mental illness, Geriatric Sexuality Breakdown Syndrome. Kaas described it as an outcropping of Social Breakdown Syndrome (Kuypers and Bengston 1973), in which older people become trapped in a vicious cycle in which they internalize and then conform to agist societal stereotypes in a negative, self-fulfilling prophesy. As a result, the older person begins to feel and act useless and helpless. Such apathy leads to further "confirmation" for that individual that society perceives older adults as mere hindrances, who require care and drain valuable social resources without contributing anything in return. Unlike many Eastern societies, our individualized, Westernized society places high value upon independence, autonomy, and the related goods or services that an individual member of society can provide. Older adults, who typically retire from the full-time workplace when financially able, are viewed as liabilities by many younger segments of society, rather than as a store of wisdom, experience, and opportunity.

Kaas (1981) outlined a series of steps that take place when an older person undergoes this emotional and subsequent sexual breakdown. This series of events is presented in Table 3.1. With the recent medicalization of sexuality and aging, as well as increased portrayals of "sexy seniors" in the media (Marshall 2010; Walz 2002; although "cougars" are typically portrayed as middle-aged and not older women), significantly fewer older adults may be impacted by Geriatric Sexuality Breakdown Syndrome than when it was first described by Kaas in the 1980s. However, it is important to note that this risk may be increased particularly for older adults with physical disabilities or limitations, individuals in long-term or other institutionalized

Table 3.1 Geriatric sexuality breakdown syndrome (Kaas 1981)

Step	Label	Manifestations
1	Preconditional susceptibility	Identity problems; diminished ego strength; changes in physiological sexual response
2	Dependence on available cues	Cues from sexual taboos and myths of asexual older adults; limited role models
3	Social labeling	"Dirty old man" and "indecent old woman" perceived as their own label if they engage in sexual behavior
4	Sick role	Role of perverse older person; perception of "sick," abnormal sexual desires
5	Learning of behaviors	Verbal disavowal of any sexual desire or activity; decrease in sexual activity
6	Atrophy of skills	Loss of sexual performance skills for sexual excitement and enjoyment
7	Self-labeling/internalization	Self-perception as "dirty old" man or woman; identification and acceptance of asexual status

care, older women without partners (c.f., Hinchliff & Gott, 20008), and any other aging adult who does not fit society's image of the vivacious, physically attractive, upper-middle class, typically White sexy senior. Similarly, the first condition that must be satisfied in order to initiate Geriatric Sexuality Breakdown Syndrome is some kind of susceptibility to negative, societal influences. These factors may include identity problems or diminished ego strength, either through difficult transitions into retirement or a difficult adjustment to a chronic illness. Sometimes a change in sexual physiology (e.g., a reduction in response time or ED related to a medical condition) can preclude the emergence of Geriatric Sexuality Breakdown Syndrome.

In the second step of the syndrome, older adults rely on cues from their environment to rationalize their feelings and behavior. They receive messages from society at large in which older adults are asexual beings either incapable or undeserving of engaging in satisfying sexual activity. Few older adults in their peer group, or those who look like them in the media, are available as role models. In the third and fourth stages, older adults begin to accept agist labels, and regard themselves as a "dirty old man" or "indecent old woman" if they seek out sexual contact or engage in sexual behavior. Society also introduces the role of a "sick elder," where sexual conduct among older adults is not just seen as unusual, it is viewed as perverse and pathological. In the fifth stage of the syndrome, between their own internalized beliefs and real or imagined pressure from their immediate social environment, the older person reports having no interest in sexuality. Such individuals literally talk themselves out of their needs and desires for intimacy and sexual expression. A decrease in the frequency and types of sexual behavior ensues.

In the sixth stage of this breakdown, the underlying skills for sexual enjoyment and arousal become diminished and often lost. The social skills required to maintain sexual relations may wither and become obsolete, and the older person may manifest symptoms of apathy, depression, and even hostility and guilt. In the final stage of the syndrome, the older adult has internalized all of the negative attitudes espoused by society and has reinforced these beliefs with his or her own decline in sexual activity, desire, and pleasure. The older adult has accepted the asexual status and has, in effect, confirmed society's negative and restrictive attitudes. The most critical aspect of intervention is assisting an older person in resisting negative social attitudes and cues.

Jayne

Jayne was a 78-year-old woman who had been divorced for more than 25 years. She was cognitively intact, and her only chronic illness was osteoarthritis. Jayne required a walker to move about, but otherwise was able to drive, shop, tend to her apartment, and prepare her own meals. She had a small cadre of single women and married couples with whom she played bingo and attended church weekly. Jayne had begun psychotherapy during the past 2 months because she had begun to feel depressed about her increasing lack of mobility; she feared that her worsening arthritis might force her to give up driving soon. Jayne was most upset about not driving because it would increase her dependence on others, and because she worried it would severely limit her ability to socialize. When asked about those in her social circle, she admitted that one man from her church expressed interest in taking her to movies, going to a local diner, and taking a cruise. Jayne responded that there was no way she could negotiate a cruise ship, although she would certainly not become involved with this gentleman anyway.

When asked what she did not like about this man, Jayne responded that there was "nothing wrong with him." She described him as funny, vibrant, and active. She remarked that since he also was divorced for many years, she would not feel as though she were in "any type of competition" with his ex-wife. Puzzled by Jayne's resistance against spending time with this man for both fun and companionship, her therapist asked what was keeping her from enjoying this man's company. Jayne responded that it had been "a long time" since she had been with a man, and that she was afraid he would make sexual advances. When asked what exactly she was afraid of, Jayne reported that she was not so concerned about her arthritis inhibiting her sexuality, but was instead preoccupied with "what people would think." "After all," she said, "I am over 70 years old .... What kind of example would I be setting for the children at church? Old women don't prance around holding hands or kissing somebody much less. I don't know what I'm thinking about!"

Jayne clearly maintained a need for companionship and sexual intimacy (e.g., she certainly had given thought to engaging in sexual activities with this man), but felt compelled by unspoken societal taboos to stop the relationship before it even got started. Only after Jayne was able to explore and compare and contrast her own needs and desires with the inaccurate and limiting views of society, as well as her fears about how she would be received by her friends at church, did Jayne have enough confidence to accept a date with this older man. In therapy, she also was open to learning about some of the typical physiological changes that often alter the sexual response with aging. At last report, Jayne's relationship with her boyfriend had blossomed slowly both emotionally and sexually, and Jayne was able to enjoy a renewed sense of love and acceptance of herself.

Attitudes in the Therapeutic Relationship

Regardless of theoretical orientation, virtually all clinicians accept that the quality of the relationship between the patient and practitioner (also identified as the therapeutic alliance, working alliance, therapeutic bond, and helping alliance) represents a common factor in psychotherapy for older as well as younger adults (e.g., Hyer et al. 2004) and prepares the foundation for all therapeutic work that follows. Trust, open communication, positive regard, and empathy are essential elements for the unfolding and working through of issues in the relationship and active participation in homework assignments, role plays, psychoeducation, identification of faulty schemas, analysis of resistance, and any other therapeutic intervention. The quality of the client–therapist relationship is especially important in relation to issues of sexuality and aging. In order to generate such a positive relationship, a number of potential stumbling blocks in terms of both potential countertransference and transference must be recognized and addressed.

Countertransference

Many clinicians have strong and salient emotional reactions to working with older clients, much less to working with older clients who are dealing with sexual issues. As practitioners, our own emotional reactions and attitudes toward our older patients can serve as a source of enhanced insight and a guide to diagnostic assessment, as well as the basis for potential therapeutic impasse or resistance in treatment. Such strong countertransference also must be examined and worked through within the context of individual self-study, psychotherapy, or supervision in order to prevent improperly imposing the clinician's values upon aging clients and engaging in repetitive cycles of unproductive behavior (Hillman 2008a, b; Knight 1996; Malamud 1996). Unfortunately, discussion of this important process is generally limited in the literature (Altschuler and Katz 2010; Watters and Boyd 2009).

One method used successfully to help both novice and expert clinicians, including supervisors, identify and begin to process their countertransferential reactions to aging, and to aging clients is through the use of a sentence-completion exercise developed by Altschuler and Katz (1999, 2002). To be most effective, an individual is asked to allot at least 10 min to complete the task, and the responses are typically shared and processed with at least one other individual, who preferably can serve in a peer consultation or supervisory capacity. This activity also can be employed in small to large groups of lay persons, as long as the group leader engages in appropriate delivery and debriefing, which includes creating an atmosphere of trust and nonjudgment. (See Altschuler and Katz 1999 for specific instructions for group delivery.) Whether completed by one individual or by many in a large group setting, the primary instructions are to allot the appropriate amount of time so that no one feels rushed and to remember that there are no "right" or "wrong" answers.

Attitudinal Exercise #2
Sentence Completion

The following sentence stems include items that can be used to elicit and then examine one's countertransference to aging:
  1. When I am with an older person, I…
  2. To me, old means…
  3. The thing I like best about old people is…
  4. The thing I like least about old people is…
  5. The secret to successful old age is…
  6. When an older person moves into a retirement home…
  7. My grandparents…
  8. When I think of being old I…
  9. Having brothers or sisters makes being responsible for an aging parent…
  10. Older people who have been emotionally abusive in the past…
  11. When I think of an older alcoholic I….
  12. When I think of an extremely self-centered older person…
  13. An older person with a chronic mental illness…

Upon completion, individuals are asked to identify any common themes or stereotypes that emerge from their responses. Countertransference to aging clients can take many forms and some of the more common themes that emerge in response to working with older adults include:

- Fear of aging
- Fear of dying
- Counterphobic reactions to aging (e.g., Aging is so wonderful!)
- Desires to heal or rescue
- Curiosity
- Impatience
- Disgust
- Awe
- Reverence
- Wishes to be parented
- Wishes to be a parent or caretaker
- Desires to rescue
- Exhibitionistic tendencies to appear vigorous and youthful
- Fear of intimacy
- Fear of loss
- Desires to experience history or another halcyon era

It also is typical for clinicians to experience one or any combination of these feelings at any time during the course of treatment. It also can be expected for practitioners to possess ambivalent or contradictory feelings when working with older patients. As noted, the importance of acknowledging and assessing one's countertransference, rather than acting upon it is essential. As we often tell our clients, having a variety of feelings is normal and to be expected, and to some extent beyond our control. However, what we can control is how we process and respond to those feelings. We need to consider whether we act appropriately or inappropriately with our clients when such powerful countertransference arises.

Lorraine

To complicate matters, countertransference to an older patient (as with any patient) can manifest itself both explicitly and implicitly. A therapist in training was working closely with a 76-year-old widow named Lorraine. Lorraine inspired feelings of awe and reverence in her therapist; Lorraine was feisty, outgoing, and determined to enjoy her life in spite of her arthritis, recent loss of her husband, a dwindling income, and rapidly failing eyesight. She always seemed to have a joke, a funny story, or tales of adventures to new places with friends each week. Yet, Lorraine still was able to talk openly about the loss of her husband and her range of new feelings in response to becoming a widow.

Approximately 6 months into therapy, Lorraine had to schedule her weekly appointments at different times in order to accommodate numerous trips to the hospital and to her physician's office for diagnostic tests. The medical team feared that Lorraine's glaucoma was worsening as a result of her high blood pressure. Lorraine's therapist also began to notice some subtle changes in her mental status that brought up a discussion of initiating neuropsychological testing in order to rule out multi-infarct dementia related to her high blood pressure. Lorraine's therapist's primary response was that her brave, outgoing, resilient patient could handle just about anything, and that surely everything would turn out just fine.

Approximately 2 weeks later, Lorraine called her therapist one weekday afternoon and asked, "What happened to you yesterday? I started to get worried." What had happened was that Lorraine's therapist had forgotten to write down their rescheduled appointment time from the previous day and had failed to show up for their session. Lorraine laughed about the incident and said, "Well, that's all right. You know me, I always like to get out of the house." When she discussed it with her supervisor, Lorraine's therapist realized that she had some unconscious motivation for "forgetting" about their appointment. Lorraine's apparent chink in her lively armor forced her therapist to consider that Lorraine, despite her positive mental outlook, was an older woman with serious medical problems.

Lorraine's therapist did not want to acknowledge that her patient's increasing physician visits (1) interfered with the structure and intimacy that they had worked toward for months to foster and maintain and (2) provided a cruel reminder that

Lorraine may not have many years of independent living left as she had once believed. Lorraine's therapist also was able to acknowledge her own feelings of loss of control. She was concerned about losing control over the treatment schedule and, more importantly, losing Lorraine both literally and figuratively if she were to have a more serious stroke.

Lorraine's therapist also was able to admit to herself that Lorraine's sudden change in appearance, from a tanned, fit, and trim older woman who wore tasteful clothing and makeup to a pale, hunched figure, who had little energy to shop for trendy clothes or to see well enough to apply her makeup, quite simply, frightened her. These changes made her think about her own aging. Lorraine's therapist often thought that when she grew old herself, she would choose to be youthful and active, just like Lorraine. Lorraine had been the healthiest patient in her caseload, compared to numerous other older adults who used walkers, had speech impediments, were housebound, or suffered from Alzheimer's disease. Lorraine's therapist had placed her own high expectations of Lorraine and her excellent physical and emotional health above her own patient's need for expression, change, and acceptance.

Supervision was helpful in allowing Lorraine's therapist to accept that her countertransferential feelings were a normal and expected part of the treatment process. She was able to realize that glamorizing Lorraine as a function of her own phobic reaction to aging was preventing a vital discussion of aging, physical decline, and coping skills with her own client when she needed it most. Lorraine's therapist also was able to stop feeling guilty about her own lapse in memory, and to use this incident as a catalyst in their next session to address issues of trust, loss, and change. In other words, acting out on the part of the therapist does not mean that the therapeutic alliance is ruptured permanently. Often, such "mistakes" can be interpreted, addressed, and used as an important tool within the therapeutic relationship.

In their next session, the therapist was able to open a successful dialog with Lorraine about her feelings of fear and loss related to her deteriorating health and independence. The therapist encouraged Lorraine to be upset with her for missing their appointment, and to also consider Lorraine's comment that she "was worried about" her therapist. Lorraine recognized that her initial worry about her therapist represented her own worries about ending treatment because of her own failing health. In response to this revelation, Lorraine remarked that it was a relief to talk about "growing old and fat and unfashionable."

Lorraine observed that her friends and relatives referred to her as the "always young, always fun one" who was never encouraged to discuss her own problems or fears. No one seemed to want to hear about her most recent visit to the eye doctor; they would rather hear stories about the cruise she took last month or the new earrings she just bought. Lorraine's therapist provided her with a unique opportunity to discuss her thoughts and fears about aging, and to devise ways to deal with the resistance to aging she encountered within her own peer group. When Lorraine later moved to a rehabilitation hospital 3 months later, her therapist also felt more prepared and accepting of this change, and the two were able to continue their sessions in this different setting.

Additional Reactions to Sexuality and Aging

When working with an older client for whom sexual issues become the focus of treatment, an even greater variety of feelings can ensue (c.f., Altschuler, and Katz. 2010; Hillman 2008a, b; Hillman and Stricker 2001; Watters and Boyd 2009). These reactions are quite similar to those experienced when working with sexually sensitive issues with younger patients, but these feelings may be magnified by the unusualness of the situation for the therapist. Pervasive agist beliefs that sexuality among older adults is inappropriate or virtually unheard of also lend themselves to more intense and ambivalent reactions. Feelings that older clients also lead them to think about sex among their own parents or grandparents also can pose a challenge for therapists of all ages and level of experience.

The use of a mental imagery exercise by Hammond (1987) allows individuals to examine their own countertransference in a safe and nonjudgmental manner. Please refer to Exercise #3. The use of additional sentence completion items (see Altschuler and Katz 1999, 2002) also allows for an examination of underlying countertransference in relation to sexuality and aging. Please refer to Exercise #4.

Attitudinal Exercise #3

Mental Imagery

Take a comfortable position and read this exercise slowly as if you would be reading a hypnotic suggestion. As soon as you finish, give the first answer that comes to your mind. Ready?

Imagine that you are walking through a park during the daylight hours when you come across two people sitting on a bench. These two people appear to see only one another; they are hugging and kissing and seem to be a perfect, loving couple. When you come closer you can see that they are…Now stop reading and imagine the couple. Take a moment to really picture them before going on to the next part of the exercise.

1. What is the first image that comes to your mind?
2. Are they a young attractive couple, a middle-aged man and women, or two young men, etc.?
3. What would be your reaction if this were an older woman and an older man?
4. What would your reaction be if this were an older woman and a younger man?
5. What if this couple were your parents or grandparents?

Attitudinal Exercise #4

Sentence Completion

Generate responses to the following sentence stems. Write down your first impressions, without any form of self-judgment or censorship.

1. Sex over 65 is…
2. When an older person remarries…
3. A lesbian, gay, bisexual, or transgender older adult is…
4. Sex among people with physical disabilities is…
5. An older adult with HIV is…
6. When people over 65 date…
7. Sex among residents of a nursing home is…

As noted previously, a thoughtful, nonjudgmental examination of one's responses to these exercises is essential in order to confront judgmental or otherwise non-objective attitudes. An examination of previous personal experiences also can be useful. Typical countertransferential reactions that emerge in relation to issues of sexuality and aging include:

1. Embarrassment
2. Avoidance (Old people don't have sex)
3. Disgust or disdain (It's repulsive to even think about)
4. Disgust, accompanied by subsequent feelings of shame and guilt about that initial, negative response
5. Titillation and excitement
6. Voyeuristic tendencies
7. Competitiveness
8. Sexual attraction
9. Denial of sexual attraction
10. Disorientation and confusion
11. Homophobia
12. Intrusive thoughts of parents or grandparents
13. Acute awareness of body image
14. Acute awareness of one's own sexuality
15. Fears of decreased perceptions of attractiveness with age
16. Hostility
17. Infantilizing or treating the older adult like a child

Allen

Many therapists also are shocked when they find themselves sexually attracted to an older client. Poggi and Berland (1985) provide one of the few discussions of this issue in the literature. They describe group therapists' positive reactions to an attractive woman in her 70s. They further note that since these therapists were involved in such work for nearly a decade, this experience probably was novel for them because any sexual feelings for other older patients were categorically repressed and denied. Most clinicians find such sexual attraction to their older patients unsettling and sometimes, even alarming. Once a practitioner realizes that this experience is within the normal range of experience, anxiety about this reaction typically subsides and it becomes easier to interpret the basis of the attraction.

As with younger patients, feelings of sexual attraction often provide useful, diagnostic information. Allen was a soldier during World War II. Before he went off to war, he was married and had two children. After his return from Europe, things began to change for Allen. His marriage dissolved, and he began to drink heavily. As he aged, he divorced his first wife, married another woman, divorced this second wife, became estranged from his children, and then married for the third time. By the time he was 68, he was in the hospital for a bilateral, above-the-knee amputation. Allen had developed diabetes, and between his diabetes, substance abuse, and poor attention to medical care, both of his legs had to be amputated. Soon after this traumatic event, Allen suffered a stroke in which he lost movement on his left side. Since he was left-handed, he struggled to even feed himself. His wife, who lived more than 40 min away from the hospital, found it difficult to take three different bus and train lines to visit him very often.

When his therapist first met Allen, she could not help but be shocked by his appearance. Allen appeared to be a thin man, even when he was in good physical shape. His face was pale and gaunt, and what little hair he had left was thin and pure white. He always lay under covers, propped up by some pillows. The thin blankets left nothing to the imagination; Allen's body appeared to be a short stump, and his left arm hung limp at his side. His therapist felt anxious and uncomfortable when she thought about his missing limbs and felt helpless thinking about being in his position. Allen was quite depressed and expressed wishes to hang himself with the phone cord by his bed. He cried when all of the cords were moved away from the reach of his right hand.

Despite the depth of his depression, Allen was a willing and active participant in therapy. He spoke at length about his service in the war and about his one grandson who lived in London and worked as a journalist. Allen also lamented that all he did all day was watch TV, and that it was like watching everyone else through a big window. No one could see him or hear him on his side. Although it was tempting to fall into Allen's passive-dependent stance, his therapist helped Allen learn to strike up conversation with his nurses, to ask to be moved outside during the day in a karol chair, and to have special remote controls and forks provided for him so that he could feed himself and change the channels easily on his own television. Within a few months, he denied any suicidal ideation and focused on renewing his relationship with his children and grandson. He was not yet able to cope with the ambivalent nature of the relationship with his third wife, however.

Suddenly, his therapist began to notice things about Allen that she had not noticed before. She noticed his bright green eyes, the colors used in the tattoos on his upper arms, and his ability to joke with the nursing staff. One day, Allen asked her to look at a card that his daughter sent to him, and his hand brushed against her as he clumsily handed her the card. For a brief moment, his therapist noticed the warmth of his hand, and for a moment felt physically attracted to Allen. Once the moment had passed, his therapist felt baffled and somewhat bemused. While she previously felt very protective and maternal toward her patient (and had worked hard not to act on those feelings), she began to regard him with more respect and interest once she was able to rid herself of feeling embarrassed about the exchange.

Deciding to investigate this countertransferential reaction further, Allen's therapist decided to broach the subject of sexuality and Allen's relationship to his wife. Allen responded that he had not had sex with his wife since his legs were amputated. He added, "These nurses are wonderful to me and everything, but there are times when I wish I didn't have that tube (catheter) stuck up my there [in my penis]. I guess I'm more used to it now." Allen added that while he did not think he would ever have sex again, "the whole thing," he still wished his wife would come to visit him and sit with him in bed, hold his hand, or just kiss him. While he previously expressed remorse and sadness about his condition, and felt that his wife had every reason to feel revolted by him, he began to assert his own needs and wants and to express anger toward her. "In sickness and health, my ass .... Even my own kids are starting to call me and visit me. What about her?"

Apparently, the therapist's attraction to Allen represented an unconscious fulfill-ment of his desire to be seen as attractive by another member of the opposite sex, and a recognition on the part of his therapist that Allen was ready for more insight-oriented work and a greater sense of entitlement. Allen's therapist also gained insight into the relationship between sex, power, and self-esteem. This relationship can be particularly salient for an older adult who feels that his or her body, as well as body image, is distorted or damaged.

Transference

Just as it is vital that clinicians evaluate their own emotional reactions to their elderly patients, elderly patients often have strong conscious and unconscious emotional responses to their typically younger therapists. Because transference is often regarded as an unconscious process, it is important to note that an elderly person who has a therapist 45 years his or her junior may still manifest a transference in which the elderly patient regards the younger therapist as a maternal figure. The age difference between therapist and patient initially may prevent therapists from con-sidering the wide range of transference configurations available in their interaction. In sum, clinicians must remember that transference is not age or reality based; patients and therapists of any age combination can provide the backdrop for any number of transferential interactions.

Eroticized Transference

A particularly difficult transference reaction for clinicians to manage both person-ally and professionally is that of an eroticized transference. Patients may make bla-tant or subtle sexualized comments toward the therapist, or may act out sexual urges in the form of inappropriate attempts to touch their therapist, to sit too close to them, or to give suggestive gifts. Other patients may make inappropriate statements per-taining directly to sexual attributes of the therapist or to wishes for sexual liaisons. Although it is more common for female patients to develop eroticized transferences to their male therapists, male patients also have been reported to develop such eroti-cized transferences to their female therapists (Lester 1985). Regardless of the sex of the patient or therapist, these interactions often are among the most disturbing events cited by both novice and experienced therapists. However, these challenges in ther-apy can lead to the uncovering of significant conflicts and to beneficial changes in the therapeutic relationship and the patient's life. As with younger patients, such sexualized interactions directed by the patient toward the therapist often mask deep-seated patient issues involving interpersonal power, boundary problems, a resistance to change (Blum 1973; Rappaport 1956), and both fears of and needs for intimacy (Peterson et al. 1989) in their own relationships.

Mario

Mario, a 67-year-old retired husband of 43 years and father of three, sought treatment for major depression. He had become so depressed that he lay in bed for most of the day, and it took significant effort on the part of his wife to get him out of bed to eat and shower. Mario remarked that he knew he needed treatment when he realized that for the first time, he could not stand to hear his grandchildren running and playing through the house. His wife also became very frustrated that her husband was no longer interested in attending social functions or going out to restaurants with friends, and insisted that he seek treatment. Mario had always assumed the role of provider, protector, and entertainer. He lamented that before he became depressed, he was "always the center of the party" and that he was always the one to make the toast or command an audience for an impromptu speech. Mario's depression emerged at approximately the same time he retired from his position as head salesman at a local construction company. His depression worsened when he learned that his son was going to declare bankruptcy; Mario was despondent that he did not have the money to get his son out of financial trouble.

During the course of his treatment, Mario worked hard to schedule interesting activities for himself and to become educated about depression. He became less frightened that any negative feeling would grow into an overwhelming, incapacitating experience of depression. He had begun to enjoy his grandchildren and had successfully ventured out on a few evenings for dinners with his wife and other couples. He also had begun to accept that his son had made some bad financial decisions, and that he did not have to feel guilty about not having the money to prevent his son from filing for bankruptcy. Once his depression had begun to lift, Mario began to spend more time in treatment upon the conflicted relationship he had with his wife (i.e., women).

A discussion of his marriage revealed that Mario maintained very traditional sex roles. He expected that his wife cook, clean, and take care of the grandchildren whenever they came to visit. He also expressed anger and concern that his youngest daughter was working outside of the home. When asked what about her job made him so angry, he said that if she were any kind of self-respecting woman and mother, she would stay at home to care for her children and let her husband be the breadwinner in the family. When asked what it was like for him to have a female therapist, who obviously worked outside of the home and who was even a bit younger than his own daughter, Mario responded, "It doesn't matter because you are a professional, and you are here to help me. How you decide to spend your personal time is not up to me."

When his therapist added that it sometimes can be difficult for clients to work closely with a therapist who has different values, Mario retorted, "Well, Doc, it's OK with me. You are helping me out. You have a Ph.D. That is a special deal, for a man or a woman. I don't mind, really." Mario went on further to discuss that he had known one other Ph.D. in his life, and she was "a terrific lady" who was his boss for the last 15 years. He became resistant to discussing the issue further, either from the perspective of sex roles, his own career expectations, or his own competitiveness. Mario's therapist decided that at least she was able to broach the issue even if he were not prepared to discuss it at that particular time.

The next week, Mario canceled his session without notice. The following week, he presented his therapist with a bottle of cheap perfume. (His therapist's policy on gifts was that she would take gifts only if they were inexpensive, and only on the condition that her patient was willing to process the interaction.) Mario's therapist attempted to hide her discomfort and thanked him for the gift. She also asked Mario if he were comfortable discussing it because it could be helpful to him in his therapy, in terms of better understanding their relationship and his relationship with other people in his life. Although Mario had been open to such self-exploration previously, he replied, "But, I don't understand. It's only a gift I got at a wholesale price from my nephew. He's got a job at [a store] and always had a big discount, you know." He mentioned that if his therapist were a man, he would have selected a "man's gift," like a wallet or a coupon toward new tires on a car. Mario denied that the choice of perfume was meaningful, or that his gift could be related to anything that had happened in a prior session, such as mixed feelings over his therapist's credentials, sex, and values.

During the next week's session, Mario focused on his relationship with his old boss, "the other lady with the Ph.D." He mentioned that "even though she was a woman," she could get along with any man, and tell any joke. She also appeared to guide Mario's rise through the ranks before he moved to his current position at a different branch of the company. He began to tell a graphic story about a friend of his who sold "big brassieres." Mario gave his therapist a wink and said, "You ladies know all about them, right, Doc?" Toward the end of the session, Mario's therapist coughed (she had a cold) and Mario retorted, "Hey, Doc, I know CPR. Do you think you need some mouth to mouth?" Mario's therapist began to feel embarrassed and uncomfortable by this inappropriate attention and innuendo. When she attempted to process these points with Mario, and asked him how he thought she might respond to those questions, he changed the subject or denied what he had said.

When the session had concluded, Mario's therapist's response to his inappropriate sexual comments began to change from feelings of discomfort and anxiety to feelings of anger and oppression. She recognized that these advances were not motivated by a desire for intimacy, but were motivated by Mario's conscious or unconscious desire to assert control through the use of sexual domination. He did not have a wish to "save her" from a coughing spell, but instead desired to, in effect, castrate her with his own sexual innuendo in order to overwhelm her and make her feel less like a professional. During the following session, Mario's therapist informed him that it was an important part of his treatment that they discuss, as openly as they could, what had happened during the previous week. Mario's therapist also suggested that if they could understand what was happening between the two of them, it might give Mario some insight into the problems he was having in his own marriage and in his feelings toward his prior boss.

Mario sat quietly for a moment and agreed to attempt to talk about what had happened. He said that he never had an affair with his old boss, but that "it sure had been tempting." He was confused but titillated and excited by his old boss's ability to move easily among both men and women. Mario recounted a time in which he drove her to a formal dinner party, and she wore a beaded gown that fit her "just right in all the right places." He found it hard to believe that someone who could "swear a blue streak with the rest of us" could look that way in a dress. He felt proud to be singled out as her special assistant; she made sure that he was placed on the best jobs with the best overtime rates and chances for advancement.

When asked if he ever wanted to give his boss CPR or mouth-to-mouth, he said emphatically that their relationship was never sexual. However, he added that once or twice, when he drove her to business dinners, she often would ask him to come inside the ladies' room and make sure that no one would come inside while she was changing. When asked how she looked while she was changing, Mario responded, "Oh, she knew and I knew that I would keep my eyes on the door while she was in there changing." Only then did Mario acknowledge that he might have had mixed feelings about his boss. He also was able to see that while unusual requests played into his traditional expectations about sex roles and his desire to protect her and assert his sexuality with her, they were anything but professional. She was, in fact, coming on to him the way he typically imagined a man would come on to a woman.

Mario also recognized that his primary experience with professional women was with this woman, who was very different than his wife, his mother, his friends' wives, and his daughter. His boss, the woman with the Ph.D., seemed to marry power with sex. When asked if his boss reminded him of anyone, he responded sheepishly that he had placed a lot of emphasis on the fact that his therapist had a Ph.D. After gentle questioning, he also admitted that, just like with his boss, he had thought about what it would be like to go on a date, or to go out to dinner. At this point, his therapist had to take a moment to collect herself and to put aside her discomfort at discussing the topic. Mario's therapist reinforced that while their relationship was professional, and would always remain professional, discussing such feelings could be important in helping him understand how he relates to women in general, and to his wife in particular.

Mario said that since it "was OK just to talk about it, then," he would enjoy going out to dinner with his therapist, talking about interesting things, and "showing her off" to his friends. He added quickly, "I wouldn't really even think about having sex with you, I mean, at least on the first few dates. Talking is really important to me." Mario realized that his feelings toward his therapist (which he now said had faded quickly since they talked about them) mirrored those that he had thought about his old boss, but had never really allowed himself to think about. He also recalled that the first Christmas after his boss had asked him to "watch out for her" in the ladies' room, he gave her a small gift of perfume at the office party.

Now that Mario recognized his underlying feelings of sexual attraction for his old boss, coupled with his feelings of impotence about being her subordinate, his eroticized transference toward his therapist dissipated quickly and allowed for a previously buried conflict to surface in its place. Mario was now able to compare his feelings for his old boss and his wife, and process his own ambivalence about having a woman who could both excite him sexually and be in a position of power at the same time. Mario traditionally viewed his wife as his subordinate, and recognized that he could still feel aroused by her. He also recognized that he had fallen victim to the traditional "Madonna/whore conflict" (although his therapist did not discuss this conflict in these specific terms) in which he viewed powerful, freewheeling women such as his boss as sexual beings, and he viewed subordinate, material women such as his wife as nonsexual beings. The enactment of an erotic transference to his female therapist also is suggestive of wishes to merge with a "vengeful, overpowering, phallic mother" (p. 284) as manifested via issues of dominance and submission (Lester 1985).

As the transference was interpreted and dispelled, Mario also felt less guilty about his romantic feelings for his boss because he learned that she actively contributed to the sexualized nature of their relationship. He also was reassured when his therapist pointed out that even in such enticing situations, Mario placed his faithfulness to his wife, his marriage, and his family first in his life; he could experience a variety of feelings without acting on them. Toward the end of his therapy, Mario's depression faded significantly, and he noted that he had begun to enjoy his wife's company more, both in and out of the bedroom. He began to enjoy his retirement more, and became intrigued by the fact that his wife was somewhat like the new "boss" of domestic chores at home. He began to learn to untie his relationship between sex and control, and understand that both he and his wife could be powerful and sexual.

June

June was an 83-year-old white, divorced woman who lived in a nursing home. She had been married and divorced twice, and had no children. She lost her brother and two sisters to cancer 5 years previously. Her only family contact was with a nephew who lived in another state. He visited once in the last 5 years, and would only occasionally telephone June on major holidays. Essentially, June had no family or friends to visit her. She suffered from bipolar disorder, and during her manic periods she could become bitter and cruel to others. Most of her friends appeared to shy away from her once they were aware of what she was capable of doing and saying. June also had physical problems that made it rare for her to get out of bed, and she needed a wheelchair to move around the home.

June was a willing participant in psychotherapy. She had been stabilized on her medication for about a year and was able to engage in meaningful conversation with some degree of introspection. Most of her treatment revolved around accepting her physical limitations and making attempts to develop friendships within the nursing home. June often felt that the other residents were "beneath" her since many of them suffered from cognitive impairment. Other residents' physical deterioration represented an unpleasant reflection of her own narcissistic injury, and she felt uneasy in their presence.

Knowing her history of poor interpersonal relations, even when she was younger and physically healthy, her therapist had guarded expectations about June's ability to have intimate peer relationships. During her treatment, June decided to pursue craft classes and music hours. Her mood brightened and she began to eat in the community dining room instead of her room. (She often ate by herself, but at least made the effort to socialize.) June also began to remark to her therapist about a "wonderful young man who comes to visit." Her therapist was surprised to learn that this visitor was not her nephew, but a janitorial aid who stopped in to talk to June during his night shift. He did not appear to be motivated by money or other favors (June was practically indigent) and visited her at least once a week. June enjoyed these visits very much and said that it was nice to have someone who was smart enough to play cards with her and to talk about things she heard on the news.

Approximately a month later, June became insolent and angry. She threw a food tray at a staff member and refused to come out of her room for hours at a time. When her therapist asked her what was going on, she said that her special friend had not been to visit for a few days. When asked if he might be sick or on vacation, she muttered, "Oh, I think I know what happened." Apparently, June had begun to "joke around" with her visitor. She said she thought it was funny that she began to refer to him as her "boyfriend," and she teased him about his nice eyes and dark hair. She also said that she told him that if she were his age, "I wouldn't let you get past my bedroom." She also admitted that the last time she saw him, she reached over and pinched him on the waist. Without a formal or professional relationship, this kindly visitor probably became overwhelmed by such sexualized responses from an elderly, physically disabled woman, and simply stopped coming by to visit. Her violation of his personal, physical boundaries appeared to be the most disturbing episode. Perhaps understandably, the eroticized transference she projected onto this young man simply was too much for him to tolerate.

June's therapist was sure to check on her mental status; she was not becoming manic nor did she deviate from taking her psychotropic medication. Apparently, June's emerging feelings of intimacy with this visitor and her therapist were too threatening; she had little experience with true intimacy and her underlying personality issues had prevented her from getting close to others in the past. To continue in this trend, she found some way to become too loathsome for her visitor to tolerate. He was accepting of her age and her physical disability; it is as though she had to generate some other means to thwart his friendship. June's therapist was only somewhat successful in guiding her in processing her actions. June maintained that "I'm no spring chicken, but I still find young men very attractive." June's eroticized transference to this visitor represented an attempt to use sex in order to thwart genuine, emotional intimacy.

Guides to Managing the Transference

The first step in managing an eroticized transference is to recognize that it is diagnostic and acts as an indicator that either consciously or unconsciously, enough trust and rapport have been established in the therapeutic relationship to allow it to enfold (Peterson et al. 1989). The second step is for the therapist to accept that any emotional reaction to such a transference, whether it is anger, embarrassment, arousal, confusion, oppression, pleasure, anxiety, or any other emotion, is entirely normal. In short, clinicians must be prepared for an onslaught of powerful and often disconcerting feelings. Clinicians also must remember that while it is normal to experience any of these emotions, they must proceed carefully in order to avoid acting on them.

The third step in managing the transference is often the most difficult for clinicians. The patient's fantasies about the therapist must be discussed at length, in order for appropriate reality testing to take place (Rappaport 1956). Patients should be encouraged to recognize that this sexualized pattern of relating that they have created with their therapist parallels an important relationship or conflict in their own lives. In discussing such an eroticized transference, both patients and clinicians are reassured by clear, unambiguous messages that the relationship between them is, always will be, and must always remain purely professional. Talking about feelings will always remain separate from acting on them in the therapeutic relationship, and the therapist is well prepared to handle any feelings that the patient brings to the session (also see Rappaport 1956). Patients are often comforted by such demonstrative statements that appropriate boundaries will be maintained, and that firm limits are in place for what is considered safe, acceptable behavior.

Ultimately, patients recognize that sexual interactions do not necessarily arise from benevolent, loving feelings; they may indeed underlie aggressive, hostile feelings or intentions (Blum 1973; Lester 1985; Rappaport 1956). When the therapist can maintain neutrality and objectivity, patients begin to learn that their illusions within the therapeutic relationship reflect real, but often alterable conflicts within their own interpersonal relationships. As difficult as it is for therapists to manage these eroticized transferences, they often provide one of the best means of assessing our patients' underlying character structure and core conflicts.

Summary

Attitudes toward sexuality and aging play a powerful and important role in the sexual identity and expression of both middle-aged and older adults. The attitudes maintained by society at large, aging adults themselves, and the health care providers who treat them are equally important items for exploration. Self-report measures such as the attitude subscale of the ASKAS (White 1982), participation in the IAT (Greenwald et al. 2003), participation in specialized exercises in guided imagery (Hammond 1987), and sentence completion (e.g., Holaday et al. 2000)

can be used to examine both explicit and implicit attitudes about aging, and sexuality and aging. These exercises can also initiate an essential examination of a clinician's potentially underlying, undisclosed thoughts and feelings, as the therapeutic relationship also can be rife with powerful countertransference and transference issues.

Clinicians must be aware of their own, potentially sexualized feelings toward their older patients and also be aware of the presence of any sexualized attitudes directed at them by their patients. Although these sexualized attitudes are one of the most difficult subjects to broach, an analysis of such transference and countertransference can provide some of the most valuable information regarding older patients' attitudes toward sexuality, their sense of intrinsic value, sex-role conflicts, and difficulties with or needs for intimacy. As in work with younger patients, having a variety of countertransferential feelings is acceptable and typically unavoidable. It is how we decide to process and act upon these feelings that makes the therapeutic relationship so beneficial.

Older adults who seek treatment for sexual issues, or for whom sexual issues emerge in treatment, often need reassurance from their health care provider that talking about any feeling in relation to this topic is welcomed and permitted. Because society provides a closed door in relation to sexuality and aging, we as clinicians must ensure that we offer the proverbial open door to an objective and thoughtful discussion of such issues. Clinicians have the unique opportunity to provide a corrective educational and emotional experience for aging clients, as well as the opportunity for personal examination and growth.

Chapter 4
Sexuality in Long-Term Care

A consistent challenge in long-term care is the need to protect
an individual resident's rights to privacy and sexual expression
with the duty to protect residents from harm.

Acknowledging the sexuality of older adults in institutional settings is vital. In the USA alone, more than one million older adults live in nursing homes (Kasper and O'Malley 2007) and an additional million older adults live in assisted-living facilities (National Center for Assisted Living 2009). By 2040, more than 81 million older adults are expected to suffer from dementia (Ferri et al. 2005), and many of these individuals will require care in nursing homes or assisted-living facilities. More than 120,000 adults over the age of 50, generally regarded as older or geriatric prisoners, are incarcerated in a variety of state and federal forensic settings (Harrison and Beck 2004). These numbers of institutionalized older adults in the USA are only expected to increase with the burgeoning older population.

The issues surrounding the expression of sexuality in institutional settings are complex, especially for older adults. The presence of sexual activity among aging residents in nursing homes, assisted-living facilities, and even jails may come as a surprise to some, and mental health providers often find themselves in the role of educator, arbitrator, legal advisor, support group leader, and case manager. Clinicians working within nursing homes often find that they provide as much education to staff members about sexuality and aging as they do actually working therapeutically with older residents themselves. Clinicians must take into account issues of competence, prejudice, and medical, legal, ethical, familial, religious, and practical concerns to make appropriate recommendations.

Certainly, an overarching theme regarding sexuality within an institutional context is that of stigma and discrimination, in which older adults, particularly those who may suffer from dementia or who are wheelchair or bed bound, are assumed and often expected to be asexual. In fact, the majority of nursing home residents who responded in a recent survey report that sex is at least "moderately important" in their lives (Aizenberg et al. 2002), and significant value was placed upon a variety of activities ranging from touching and kissing to masturbation and intercourse

J. Hillman, *Sexuality and Aging: Clinical Perspectives*,
DOI 10.1007/978-1-4614-3399-6_4, © Springer Science+Business Media New York 2012

(Ginsberg et al. 2005). Because nursing homes are one of the institutions believed to be most devoid of human contact and sensuality, psychologists and other mental health care providers are in a unique position to foster appropriate expression in an often otherwise sterile environment.

Primary considerations in the expression of sexuality in long-term care and other institutional settings are that of privacy, capacity for sexual consent, and competence among aging residents, attitudes among staff and adult family members, and policies of the institutions themselves. Fortunately, the American Psychological and the American Bar Associations now provide some guidelines regarding the assessment of "sexual competence" among older institutionalized residents (2008). This chapter is designed to review issues of privacy and attitudes toward sexuality among various individuals including staff members, psychologists, and other mental health care providers as well as family members and older residents themselves. Additional information will be provided regarding policy considerations for residents' rights and the assessment of sexual consent capacity.

Various case examples will be offered to illustrate some issues commonly encountered in long-term care settings related to sexuality among cognitively intact residents including room sharing, dating, masturbation, pornography, and the impact of conditions including stroke and urinary tract infections. Recommendations for the management of clearly inappropriate resident behavior (e.g., sexual abuse of non-consenting or non-competent residents or visitors; masturbation in public area; inappropriate sexual comments and behavior toward staff) will be addressed. Other topics to be reviewed include benchmark programs and policies that may lead to a more ethical balance of residents' individual rights to privacy and sexual expression with the rights and safety of other residents and staff. Additional concerns to be addressed in long-term care include many LGBT elders' concerns about placement, in which they fear significant disrespect and discrimination if their sexual orientation is revealed (e.g., MetLife 2010). The influence of adult children and spouses upon their institutionalized family members will also be discussed.

Differences Among Settings

It is important to differentiate among nursing homes, assisted-living facilities, and forensic settings in terms of their resident and staff characteristics, conditions of privacy, and federal and state oversight. According to a national study sponsored by the Kaiser Family Foundation (Kasper and O'Malley 2007), the average nursing home resident is a non-ambulatory 86-year-old woman who requires assistance or supervision with at least five activities of daily living (ADLs) including bathing, dressing, and eating. Nearly 70% of all residents, including those who are newly admitted, suffer from some combination of heart disease, stroke, or hip fracture. More than three quarters of nursing home residents are women, the majority of whom are widowed. However, it also is important to note that nearly half of all nursing home residents are under the age of 65, and that nearly 40% of all nursing home residents are ambulatory, without cognitive impairment.

Many lay people assume that nursing home residents are severely medically ill, requiring round the clock care, and that assisted-living residents are generally healthy, with their ability to attend to their ADLs essentially intact. Even though assisted-living facilities do tend to employ a social model to care, whereas nursing homes typically employ a medical model (Frankowski and Clark 2009), up to 70% of residents in assisted living are estimated to have some form of dementia (Rosenblatt et al. 2004; Spillman and Black 2006) and heart disease (Hyde 2001). Nearly an additional 33% and 15% of assisted-living residents are estimated to suffer from arthritis and diabetes, respectively (Hyde 2001). The average resident in assisted living is an ambulatory, widowed, or single 85-year-old woman who needs assistance with at least two ADLs, most likely bathing and dressing. In other words, many residents in assisted living have similar health care needs as those in nursing homes care. In fact, after an average stay of approximately 2 years, more than half of all assisted-living residents move into nursing homes care. The majority of assisted-living residents (74%) are women (National Center for Assisted Living 2009).

Although the federally enforced Nursing Home Bill of Rights (see Medicare 2008) dictates that residents have the right to dignity, privacy, and to keep and use personal belongings as long as they doesn't interfere with the rights, health, or safety of others, as well as to be free from the use of physical and chemical restraints unless prescribed by a physician, many residents share a room due to space or financial constraints and must keep their doors unlocked as residents are frequently checked upon by staff every 1 or 2 h, even at night time. Although nursing home staff are instructed to knock and seek permission before entering a resident's room, this process may be cursory. Structurally, few nursing homes provide private spaces for residents beyond individual or semi-private rooms, and most areas for socializing are established lounges or the cafeteria.

The range of sexual behaviors manifested by older residents in nursing homes, assisted living, and other long-term care institutions can vary widely. The expression of sexual feelings and desires may range from more subtle and often publically acceptable expressions such as taking evening walks, holding hands, and hugging to more overt and private expressions such as kissing, petting, fondling, individual or mutual masturbation, and intercourse. Along this continuum of behaviors, it is notable that these more overt expressions of sexuality, in which personal boundaries are often traversed among consenting adults, typically require an element of privacy, something in short supply in many institutional settings.

Although the US Nursing Home Bill of Rights (Medicare 2008) dictates that a non-resident spouse should be able to visit his or her partner privately in the resident's own room, and that married couples who reside in the facility should be able to share a room unless medically contraindicated, some facilities claim that they do not have the space to provide these requisite accommodations. It also is notable that these particular rights for private visitation and living arrangements extend only to legally married, ostensibly heterosexual couples. Many administrators of these facilities appear to differ significantly in the criteria they use to determine if it is medically unsafe for an older resident to engage in sexual behavior. In other words, even though nursing home residents may be granted various federal rights regarding privacy, particularly in relation to sexual expression, it remains unclear to what extent these rights are protected or exercised in individual institutions.

Similar challenges to residents' privacy emerge from studies of assisted-living facilities. Although The National Center for Assisted Living formally advocates that residents' rights should include the right to privacy, to be treated with dignity and respect, to retain and use personal possessions (which ostensibly include pornographic material, vibrators, and other devices), and to interact freely with others both within the assisted-living residence and in the community (2008), no federal laws have been passed to establish such a nationally recognized bill of rights. Currently, the rights of residents in assisted living are determined by individual states.

Although residents of assisted-living communities typically do not require around the clock medical care, impediments to residents' privacy (similar to those in nursing home settings) can be readily identified. Assisted living requires some degree of structure, oversight, rules, and monitoring (Frankowski and Clark 2009). For example, many residents have roommates, and many communities have unlocked door policies and engage in regular room checks by staff members (Eckert et al. 2009; Morgan 2009).

The rights of residents in long-term care settings can also differ significantly by country. For example the 2007 Canadian Residents' Bill of Rights in long-term care states that each resident has the right to share their room with another resident if they both so desire, as long as accommodations are available (Ministry of Long Term Care 2011). In other words, Canadian nursing homes are expected to allow couples other than legally married ones to live together in the facility. Such couples could include married couples, same-sex committed partners, other same- and opposite-sex couples who may live together without being married, and newly established couples including those who begin their relationship upon entry to the nursing home itself. In some nursing homes in Europe, residents are able to order X-rated movies and even visits from prostitutes (Engber 2007). In Denmark, nursing supervisors from all 98 municipalities in the country are allowed to solicit "call girls" for nursing home residents if they make sexually inappropriate requests of staff and have no appropriate, consenting partner available (Wienberg 2008). Compare that institutional response to those documented in the USA long-term care facilities in which physicians have been known to order "cold showers" (Engber 2007) or sedating prescription medications for residents making similar sexual requests.

Of all long-term care residents, prisoners have the least amount of privacy afforded in any of these aforementioned institutions. Some states in the USA have begun to establish separate facilities for their growing numbers of older prisoners, typically regarded as those over the age of 50. (Prisoners are considered geriatric or elderly at the age of 50 due to their typically poor health and shortened life expectancy, probably due in part to living under conditions of high levels of emotional and environmental stress; Kemper 2003). Some potential benefits identified for the age-specific housing of older prisoners include cost savings and the ability to protect older prisoners from assaults by younger, aggressive inmates (Abner 2006). However, virtually no information is available regarding official regulations or policies regarding privacy for masturbation or consensual sexual activity (e.g., conjugal visits) among older incarcerated individuals. There also is limited information regarding the prevalence of sexual abuse and STDs among this older prisoner population.

Typically Negative Countertransference

Many clinicians report having an aversion to working with institutionalized older adults, for a variety of reasons. Practitioners and professional organizations have described the following challenges to involvement with clients in long-term care settings (e.g., American Psychological Association 2011b; Council on Social Work Education 2011; James and Haley 1995):

- Confusion regarding the intersection among medical problems, medical treatment, and psychiatric care
- Discomfort at working with medically ill older adults
- Difficulties in navigating sometimes cryptic administrative and policy issues
- Antagonistic relationships between staff psychiatrists, internists, and nursing personnel
- Expectations that older adults have no hope for improvement
- Fears of becoming emotionally attached to clients who may die
- Fears of becoming depressed
- Low levels of reimbursement from federally and state-funded insurance plans including Medicare and Medicaid
- Difficulties in completing the paperwork for such insurance plans and frequent delays in receiving payment

It is notable that the majority of mental health providers do not even cite concerns regarding the expression of sexuality among older adults as a deterrent to working in an institutional setting. Instead, concerns about working with older adults, in general, appear to be superseded by concerns about the aging process, and feelings of hopelessness and expectations for inevitable decline. Consistent with negative societal expectations about sexuality and aging, thoughts of sexual expression among older adults within an institution, and particularly those with dementia or in wheelchairs, generally appear to be an afterthought, or even an implausibility.

For those who do work closely with institutionalized older adults, however, many nursing home staff members, psychologists, psychiatrists, social workers, and occupational and physical therapists, among others, find that their work with older residents is among both the most demanding and rewarding work that they have ever performed. These professional caregivers learn that older adults do have outlets for growth and change, and they often come to accept the physical decline that may accompany their residents' care. However, these same staff members also report that incidents regarding sexuality and aging are more common than uncommon in their workday, and that they represent some of the most disturbing problem behaviors that they encounter on the job (Onishi et al. 2006). In most accounts, these egregious "problem behaviors" include those in which residents engage or attempt to engage in sexual activity with, or expose themselves to, non-consenting individuals including both residents and staff members.

Assessing Attitudes

The attitude section of the Adult Sexual Knowledge and Attitudes Scale (ASKAS; White 1982) offers a self-report measure that easily allows clinicians, health care providers, adult children of older adults, and older adults themselves to assess their attitudes toward sexuality and aging, particularly within the context of an institutional setting. It is comprised of 26 items that can be answered using a seven-point Likert scale. The scale also can be useful in generating discussion about potentially sensitive sexual issues among older adults, older adult couples, and among older adults and their family members. The scale is more than 20 years old, and it remains as a primary measure used in empirical research, demonstrating appropriate levels of reliability and validity (e.g., Bouman et al. 2006; Snyder and Zweig 2010). It also is important to note that the ASKAS has some limitations. For example, it only addresses attitudes toward heterosexual activity among older adults. Currently, no empirically validated measures are available regarding attitudes about LGBT sexuality within the context of aging.

A number of topics are addressed in the attitudes scale of the ASKAS including sexuality in institutional settings, sexual mores, and masturbation. The items from the ASKAS are presented below. Persons formally taking the test are instructed to use the following scale to evaluate each question:

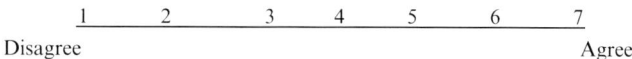

Otherwise, persons taking the subtest are simply advised to think about each question and consider their initial reaction.

1. Elderly people have little interest in sexuality (elderly or aged = 65+ years of age).
2. An aged person who shows sexual interest brings disgrace to himself or herself.
3. Institutions, such as nursing homes, ought not to encourage or support sexual activity of any sort in its residences.
4. Male and female residents of nursing homes ought to live on separate floors or separate wings of the nursing home.
5. Nursing homes have no obligation to provide adequate privacy for residents who desire to be alone, either by themselves or as a couple.
6. As one becomes older (say past 65) interest in sexuality inevitably disappears. For items 7–9: If a relative of mine, living in a nursing home, were to have a sexual relationship with another resident I would ..
7. Complain to the management.
8. Move my relative from this institution.
9. Stay out of it as it is not my concern.
10. If I knew that a particular nursing home permitted and supported sexual activity in residents who desired it, I would not place a relative in that nursing home.
11. It is immoral for older persons to engage in recreational sex.
12. I would like to know more about the changes in sexual functioning in older years.

13. I feel I have all I need to know about sexuality in the aged.
14. I would complain to the management if I knew of sexual activity between any residents of a nursing home.
15. I would support sex education courses for aged residents of nursing homes.
16. I would support sex education courses for the staff of nursing homes.
17. Masturbation is an acceptable sexual activity for older males.
18. Masturbation is an acceptable sexual activity for older females.
19. Institutions such as nursing homes ought to provide large enough beds for couples who desire to sleep together.
20. Staff of nursing homes ought to be trained or educated with regard to sexuality in the aged and disabled.
21. Residents of nursing homes ought not to engage in sexual activity of any sort.
22. Institutions such as nursing homes should provide opportunities for the social interaction of men and women.
23. Masturbation is harmful and ought to be avoided.
24. Institutions such as nursing homes should provide privacy to allow residents to engage in sexual behavior without fear of intrusion or observation.
25. If a family member objects to a widowed relative engaging in sexual relations with another resident of a nursing home, it is the obligation of the management and staff to make certain that such sexual activity is prevented.
26. Sexual relations outside the context of marriage are always wrong.
    A series of additional attitudinal questions have been employed in conjunction with the attitude subscale of the ASKAS (Hillman and Stricker 1996). These were designed to spur additional interest and conversation among its test takers. These few, additional items are to be addressed with the same seven-point Likert scale employed by the ASKAS.
27. If I found out that an older relative of mine were engaging in sexual activity, I would be delighted.
28. It is easy for me to imagine elderly couples in loving, romantic sexual relationships.
29. Masturbation among older people without partners should be encouraged, so that they have a sexual outlet.
30. Sexuality is a wonderful thing to express at any age.
31. Physicians should talk to older people about sexual relations on a regular basis.
32. I would feel very comfortable having a conversation about elderly sexuality.

The Relationship Between Knowledge and Attitudes

The attitude subscale of the ASKAS has provided the majority of empirical data available regarding individuals' attitudes toward sexuality and aging. General research findings suggest that health care providers possess moderately more permissive than restrictive attitudes (Bouman et al. 2006; Snyder and Zweig 2010). However, responses to case studies about individual residents' sexual "problem behaviors," as well as analyses of individual attitudinal items addressing specific behaviors such as masturbation or sexual activity between same-sex residents have elicited significantly more negative and restrictive responses (Hillman and Stricker 1996; Hinrichs and Vacha-Haase 2010; Wallace 1992). It is as though health care providers and other individuals intend to profess more positive and permissive attitudes until they are forced to deal with specific incidents that evoke anxiety and other unpleasant responses within the context of therapeutic, caregiving, or family relationships.

For example, it is one thing to talk abstractly about an older resident's sexual behavior with a colleague, and quite another to speak in detail, at length, with a wheelchair-bound 83-year-old woman who wants to learn how to masturbate to orgasm and who needs help buying a vibrator, or to walk into a resident's room (after first knocking and obtaining permission) to see a 67-year-old man with dementia lying on his bed with his pants around his ankles, masturbating vigorously while leering, "That's it, Honey, now come and get it!" Nursing home staff members may cope with such incidents daily in a variety of ways, including both professional and unprofessional. It often becomes the role of a mental health practitioner to help guide staff members in their management of such situations, along with the negative countertransference that typically results. Psychologists and other mental health providers in nursing homes often find that they provide as much "treatment" and support for staff members as for residents themselves when issues of sexuality are involved.

It also is important to note that many studies show that an individual's knowledge of sexuality and aging is generally unrelated to their attitudes toward sexuality and aging, particularly within an institutional context (Hillman and Stricker 1994). Although greater knowledge of issues associated with aging in general have been linked with more positive attitudes toward sexuality and aging, greater knowledge of sexuality and aging, even among medical students, was not associated with more positive or permissive attitudes toward sexuality and aging (Snyder and Zweig 2010). Among nursing home and assisted-living nursing staff, more years of experience working with older adults was associated with more positive and permissive attitudes toward sexuality and aging. There is evidence that training in specific aspects of sexuality and aging, including information about how to differentiate between appropriate and inappropriate sexual activity, can lead to more positive staff attitudes (Walker and Harrington 2002). One finding from the literature is that helping staff to view long-term care residents as impaired older adults, rather than as children, is essential to help residents maintain both their dignity and more appropriate, neutral or accepting attitudes toward their sexual rights and needs (Parker 2006).

Richard

Richard was an 81-year-old man living in a nursing home. He suffered from Alzheimer's dementia and was in the final stages of the illness. Richard required assistance in feeding, dressing, bathing, and toileting himself. He was no longer able to speak or to move easily about the unit. Richard's weight had dwindled to nearly 110 pounds, though he stood more than 6 feet tall. Although he did not receive individual psychotherapy because of his severely impaired mental status, the unit psychologist encouraged all staff members to actively involve Richard in occupational, music, and pet therapy. When engaged directly in such activities, Richard displayed a gentle character, and would sometimes smile and make gurgling noises.

When not engaged in planned activities, Richard spent most of his time during the day in the third-floor lounge. He often was placed in a karol chair (i.e., essentially a modified recliner on wheels, with an attached lap table) to help maintain his posture. The staff members took care to have Richard sit next to other male residents who also had dementia. Even though they could not talk to each other, these men would sometimes look at each other, reach out to tap each other's arms, and smile at one another. The staff reported that Richard was an easygoing man who was easy to provide care for. He was widowed, and his four children lived out of state. He generally had no visitors. Although the staff psychologist did not know Richard well, she was aware of his cognitive status and conducted periodic assessments with him.

When walking through the unit to have a therapy session with a resident who was bed bound, the psychologist heard an announcement over the public announcement system, "We have a naked man here in the lounge and we need help....Mr. N. is taking his pants off. [laughter] We need a nursing assistant right now!" The psychologist rounded the corner in time to see the unit receptionist buckled over in laughter, two of the nurses rushing to Richard's aid, and one of the occupational therapists shading his eyes and walking back to his office. Richard was standing up in front of his karol chair, with his pajama bottoms around his knees. (He was not wearing any underwear.) Richard's arms were shaking from the effort of supporting himself, and there was concern that he would fall. The psychologist grabbed the microphone from the receptionist to stop her from broadcasting over the public address system. Within a few minutes, Richard was helped back into his pajama bottoms, and was taken to the bathroom to see if he needed to go, or to see if he had some kind of rash or other previously unnoticed condition that could be causing him discomfort.

Although Richard appeared to be unscathed from this incident, a number of residents and staff members were noticeably disturbed. The receptionist was speaking loudly enough for other residents in the hallways and in adjacent rooms to hear, "I didn't want to see Richard's thing! Put that nasty thing away. He is one wrinkly, old dude .... Oh, my God! That was so disgusting." Residents who were moving about the unit appeared embarrassed to see another resident's genitals. Some seemed to laugh off the incident, while others hung their heads and slowly walked away or to their rooms. The primary concern for the psychologist was that the other, observant residents were thinking that if they were to lose control of their faculties, would they be treated with such disrespect and disdain by the people whom they depended on to care for them? Regarding the residents' concerns, the psychologist held a support group (e.g., Hodson and Skeen 1994) a few days later in which issues of privacy, sexuality, and expectations for care were discussed. Individual residents also were approached after the incident to ask them about what they saw and how they interpreted both Richard's behavior and staff members' responses to it.

After this incident was rectified and documented, the psychologist was able to speak informally with the receptionist and members of the nursing staff who were present during the incident. The receptionist was informed gently but directly that it is essential that residents are never laughed at or humiliated, particularly in front of other residents. The receptionist also was provided with emotional support and was able to admit that she was so shocked and upset about seeing this older man naked that her first reaction was to laugh and make a joke out of it. She seemed to accept that although humor can be a good way to dispel tension and discomfort, it can never be expressed at the personal expense of a resident (Bauer 1999). The receptionist was helped to devise a plan that if such an event were to happen again, she would immediately, and privately, seek out the assistance of a nurse or other staff member involved in primary care. The psychologist and nursing staff also were able to counsel the receptionist that she could expect to see more resident nudity, even if that was not in her job description. In sum, the use of a team approach, along with psychoeducation, led to the most effective resolution of a potentially serious problem.

Gregory

Gregory was a 71-year-old single man who lived in and out of a nursing home for rehabilitative needs after several heart surgeries and fractures from falls. He was moderately depressed and also presented with dependent personality disorder. Gregory had been active in hand-to-hand combat during World War II and lamented that he was now forced to use a walker to get around. He also was no longer allowed to drive because of his glaucoma and slowed reflexes. Despite these physical limitations and insults to his previous independence, Gregory faithfully attended a day hospital program on the nursing home grounds. He enjoyed talking and joking with the other residents and learned how to paint and to work with ceramics. He also participated in psychotherapy on a weekly basis in order to help him cope with the changes in his physical health and his subsequent depression.

During therapy, Gregory liked to display his newest works of art. He often made things for other residents in the nursing home who "couldn't get around as well as [he could]." Soon, however, he began to make more and more pieces for a woman whom he was dating. He reluctantly admitted to his therapist that even though he had his own apartment when he was not required to stay at the nursing home for care, he often spent the nights and weekends with his "lady friend" at her own home. He seemed relieved to learn that the psychologist had no intention of judging his behavior, but was more interested in helping him decide how he felt about his relationship with this woman. Gregory disclosed that he was concerned that his girlfriend would abandon him in favor of another man who was more physically fit. He said that even though the physician told him it was OK for him to have intercourse with her, he sometimes felt scared that he would suffer another heart attack. He also found it difficult to discuss these concerns with his girlfriend directly.

Gregory's paranoia grew when he mentioned to his psychologist that he tried to call his girlfriend a few times a day, just to find out what she was doing. His girlfriend appeared to grow tired of this constant monitoring. She told Gregory that she would only talk to him in the evening; she had errands to run and friends to visit during the day. Gregory also revealed that he gave more than half of his pension check to his girlfriend, even when he was unsure if she were being faithful to him. Things began to escalate quickly when Gregory hit another resident at the nursing home. When Gregory was waiting for an elevator, a wheelchair-bound resident with dementia ran into his injured leg. Gregory's immediate response was to strike out and hit the resident in the wheelchair with his cane. A nurse on duty documented the incident, and Gregory was asked to appear before the treatment team for a meeting to discuss his behavior.

During this treatment-team meeting in which Gregory was asked to evaluate his assaultive behavior, the unit psychologist elicited more information from Gregory that he was becoming increasingly paranoid that his girlfriend didn't love him anymore. He became obsessed that she might be having an affair with another man. He was upset that they had planned a weekend trip to the mountains (she was going to pick him up and drive them to the cottage), and that his girlfriend had canceled at the last minute. When Gregory asked her why she didn't want to go, she responded curtly that she was going on vacation, but that she was going to go "with a friend" instead. Gregory was devastated when his girlfriend would not answer his question, "So is it a male or female friend?" It also was unclear to the treatment-team members whether his girlfriend (who also appeared to have her own mental health issues) was invested in maintaining their relationship or not. The treatment team informed Gregory that while it was understandable he was upset, such abusive behavior toward other residents was unacceptable. He was encouraged to vent his feelings through his therapy sessions and his art.

When discussing the issue with the treatment team privately, Gregory's psychologist brought up the difficulties he was having with issues of power, sex, and masculinity and the potential benefits of couples counseling. Other members of the treatment team seemed nonplussed, and remarked that he was lucky to have a girlfriend of any variety at this point in his life. Others remarked that since he was not married, couple's counseling might not be effective. Another team member would not even refer to this woman as Gregory's girlfriend; she insisted on referring to her as his "paramour." It was as though calling the woman his girlfriend would elicit concerns and issues that were too realistic and personal. (The term paramour seems to carry additional, negative implications of artificial formality as well as illicit sexual relations within the context of an affair.)

When asked about the implications directly, many staff members avoided the topic and remarked that they were simply concerned for Gregory's health (e.g., "If he had a heart attack and was using a walker, how could he possibly have a sexual relationship?") Despite assurances from the team physician that Gregory could engage in sexual activity, many team members appeared unconvinced. Quite simply, the treatment team members' prejudices and fears were projected onto Gregory. It took a substantial amount of group work for many of the staff to acknowledge their own feelings and to legitimize those of Gregory. One of the team members remained unconvinced that this resident had "sufficient enough of a reason" to be so upset.

Two weeks later, Gregory's girlfriend broke off their relationship, and he was able to receive support from the majority of the team members as he mourned his loss. It had become clear that Gregory was not inappropriately paranoid; his girlfriend admitted to seeing another man while he was recovering in the nursing home. Gregory also continued to receive, both directly and indirectly from a variety of staff members, mixed messages about his right to pursue a sexual relationship. It also is notable that no one discussed safe sex or STD testing with Gregory, even after Gregory shared with the treatment team that his girlfriend had been sleeping with other men.

Sexual Abuse

Although abusive sexual behavior among older adults represents less than 1% of all verified cases in Adult Protective Service investigations (Teaster et al. 2006), the need to help prevent, report, and treat each single case is obviously essential. Although no reports from federal agencies are available, data from individual states' Adult Protective Services programs suggest that the majority or 70% of reported cases of sexual abuse among older adults occurs in nursing homes (Teaster and Roberto 2004). Findings from individual states' Adult Protective Services programs (Teaster and Roberto 2004; Teaster et al. 2007), reviews of substantiated sexual abuse cases among both older male and female residents (Teaster and Roberto 2003), and a national survey of long-term care facilities (Ramsey-Klawsnik et al. 2008) also indicate that the majority or 70% of sexually abused nursing home residents are victimized not by staff, but by other residents.

As defined by the National Center on Elder Abuse (NCEA, 2011), sexual abuse includes non-consensual sexual contact of any kind including unwanted touch, rape, sodomy, coerced nudity, and sexually explicit photography. The NCEA also considers any sexual activity with an individual who is incapable of giving their consent sexual abuse. Additional acts regarded in the literature as sexual elder abuse include sexual harassment, threats of sexual abuse, exhibitionism, voyeurism, coercion to engage in sexual activity, forcing someone to view sexually explicit images or acts, and touching of the genital area not required for nursing care (e.g., Ramsey-Klawsnik 1996).

It also is essential to note that due to cognitive impairment, many long-term care residents are unable to report or describe such sexual abuse, or hesitate to report such abuse out of embarrassment, shame, or fear of retribution (Rosen et al. 2010). Due to these challenges with reporting, coupled with the fact that some states actually do not mandate that professionals such as psychologists report suspected elder abuse (Zeranski and Halgin 2011), the prevalence of all types of elder abuse, as well as sexual abuse, is likely to be significantly underreported. Reports of long-term care staff covering up or minimizing the impact of elder abuse, for even fiscal reasons, also exist (Frankowski and Clark 2009).

Although the majority of perpetrators of elder sexual abuse in long-term care are fellow residents (Ramsey-Klawsnik et al. 2008), other perpetrators include strangers or acquaintances who may visit the facility, staff members and other care providers (who sometimes have undisclosed histories as sex offenders), community-living spouses or partners (e.g., in one case, a husband was caught raping his unconscious wife in her nursing home bed), and even incestuous adult family members (Ramsey-Klawsnik et al. 2007). The majority of nursing home residents who sexually abuse other residents tend to be older men who may or may not have a cognitive impairment. (Inappropriate sexual behavior has been reported in up to 25% of residents with dementia; Gua 2008). It also is important to note that although some perpetrators have a history of pedophilia or other sexual offenses, limited data is available to suggest that this is the only predictive factor (Kennelly et al. 2011).

More than half of all victims of sexual abuse in long-term care are believed to suffer from dementia, require assistance with at least some ADLs, require assistance for ambulation, and are unable to manage their own finances (Teaster and Roberto 2004). Although nearly 95% of all identified victims in sexual abuse cases are women, male residents can and do become sexually victimized as well. Older male victims of sexual abuse share similar characteristics with their female peers; the majority required assistance with ambulation and were unable to manage their finances and nearly half appear to be cognitively impaired (Teaster et al. 2007).

Because many victims of sexual abuse in institutional settings may be unable or unwilling to report the abuse, a number of warning signs can be observed to help alert staff members, family members, and other caregivers to any potential problems. Sample warning signs include (e.g., NCEA 2011):

- Sudden fears of bathing or dressing
- Sudden refusal to socialize or be around other residents
- Fear of physical contact of any kind
- Refusing to eat or drink
- Sucking, biting, or rocking motions
- Cuts or bruising anywhere on the body, and particularly on the genitals or breasts
- Broken bones, particularly wrists, arms, ribs, and hips
- Evidence of torn, missing, or bloody clothing and undergarments
- An increase in aggressive behavior
- Scratching or rubbing of the genitals
- The emergence of hypersexual behavior

The sudden appearance of sexually transmitted infections such as genital warts, gonorrhea, herpes, and HIV, particularly among multiple residents (e.g., Burgess et al. 2000) can present a clear warning sign.

Older female victims of sexual assault face increased risk of harm. When an older female resident is sexually assaulted, greater physical injury is likely to occur than with a younger woman due to changes in estrogen levels and subsequent thinning of the vaginal lining. More than half of the older women who are raped sustain internal injuries after the assault, and more than one-fourth sustain injuries so series that they require surgical repair (Muram et al. 1992). Bruising or swelling of the genital area should never be dismissed as a normal sign of aging. Please see the appendix at the end of the chapter for contact information for professional and national organizations that provide additional information regarding elder abuse (c.f. Zeranski and Halgin 2011).

Edward

Edward was a 73-year-old white man who lived in a long-term care home for the past 10 years. He suffered from vascular dementia, and both his long-term and short-term memory were significantly impaired. Although Edward could ambulate independently, he only spoke in sentence fragments and required assistance for ADLs including dressing and personal care. He was divorced for over 15 years and his three adult children have disowned him. Apparently, Edward had a history of perpetrating child abuse. He reportedly beat his children, and he was arrested in his 40s for exposing himself to a minor. Although it remained unclear what precipitated the breakup of his marriage, staff members assumed that his wife and children grew tired of his inappropriate and abusive behavior. It also remained unclear to what extent Edward engaged in inappropriate sexual behavior that did not lead to his arrest by the authorities. Edward's records also indicated that he never received, or was required to receive, psychological counseling of any kind.

As his dementia progressed, Edward's impulse control diminished significantly. Various staff members experienced or observed Edward's inappropriate behavior in the nursing home. Despite repeated requests not to do so, Edward would sit in the lounge or in the cafeteria and pull his pants down to his thighs, exposing himself to others. When asked to pull his pants up and cover himself, however, he would comply. After a few months of this behavior, Edward began to lower his pants and make inappropriate comments to staff members. He would instruct nursing staff to "touch me" or to "give me blow job." At other times he would look at female staff members and state loudly, "nice tits" or "I want to fuck you." Within a few weeks, his behavior escalated to asking female residents, as well as female staff members, to touch him or kiss him. Staff members were always present when Edward made such inappropriate advances to residents and were able to successfully intercede. It was unclear whether Edward would have taken more aggressive measures to have his requests for sexual activity met if he were unsupervised.

A critical incident occurred during the holiday season when a number of school-children arrived at the nursing home to sing some Christmas carols. When Edward was escorted into the common room to listen to the chorus, Edward pointed at one of the grade school girls and said, "Nice tits. Want some!…Touch me." The alarmed staff removed Edward from the room within seconds. The girl was talking with other friends at the time and, upon questioning, had not heard what Edward said to her. The staff called an emergency team meeting to decide how to assess and manage Edward's inappropriate behavior. Staff members were concerned about the safety of the other residents, and particularly for the safety of children who might visit.

During the team meeting, staff members were shocked to learn that this incident involving Edward's inappropriate comments to the young visitor were preceded by a series of other inappropriate events. Although each event was documented in Edward's file, the nursing home staff members, as a whole, were unaware of the incidents reported by each other, even when they were working during the same shift or when they worked on the same wing. The first order of business in the team meeting was to gather all available information about Edward's inappropriate behavior, and to look for any patterns in its escalation or in its precipitants. Edward's move from exposing himself to making inappropriate sexual comments and requests coincided with a mini-stroke and a further decline in his mental status. Significant time was spent discussing the fact that except for exposing himself, Edward never physically forced himself upon another resident or staff member. Staff members also reported that although Edward could be redirected for a few minutes, his impaired memory prevented him from remembering the prohibition. He could be found to engage in the same problem behavior a few minutes or hours later.

The staff decided that they had to keep lines of communication open to share vital information with one another in the future and to limit Edward's unsupervised activity to a wing of the nursing home that housed only male residents. (He had not initiated any inappropriate sexual behavior with men.) Edward also was given one-on-one observation when he went to the cafeteria or to group events. However, because of the limitations on staff members' already limited time and resources, Edward would sometimes be forced to eat meals in his own wing or private room and would not be able to attend all scheduled group events. All female visitors were informed to remain within designated areas (without revealing any personal information) in order to avoid exposing him to other young children or adults.

Staff members also debated about whether they were more disturbed by this resident's behavior because he was a man and not a woman. Many staff members noted that Edward had not actually touched anyone; he had only made verbal comments and requests. The counterpoint was raised that regardless of his gender, these verbal accosts impinged on the personal boundaries of others. Even though staff members were trained to handle such situations, other residents and visitors certainly were not. Because the best predictor of future behavior is past behavior, Edward's prior criminal record, coupled with his poor impulse control and inability to be redirected, did not bode well for his ability to manage inappropriate sexual impulses in the future. Ultimately, the staff had to make a difficult decision between protecting the comfort and safety of other residents and maintaining Edward's individual freedom and his ability to ambulate freely about the complex. Because Edward was not capable of making the choice to behave appropriately with women, the staff was forced to make that choice for him, but in the least restrictive manner available.

Management of Inappropriate Behavior

Just as it is important to remain open minded about institutionalized older adults' abilities to engage in satisfying sexual behaviors and relationships, it is equally important to stand vigilant and recognize that situations can and do arise in which sexual behavior can be inappropriate and potentially dangerous to older residents (Hodson and Skeen 1994) as well as to staff members and visitors to the institution. The impact of debilitating dementias such as Alzheimer's disease, vascular dementia, and Huntington's Chorea; psychotic disorders such as schizophrenia and bipolar disorder (whose symptoms may become less responsive to medication with age); and entrenched personality disorders can be associated with decreased impulse control and the emergence of inappropriate sexual behaviors.

It often becomes the job of the psychologist, social workers, or other mental health care professional, either as an on-site staff member or outside consultant, to ascertain the appropriateness of sexual behavior among older residents. Even when a sexual behavior is clearly inappropriate, it sometimes becomes more difficult to determine how to manage residents' behavior without imposing undue restrictions on their personal freedom. However, various guidelines and recommendations can be offered to help cope with these challenging, if not unusual or atypical, behaviors.

Clinicians and staff members can be advised to:

1. Employ a team approach. Gathering information is essential in order to ascertain the appropriateness of the resident's behavior and to generate hypotheses as to its possible origin or escalation.
2. Value all team members' input equally. Ideally, the team will include everyone in the facility with whom the resident has direct contact, including nurses, nursing aides, psychologists, psychiatrists, social workers, internists, receptionists, volunteers, nutritionists, occupational therapists, physical therapists, and even administrators. It also can be beneficial to involve food service and custodial staff as they can often provide additional, valuable input and assistance.
3. Explore the residents' sexual history (Parker 2006). Be alert for any criminal record or evidence of any potential illegal or deviant behaviors.
4. Determine if underlying stressors such as undiagnosed medical problems (e.g., a urinary tract infection), a perceived loss of independence related to sudden physical decline, the loss of a significant other, or even a room change or change in staffing preceded the emergence of the inappropriate behavior. Sexual acting out often masks older residents' responses to pain, fear, instability, or loss of control.
5. Ensure that the resident's sexual behavior is evaluated on the basis of the actual, observed problem behavior and its effect upon others, and not upon staff members' own fears, insecurities, or religious beliefs. Also make sure that both male and female instigators are evaluated without gender bias. For example, male residents who engage in aggressive inappropriate sexual behaviors are typically viewed as predatory and dangerous, whereas female residents who engage in similar behaviors are often viewed inappropriately and incorrectly as humorous, hapless, or benign.
6. Consider that the residents' behavior may not represent a sexual need. For example, inappropriate touch may simply indicate that the resident wants some form of human contact. A pat on the back, arm, or a visit from a massage therapist, if appropriate, may alleviate some inappropriate behaviors. Sometimes covering a resident with a warm blanket or providing a soft pillow or stuffed animal can provide pleasant tactile sensation. At other times, playing music or allowing the resident to walk or dance with others can either provide a distraction or meet needs for increased sensory input or contact.

7. Set firm limits with both residents and staff. If a resident engages in clearly inappropriate sexual behavior, it should be identified as such, and preventive measures should be taken. Because some staff members may find certain behaviors (e.g., masturbating in a public area) particularly alarming and offensive, whereas others may not be as egregiously offended, treatment team members should confer so that they can take a uniform and united approach to the behavior. Personal feelings should not influence the manner in which a residents' problem behavior is managed. An objective and effective plan for management of specific problem behaviors should already be in place. In other words, all staff members should provide a uniform and consistent standard of care.

8. Inform residents verbally, in a clear, direct, and calm manner when their actions are sexually inappropriate, even if those residents are severely cognitively impaired and unable to communicate verbally (e.g., "Mr. Jones, it is not appropriate to pull your pants down in the cafeteria. Let me help you pull them back up, and let's try to figure out what is going on to help you." In this scenario, the nurse then takes Mr. Jones to his room to see if he needs to use the restroom, if he has some kind of rash or skin irritation, or if he needs privacy to disrobe or masturbate in the privacy of his own room.)

9. It is helpful for staff members involved to concretize and label problem behaviors specifically, and it is helpful for other residents who may witness the behavior. Such a verbal approach lets other residents know that appropriate actions are being taken, that everyone's welfare is acknowledged, that residents with severe impairments are treated fairly and with respect, and that all residents are expected or helped to engage in behaviors within appropriate limits.

10. Consider the safety of other residents and visitors to the facility when making decisions about the management of a resident's problem behavior. If a resident only engages in inappropriate behavior toward members of the opposite sex, for example, limiting unsupervised activity to a same-sex unit, when possible, can be helpful in minimizing occurrences.

11. Use the least restrictive measures possible to prevent residents from engaging in inappropriate sexual behaviors or endangering others. For example, it is preferable to allow severely cognitively impaired, ambulatory residents with a history of repeated attempts to sexually accost other cognitively impaired residents to move freely about a monitored or locked unit instead of restraining them in a karol chair or literally drugging them into submission. Pharmaceutical approaches to behavioral management can be useful, but they should be employed only when all other attempts to manage the behavior have failed (Tsatali et al. 2011).

12. Understand that staff members who provide primary care to residents are likely to already have significant demands placed upon their time. Administrators and supervisors can be asked to provide these essential staff members with flexible scheduling, multiple breaks during the work day, and a varied work routine. Staff psychologists and other mental health providers often find themselves in a unique position to help foster such mutual cooperation and respect.

13. Monitor staff members' responses to residents who engage in sexually inappropriate behavior carefully. Encourage staff to vent their feelings appropriately

in a private forum. Provide emotional and instrumental support, training, and continuing education.

14. Mental health providers and other staff members can be encouraged to maintain their own support system both inside and outside of the institution. Because a staff therapist is likely to maintain multiple roles in the institution, it is vital that these clinicians make it a top priority to monitor and process their own counter-transference to potentially emotionally charged situations.

Antonia

Antonia was a 66-year-old female resident who suffered from paranoid schizophrenia. She had been in and out of state mental hospitals and nursing homes for most of her adult life. Both of her parents had passed away, and her only surviving relative was a niece who visited only irregularly. Antonia's psychotic symptoms also had become more resistant to treatment with psychotropic medications. Her health care providers were becoming anxious that increasing the doses of her medication would result in dangerous side effects and little improvement in her symptoms. Antonia also believed that her physician was the Devil and that her niece was Jesus incarnate. She was ambulatory and required only minimal assistance with ADLs. Antonia had been attending group therapy and music therapy regularly until she became increasingly paranoid that other residents and staff members wanted to steal her possessions and invade her thoughts.

One staff member noticed that Antonia had begun to spend a lot of her free time on the men's wing of the nursing home. Notably, however, she would not sit and talk with other men on the unit. Instead, she would lurk around corners and appear to spy on some of the men who were more disabled than she was. A male resident approached a nursing home staff member and reported that Antonia had been "sitting too close" to him and had been asking "lots of strange questions." The resident refused to elaborate further, but indicated that he did not like her spending so much time on their wing. Alerted to this potentially problematic behavior, the staff member then observed Antonia accosting other men on the unit. On one occasion, she saw Antonia rub her hand briefly against another resident's crotch while he sat in a wheelchair. He appeared shocked and bemused, and she left as quickly as she appeared.

When confronted about her behavior directly, Antonia denied that anything had happened. She was counseled that touching other residents without their permission was unacceptable, and that if she touched another resident she would lose privileges on the unit. Because medication and supportive psychotherapy were not effective in alleviating many of her symptoms, a behavioral approach (i.e., a token economy) was employed to help manage her sexual acting out. Interestingly, members of the treatment team did not find Antonia's behavior as alarming as the behavior of male residents who engaged in similarly inappropriate behavior. It was as though being a woman made Antonia's advances seem more innocent and even humorous.

When staff members were asked to consider how alarmed they would be if this type of behavior were initiated by a male resident toward a female resident, they quickly recognized their gender bias and redoubled their efforts to manage Antonia's sexual acting out. The token economy, in which Antonia received a point for every day she engaged in appropriate behavior, in order to "save up" for snacks from the food cart, led her inappropriate behavior to virtually cease.

Walter

Walter was a 94-year-old suffering from Alzheimer's dementia in its middle stages. He lived in a nursing home with a locked unit for Alzheimer's and dementia residents. (Residents on this unit were at risk for elopement.) Walter's wife had passed away and his children lived out of state, leaving him with infrequent visitors. His closest relationships were with his nurses and other nursing home staff who worked in the locked unit. It was not unusual to hear Walter making sexual comments toward his personal care nurses. For example, when one nurse requested that Walter move to the other side of a table, asking, "Would you mind getting up?" he replied, "I already am!" with a quick wink and a smile. On another occasion he told a staff member, "You should get a tattoo on your lady parts that says, 'Pay as you enter!'" He referred consistently to one caregiver as "legs," and he often asked the nurses if they would go out with him to play cards and have a drink. He was sometimes found masturbating in his room when the door was open, and after noticing if Walter was safe in his bed, staff members simply shut his door and returned to check on him an hour later.

In Walter's case, the nursing homes staff members tolerated his behavior without if negatively affecting his treatment. The staff members knew from experience that any scolding of his behavior would be forgotten within a few hours or even minutes, and that Walter might take their words as a personal attack due to his sometimes irritable nature and sensitivity to criticism (a characteristic of earlier stages of Alzheimer's). Some nurses used humor to help cope with the situation, telling Walter "Oh, my, I appreciate the compliment but we can't get involved since I work here! You will have to do that with some other lucky lady." The workers on the unit knew that Walter was experiencing a variety of confusing and hurtful circumstances in every facet of life, and that the unit was his home, so he deserved to be treated in the most accommodating and respectful way possible. Regular team meetings helped staff members share information about what techniques worked best to help redirect Walter when he engaged in inappropriate sexual behavior in public areas of the facility, and to obtain emotional support. Clearly, an empathetic approach, in which humor is used to dispel tension without denigrating a resident (e.g., Bauer 1999) can work well when dealing with challenging resident behaviors.

Staff often find it helpful to learn to judge and evaluate a resident's behavior, but not the resident. It is essential for health care providers to understand that sexual acting out does not represent a flaw in the individual, but that they often occur as a result of dementia. When residents with moderate and severe dementia engage in uninhibited sexual behavior, their behavior is typically and unfortunately beyond that residents' immediate control (Gua 2008).

Capacity for Sexual Consent

It may be necessary to turn to facility administrators and legal professionals to better delineate the roles that psychiatrists, psychologists, social workers, nursing staff, and adult children and community-living spouses play in making binding decisions regarding the sexuality of institutionalized residents. For example, although residents are endowed with rights to engage in a variety of behaviors, long-term care facilities are simultaneously legally bound to protect their residents from harm. Case law varies from state to state, and among private, public, and Veterans Affairs (military) long-term care institutions. Power of attorney and legal guardianship, which may be held by a family member, state, or federal agency, may apply to various aspects of resident care including medical decisions and financial spending. However, no "surrogate decision maker" (i.e., decisions by proxy) can be assigned in relation to a residents' sexual activity (p. 63). Sexual consent also is different from giving medical consent, as sexual consent must be made *in the moment*, and not in advance, as can be established with medical consent via advanced directives (ABA and APA 2008).

Questions typically arise about whether older male or female residents with cognitive impairment are capable of making their own decisions about whether or not to engage in a variety of sexual activities and situations. (If an older resident has some physical disability but no mental or cognitive decline in function, she or he obviously should be empowered to make their own decisions about participation in sexual activities.) Previous literature on this topic has been limited, with a primary focus upon the sexual consent capacity of individuals living in institutions due to psychiatric illnesses such as schizophrenia or of individuals with mental retardation. Fortunately, the American Psychological Association in conjunction with the American Bar Association have published guidelines regarding the assessment of older adults' competence in relation to medical, financial, testamentary (i.e., the ability to testify in court), independent living, automobile driving, and sexual consent capacity. For purposes of this chapter, focus will be placed upon the determination of capacity for sexual consent (ABA and APA 2008).

Currently, no nationally accepted criteria for sexual consent capacity is available; definitions vary from state to state. Becoming aware of individual states' legislation and case laws becomes essential (Lyden 2007). Generally accepted criteria for sexual consent, however, include that the individual be over the age of 16 to 18, depending upon the state, and an individual's (e.g., ABA and APA 2008; Grisso 2003):

1. *Knowledge*. This criterion involves an individual's knowledge of the sexual behaviors involved, any potential risks associated with the sexual relationship or behavior (e.g., STDs; pregnancy; psychological abuse; physical abuse), illegal sexual activities including rape and coercion, and how to determine whether a

potential partner wants to engage in a sexual activity. Additional risks for older residents include potential rejection or harassment from staff, other residents, and visiting family members, as well as emotional distress if one resident's placement is temporary and the relationship ends abruptly. Another factor related to this criterion is knowledge of appropriate times and locations for sexual activity (e.g., engaging in sex with the door closed versus in the lobby.) The cognitive abilities typically required to meet this knowledge criterion include attention, semantic memory, autobiographical or episodic memory, and executive functioning.

2. *Capacity.* This criterion is associated with the ability to make a decision about engaging in a sexual behavior based upon a logical, rational understanding of the situation, the likely consequences of various actions including risks, as well as the ability to make a decision consistent with one's personal values. Care should be taken to acknowledge an older resident's sexual orientation and related values, particularly if they are LGBT. Even if a staff member or community-living family member holds different personal values, the values of the older resident must be respected and upheld. Cognitive abilities that allow an older individual to meet the capacity criterion include attention, verbal comprehension, semantic memory, abstraction, executive function, auditory naming, and autobiographical or episodic memory for related personal experiences, values, and preferences.

3. *Voluntariness.* For this criterion, an individual is able to make a conscious decision about their participation in sexual activity free from coercion, unfair persuasion, inducements, threats, and other inappropriate influences (Moye 2003). Legal definitions of influence also vary across state and individual jurisdictions (Wertheimer 2003). Cognitive skills associated with this criterion include attention, abstraction, executive functioning, and sematic and episodic or autobiographical memory.

Geropsychologists (Lichtenberg and Strzepek 1990) and clinicians familiar with consent among individuals with developmental disabilities (Ames and Samowitz 1995; Stavis and Walker-Hirsch 1999) offer additional questions that can be posed to help evaluate a cognitively impaired resident's capacity to provider sexual consent. These questions and considerations include:

- Is the resident aware of *who* wishes to engage with them in sexual contact? For example, does the resident believe incorrectly that the person who lives next door to them in the assisted-living facility is their deceased spouse or partner?
- Can the resident communicate "no" (either verbally or nonverbally) to unwanted sexual advances, or put limits on their type of sexual activity if desired?
- Can the resident describe how they will feel when the relationship ends, particularly if the other party dies or is suddenly moved to another unit or facility?
- Can the resident identify when another individual says or nonverbally communicates "no" to their sexual advances and is able to stop themselves from pursuing it?
- Can the resident keep others from exploiting them, particularly if they are cognitively impaired?
- Is the resident knowledgeable about how, when, and why to use condoms and lubricants?

The use of carefully documented clinical interviews and observations with the resident, staff interviews, and a review of the resident's social history, medical status, and current medications can guide a clinician in the assessment of a resident's sexual consent capacity. Input from community-living partners and spouses and adult children can also be included, particularly if the resident is unable to communicate. Probably the most important part of the assessment includes the clinical interview and functional assessment. With any psychological evaluation, the clinician should obtain consent from the resident, explaining the purpose of the discussion and clinical assessment. The examiner should be as direct and frank as possible when discussing the specifics of sexual knowledge, sexual activities, and potential resident risks.

Some empirical measures that can be included in the assessment include:

- Mini-mental state exam (MMSE; Folstein et al. 1975). The MMSE serves as a useful thumbnail sketch of a client's mental abilities, and provides an objective measure recognized across a variety of disciplines. Although Lichtenberg and Strzepek (1990) suggest that a MMSE score of 14 or less represents a formal cutoff in which an individual should be considered unable to consent to sexual activity, whereas others may argue that this represents an arbitrary cutoff that can automatically preclude additional investigation of competence, and prevent an older resident from engaging in life fulfilling sexual activities as well as individual activities such as masturbation.

- Assessment of activities of daily living or ADLs (e.g., Lawton's Instrumental Activities of Daily Living Scale; Lawton and Brody 1969). It is helpful to gather information from the resident's significant others as well as staff members to gather an appropriate appraisal of the resident's abilities to care for him or herself. The use of a team approach, and good relationships with primary care workers, will allow for the most accurate assessment.

- Mattis Dementia Rating Scale (DRS; Mattis 1973). The DRS is unique in that it allows for an assessment of different aspects of intellectual and cognitive functioning. It also provides an objective overall score that may be recognized readily among professionals. The caveat with this instrument is that it is best used as an overall screening measure, particularly if a client scores along the cutoff point for dementia.

- Other neuropsychological testing where appropriate (e.g., if the DRS score is at the cutoff point for dementia and little is known about the dementia's origin or rate of progression), such as the California Verbal Learning Test (CVLT). The CVLT can be useful in determining dementia etiology, as well as capacities for short- and long-term memory.

- The Cognistat, formerly the Neurobehavioral Cognitive Status Examination, can be used as a general cognitive measure, and it can help assess five different cognitive skills including orientation, memory, reasoning, and social judgment (e.g., Starratt et al. 1992). The Cognistat also is available in Spanish.

- The WAIS-IV also can be used to help assess a resident's cognitive functioning. The digit span score forward can be used to assess attention, and Trails B can be used to help assess executive function.

- The Geriatric Depression Scale, Short Form (e.g., Sheikh and Yesavage 1986) can help evaluate an older resident's emotional state.
- A mental status examination that includes an assessment of the reality testing, judgment, and orientation to person, place, and time.

In relation to medical issues, underlying, undiagnosed problems such as urinary tract infections can be responsible for impaired mental status among older adults (Lerner et al. 1997). It is vital to separate the effects of long-term decline in mental status caused by dementia or chronic mental illness from the short-term effects of delirium, which often calls for immediate medical intervention. The impact of prescription and over-the-counter medications upon sexual interest and function also must be considered. For example, some medications used in the treatment of Parkinson's disease have been associated with hypersexual behavior (Bostwick et al. 2009).

A critical factor to consider in any evaluation is that the capacity of an older resident to engage in sexual consent is not necessarily dichotomous. Because sexual behavior exists along a continuum, with various activities carrying different risks than others (e.g., holding hands, hugging, and dancing accord different risks than kissing and fondling with clothing on, than petting with clothing, than oral, vaginal, and anal sex) an older resident with cognitive impairment may maintain the ability to consent to certain types of sexual activity but not others. For example, the treatment team may determine that a cognitively impaired male resident may be competent to decide whether he wants to hug and kiss another resident, but not engage in intercourse due to his apparent inability (as identified via clinical interview with the resident and a review of recent incidents in the facility) to recognize when a partner wants him to "stop" at fondling rather than continue on to intercourse.

Emma and Paul

Emma was an 82-year-old resident of a nursing home. She suffered from advanced Alzheimer's dementia. About 2 months into her stay at the nursing home, she began to hold hands and hug a male resident, Paul, who lived on her floor and who suffered from the same disorder. Although Emma and Paul were unable to communicate any longer to anyone, they would shuffle toward each other each morning after breakfast, and sit quietly together in the lobby for a few hours each day. Neither resident attempted to touch other residents or staff members.

The sexual consent capacity of both Emma and Paul were examined by a consulting psychologist, who employed a team approach. Both residents were found to have an impaired capacity to engage in sexual activities such as mutual masturbation and intercourse due to their impaired memory and inability to nonverbally recognize the other partner's cues to stop or protest. However, both residents appeared competent to consent to sexual behaviors such as hugging, dancing, and hand holding. In this particular case, staff members and family members decided that neither resident was at risk for harm as both residents were single, and the behaviors they exhibited in did not pose risk of psychological or physical harm such as STDs. The residents' adult children agreed with staff members and felt strongly that this non-verbal connection was beneficial (e.g., Walz and Blum 1987) for their otherwise socially isolated, nonverbal parents.

Familial Concerns and Influence

Adult children and spouses of institutionalized older residents often become inti-mately involved in their family members' care. Yet, few adult children have ever had to examine their parents' sexuality at any other time in their lives. While serving as legal guardian, situations sometimes arise in which adult children are asked to make difficult decisions about their cognitively impaired older parent's room assignments, to "give permission" for their parent to engage in romantic relationships with other residents, to give their opinions as to whether their parent is competent to make his or her own decisions or not, and to provide any information available about their parents' sexual history or religious convictions. The spouses, often referred to as community spouses, of cognitively impaired, institutionalized residents may also face challenging decisions about whether they are comfortable with their cogni-tively impaired partner engaging in sexual activity with another resident.

Consider the situation in which an older resident who is significantly cognitively impaired seeks out hugging and kissing with another resident who is mildly cogni-tively impaired. Although the interaction at first may appear to staff members as mutual and beneficial for both parties, they realize quickly that the older resident who initiated the relationship is married, with a physically disabled community spouse who lives in a neighboring state. Many adult children and spouses of such residents do the very best they can to manage such difficult emotional issues. Many take advantage of help from peers and professionals to make sometimes more lenient, and sometimes more restrictive, decisions in hopefully the best interest of their family member.

Other adult children can be observed to become overly involved and controlling in their parent's life in the institution, while others opt to remove themselves com-pletely from any decision making and deprive staff members from their typically insightful input. This issue can become particularly problematic within the context of assisted living. Unlike nursing home placements, in which many older residents' stays are funded by federal programs such as Medicaid and Medicare, assisted liv-ing is generally based upon private pay. In many circumstances, particularly if an older resident has some degree of cognitive impairment, adult children may fund their living arrangement.

Administrators of assisted-living facilities may find themselves in situations in which adult children threaten to transfer their parent from the facility if they date or have sexual relations with another resident. Having the resident leave would then cause significant potential financial problems if the bed could not be filled quickly. Some researchers have documented incidents in which administrators and staff made restrictive decisions about a resident's ability to engage in sexual activity and romantic relationships that were clearly based upon the financial interests of the assisted-living facility rather than those of that individual resident (Frankowski and Clark 2009). Discussing matters openly with adult children, and normalizing the discomfort that can arise when discussing an older adult's sexuality can sometimes help adult family members reaching a decision consistent with their relative's ability (and rights) to exercise their own sexual autonomy.

Some common issues can influence the judgment of adult children regarding the sexual activity of their parents in a nursing home facility. These may include:

- Fear that their parent will be taken advantage of emotionally by a new partner. Many adult children project their own fears of loss for their aging parent (through death or decline in mental state) onto the situation, and reason that their mother or father will not be able to withstand the loss of a potential suitor in the immediate future.
- Strong religious views that may or may not be consistent with those of their parent regarding sexual activities. Religious beliefs may be particularly strong regarding sexuality outside of marriage and masturbation. While the religious beliefs held by the resident's children must be respected, it often is helpful to help these adult children base decisions for their parent on their understanding of, and respect for, their *parent's* religious beliefs.
- Financial concerns that if their mother or father becomes involved with and subsequently marries a visitor or resident, they may lose some or all of their inheritance. Adult children also may fear that if their parent is cognitively impaired, he or she may be taken advantage of financially by more cognitively intact suitors.
- Unresolved issues regarding the death of a parent. If the older parent in the nursing home is a widow, the adult child may cling to the belief that the surviving parent should not seek out another relationship out of respect for the marriage vows to the deceased spouse. This is commonly observed when the adult child has not fully mourned the death of the parent or when he or she somehow blames the widowed parent for the other parent's death.
- Personality disorders or deep-seated conflicts in which the adult children may have difficulty establishing their own romantic relationship or may have dissatisfaction in their current romantic relationship. Because of conscious or unconscious feelings of competition, the adult children do not want their elderly parent to "be happy" when they themselves are not.

Managing the conflicts that such adult children bring to the treatment setting can sometimes be as difficult as managing the practical issues involved in their own parent's sexual activities. Adult children often respond well when they are encouraged to participate in long-term planning and to discuss their thoughts and concerns with treatment team members and with other adult children of older residents. Thanking these often reluctant caregivers for their time and participation can allay many of their anxieties and allow these adult children to focus more clearly on the issues at hand.

A clear example of such a situation is that of retired Supreme Court Justice Sandra Day O'Conner. When her husband, John, first developed Alzheimer's disease, he reportedly became restless and depressed when his wife was not near. So, Justice O'Conner began to bring John along with her on part of her 12-h work days. After a driver dropped the couple off in the building's basement and they took a private elevator upstairs, John spent the day in Sandra's chambers, looking at magazines and pictures while O'Conner's clerks would watch over him. The two ate lunch together and John would even sit in on some meetings with her clerks.

In time, John's dementia worsened, and he was no longer content to sit in O'Conner's chambers waiting to see her between briefs, meetings, and cases. He began to wander in the hallways and conference rooms, and the clerks were no longer able to placate him. John also lost the ability to recognize most of his family members, including his wife of more than 50 years. Reluctantly, Justice O'Conner placed her husband in an assisted-living facility. With her heavy work schedule it was difficult to spend much time visiting John, and O'Conner observed her husband becoming more introverted and depressed. O'Conner then retired from her position as Supreme Court Justice to spend more time with her husband.

Much to the surprise of many members of the public, O'Conner described how her husband seemed to rebound from his depression after he started having a romantic affair with another resident, "Kay." Apparently, John and Kay met in the assisted-living facility and began to sit next to each other at meals and in the lobby. They could often be found sitting together, holding hands on the porch swing. Staff members discussed the situation with O'Conner, and she opted to permit John to continue with his new relationship. Her son has commented that Saundra is "thrilled" that John does not complain about living there as a resident and seems happy again (Biskupic 2007). O'Connor is reportedly not jealous of the affair as John no longer has memories of their marriage, nor recognizes O'Conner when he sees her (Zerninke 2007). O'Conner's willingness to discuss her family's situation publicly provides staff members who encounter residents and family members in similar situations with a springboard for discussion.

For family members who must face challenging issues about whether or not to "allow" a resident with dementia to engage in a romantic relationship (particularly when there in a community-living, cognitively intact spouse), some points of discussion that can be included in talks with a mental health provider and the treatment team include (e.g., Anderson 2007):

- It is important to acknowledge the situation, along with the often painful feelings of loss, anger, betrayal, and grief that may result. Validation of the family members' feelings is essential.
- Sometimes it helps to depersonalize the situation, in which the focus becomes the resident and his or her life (and cognitive ability) as it is now, and not their life before entering the institution.
- Recognizing that any relationship that the cognitively impaired resident has within the long-term care facility does not negate or minimize the quality of a previous relationship with the community-based partner or spouse; it may simply be different, and an attempt to regain something that was lost.
- Community-living family members can be encouraged to establish new support systems, friendships, and other relationships in order to help alleviate stress.

- Ask the family members to consider that there may be few options for comfort and happiness for the resident, and that the new relationship may provide a sense of comfort and security.
- Consider that relationships can include a variety of sexualized behaviors, including ones that may appear more benign such as holding hands, hugging, and dancing.
- It is important to help honor the past values and wishes of the resident. For example, if a resident often spoke of strong religious convictions that sex outside of marriage was sinful or that she would never become involved with another person romantically if they lost their partner were to become cognitively impaired, the treatment team and family members may consider that a new sexual relationship outside of marriage may not be in that resident's best interest.

Practical Considerations

Many researchers maintain that nursing homes are devoid of opportunities for sexual expression. Effective comparisons have been made between prisons and their allowances for conjugal visits, and nursing homes and their outright lack of allowances for such visits (Hodson and Skeen 1994). Most believe that these restrictive environments primarily reflect the negative attitudes held by staff members and the oppressive administrative policies enacted simply to make things easier for staff members (McCartney et al. 1987). Others suggest that these problems could be ameliorated simply by instituting privacy rooms (Wallace 1992) or minimizing the frequency of nightly bed checks. Comments also have been made that traditional hospital beds are simply too narrow to permit two people to comfortably engage in sexual activity (Waslow and Loeb 1979). Thus, in an ideal setting, institutions would engender more permissive administrative policies that do not label any sexual behaviors as "problem behaviors," provide double beds for all residents, offer a privacy room for residents to engage in masturbation or intercourse, provide less frequent bed checks for residents who seek privacy in the evenings, and foster interaction between male and female residents. However, it appears that as much as some administrators may be guilty of enacting restrictive policies to simplify their duties and concerns, many gerontologists wish to provide an equally "simple" solution to a very complicated problem.

A number of practical issues detract significantly from the utility of the aforementioned suggestions. Because the need for bed space is increasing rapidly in nursing home settings, many institutions are not equipped to provide double beds for its residents, much less single, private rooms. Many nursing homes offer a combination of accommodations ranging from single rooms and double rooms, to rooms housing four residents. Some administrators provide single rooms to residents who require the most acute care, others provide single rooms to the residents who can afford to pay the associated higher price, and others provide single rooms based on availability and family request. Certainly no national standards or guidelines are set regarding such decisions or policies. Other residents are advised to sleep in the proverbial, narrow hospital bed because they need support from bed rails or from the ability of the bed to incline.

The use of a privacy room poses other challenges for nursing home staff members. One limitation is that of space; one private room must be equipped with a double bed and a locked door. This also represents a space that cannot be used to store supplies, function as a recreation area for all residents, or to serve as living quarters for residents. Although some administrators are opposed to the idea simply on its principle, others fear for the loss of revenue, whereas others fear that the legal liabilities related to the use of the privacy room are simply too great. Some have posed questions such as, "Who would decide who was capable of using the privacy room? How will mutual consent be determined? How long could a person or a couple use the room? What kind of safety risks are posed if the residents are behind a locked door and there is some kind of medical or psychiatric emergency?" More pointedly, many primary care nurses ask, "What if they need help taking their clothes on and off? What if they fall out of bed? Who will be left to clean up after a resident masturbates or has intercourse?…We will." When asked about less frequent bed checks, both administrators and nurses balk for reasons of health and safety. In sum, the ideas proposed are theoretically sound, but are very difficult to put into practice for a variety of practical, as well as some prejudicial, reasons. There appear to be no simple answers.

As noted previously, in some European countries it is considered normal, legal practice to provide prostitutes for nursing home residents. Although documented cases exist in which family members of US nursing home residents have requested escort services for an aging parent, this action would certainly be illegal (Frankowski and Clark 2009). In contrast, long-term care administrators could consider providing such residents with pornographic materials, vibrators, or other sex toys for masturbation. Possession and use of these materials in private would be in accord with the US Nursing Home Bill of Rights. However, practical considerations often arise when staff members note, "What if the resident forgets to put the [pornographic materials] away," "Does this mean I have to change the sheets even more often when I'm already busy?," and "So now I'm going to have to probably put up with even more 'come-ons' from this resident after they get riled up from looking at porn." Similar concerns exist regarding the distribution of Viagra, condoms, and personal lubricants for long-term care residents. Who pays for them? Who is allowed to have them? Who monitors this process, and how can the safety of residents (e.g., from HIV and other STDs; falling out of bed) be maintained? Again, there are no easy answers.

LGBT Issues

Limited empirical data is available regarding the sexuality of older LGBT residents in long-term care. Unfortunately, the available findings suggest that LGBT older adults maintain significant concerns and reservations about living in long-term care facilities. In a recent large-scale Metlife (2010) study, older lesbians, in particular, reported that they had no expectations that staff members would treat older non-heterosexual residents with any respect. Many lesbian and gay study participants indicated that they would not reveal their sexual orientation to anyone in the facility out of fears of hostility and social isolation from other residents and staff.

Also unfortunately, the fears that many LGBT elders have expressed regarding likely stigma and discrimination in institutional living may be well-founded. An empirical study of assisted-living staff members' attitudes toward residents' same-sex versus heterosexual sexual activity revealed that staff members expressed significantly more negative and restrictive attitudes in relation to sexuality among lesbian and gay residents, regardless of the level of knowledge about sexuality and aging individual staff members possessed (Hinrichs and vacha-Haase 2010). In addition, qualitative analyses of ethnographic data from three different NIH studies in long-term care revealed that a number of staff members openly made derisive comments about residents who were even suspected of having an LGBT orientation.

Although some nursing homes and assisted-living facilities exist that cater specifically to LGBT residents (see www.gayretirementguide.com for sample listings), these facilities certainly represent the exception rather than the rule. Concerns also exist that these LGBT friendly residential communities typically have a long waiting list, are expensive, and are located in geographic areas with a high cost of living (e.g., Palm Springs and San Francisco, CA). Some potential resources for older LGBT individuals and staff members in long-term care who wish to better serve older LGBT residents include The US Department of Health and Human Services' funding of the National Technical Assistance Resource Center for LGBT Elders (Services and Advocacy for Gay, Lesbian, Bisexual, and Transgender Elders; SAGE, 2010), as well as the National Center for Assisted Living's (2011) educational training tool, "Better serving lesbian, gay, bisexual, and transgender populations in assisted living communities." In essence, staff members can be helped to provide person-centered support, in which the individual residents' histories, preferences, needs, and expectations are known, and to appreciate the cultural issues and concerns of LGBT populations.

Benchmark Policies and Training

Although many institutions do not appear to be moving toward more permissive policies toward sexuality and aging, a number of nursing homes promote an open discussion of sexual activity among their residents, staff members, and community-living spouses, partners, and adult family members. The Hebrew Home at Riverdale, New York represents one of the first long-term care facilities of its kind to establish a formal policy regarding residents' sexual expression (Hebrew Home 2010). The administrators of the Hebrew Home crafted their policy, designed to "recognize and protect the sexual rights of nursing home residents, while distinguishing between intimacy and sexually inappropriate relations," with the help of a workgroup composed of nursing home staff, researchers, residents, residents' family members, and religious leaders.

The Hebrew Home at Riverdale's policy places emphasis upon the consensual nature of sexual expression between residents, and a careful examination of sexual expression between residents who may have dementia. For example, care is taken to examine both verbal and nonverbal means of expressing consent, and whether other residents may be at risk for having their individual rights or safety compromised. The New York State Department of Health provided a grant to the Hebrew Home and one outcome is a staff training video titled, "Freedom of Sexual Expression: Dementia and Resident Rights in Long Term Care Facilities." This training video was provided to all nursing home administrators in the state of New York, and it is available by contacting the Hebrew Home directly. Please see the appendix for additional information about this and other training options for long-term care staff and other providers.

Family members and older adults themselves also can select long-term care options based upon certain criteria that allow for greater freedom of sexual expression. Some questions to consider asking when selecting a residential care home include:

- Do staff members receive education and training on how to interpret and discuss issues related to sexuality and aging?
- Is there a written policy for the institution regarding sexual activity and expression?
- Are men and women relegated to discrete aspects (i.e., wings or floors) of the facility? What opportunities are there for social interaction between men and women (e.g., dances, dining, concerts, exercise)?
- Does the facility offer different kinds of seating in both pubic areas and in individual rooms? Even environmental factors such as seating can encourage residents to either sit alone (e.g., with arm chairs) or with others (e.g., on couches; Parker 2006).
- Can husbands and wives share a room (Roach 2004)? Will they have two double beds or could arrangements be made to share a double bed? What happens if one spouse becomes ill and requires more intensive care?
- Is a privacy room available for residents' use (Wallace 1992)? Is it available for masturbation, intercourse, and even for dates or card games? Do residents have to sign up in advance? Is medical assistance available? Is there a lock on the door, and if so, what requirements are needed to use it?
- Do residents have opportunities to openly and objectively discuss a range of issues such as dating and sexual dysfunction (e.g., erectile dysfunction)?
- Do residents receive education about possible changes in sexual physiology with aging (e.g., vaginal dryness), including possible treatments?
- Do residents receive education about HIV and its prevention (Kendig and Adler 1990)? Are condoms and water-based lubricants available or can residents readily purchase them?
- Does the facility have a policy about pornography, vibrators, or other sex toys if residents want to use them?
- If problems or concerns do arise in relation to a resident's sexual expression, is the issue addressed via an interdisciplinary team meeting including nursing home administrators, nursing home staff, a psychologist, social worker, or other mental

health care provider, the residents' spouse, partner, or adult family members, and the resident, when possible?

- Are opportunities available to increase self-esteem and experience sensual pleasures through frequent visits from hairdressers, barbers (Malatesta et al. 1988), cosmetologists, massage, and music therapists, among others?
- Can residents bring their own furniture and belongings, including beds, pillows, and bed linens to increase a sense of home, familiarity, and sensuality?

Summary

Issues related to sexuality and aging within an institutional setting are multifaceted. Various challenges are posed to older residents, long-term care administrators, nursing home staff members, and their community-living spouses, partners, and family members. Psychologists, social workers, and other mental health professionals are in a unique position to assist these long-term care facilities in navigating the sometimes complicated integration of individual resident rights to privacy and freedom of sexual expression with the duty to protect other residents from harm. Fortunately, the American Bar Association and the American Psychological Association have published some general guidelines in relation to assessing an older resident's capacity to sexual consent, particularly if there is some degree of cognitive impairment. For residents who engage in clearly inappropriate sexual behavior (e.g., masturbating in public areas; assaulting staff), various behavioral and environmental interventions can be used to help stop or minimize the frequency of those behaviors; drug therapy should be considered only as a last resort. The Hebrew Home in Riverdale, New York offers a benchmark Sexual Expression Policy and training video that can be used to assist psychologists and other health care providers as they interface with staff and residents in long-term care facilities. Additional research is needed, particularly in relation to LGBT and incarcerated elders.

Appendix

The following are organizations that provide information about sexuality in long-term care. Many of these organizations offer assistance specifically in relation to elder abuse and older adults' rights (also see Zeranski and Halgin 2011).

Administration on Aging
www.aoa.gov

Alzheimer's Society, Ethical guidelines for intimacy and sexuality
http://www.alzheimer.ca/english/care/ethics-intimacy.htm

American Bar Association Commission on Law & aging
http://new.abanet.org/aging/Pages/elderabuse.aspx

Gerontological Society of America (GSA)
www.geron.org

National Adult Protective Services Network
www.aspsnetwork.org

National Center on Elder Abuse (NCEA)
www.ncea.aoa.gov/main_site/library/cane/cane.aspx

National Clearinghouse on Abuse in Later Life
www.ncall.us

National Committee for the Prevention of Elder Abuse
www.preventelderabuse.org

Psychologists in Long Term Care (PLTC)
www.ptcweb.org/index.php

National Adult Protective Services Network
www.aspsnetwork.org

National Sexuality Resource Center
http://nsrc.sfsu.edu/

The following groups and organizations offer materials for staff training or continuing education for professional caregivers in relation to sexuality in long-term care.

Center for Medicaid, CHIP & Survey & Certification
Webcast, Sexual abuse in long-term care
www.cmstraining.info/pubs

The Hebrew Home at Riverdale (staff training DVD)
5901 Palisade Avenue
Riverdale, NY 10471
www.hebrewhome.org
800.567.3646 or 718.581.1000

McMaster University, Canada
Intimacy, sexuality, and sexual behavior in dementia: How to develop practice guidelines and policy for long term care facilities
http://fhs.mcmaster.ca/mcah/cgec/toolkit.pdf

National Center for Assisted Living
Educational training tool, "Better serving lesbian, gay, bisexual, and transgender populations in assisted living communities."
http://www.achca.org/content/pdf/LGBT-%20Final%20Draft-%20Slides%20Only-Formatted.pdf

The University of Kansas
Pioneering change: Sexuality in nursing homes education module
http://www.agingkansas.org/LongTermCare/PEAK/Modules/sexualitymodulef.pdf

Chapter 5
Coping with Chronic Illness and Disability

More than 85% of older adults suffer from at least one chronic
illness, and 50% must cope with at least two chronic illnesses.
Unfortunately, most health care providers fail to discuss the
impact of those illnesses upon sexuality as part of treatment.

Prevalence of Chronic Illness

The presence of chronic illness among older adults is pervasive, even among Western, Industrialized countries that boast high standards of living and advanced health care. The vast majority (85%) of older Americans suffer from at least once chronic illness, and nearly 50% cope with at least two chronic illnesses. Chronic illnesses include a range of physical (e.g., heart disease; arthritis; sleep apnea; diabetes; Parkinson's disease) and mental disorders (e.g., depression; schizophrenia; substance abuse). Some of the more commonly reported physical chronic illnesses in the US include arthritis (50%), high blood pressure (40%), heart disease (30%), diabetes (12%), cataracts (15%), and stroke (10%; Speer and Schneider 2003). In relation to the aforementioned physical disorders, nearly 20% of independently living older adults suffer from chronic mental disorders such as depression, and approximately 5% of older men and 1% of older women suffer from alcohol dependence or abuse (Administration on Aging 2001). The World Health Organization recently reported that depression is the primary cause of disability in the world (Lopez et al. 2006). Understanding the impact of these chronic ailments upon older adults appears vital as the number of older adults coping with functional disabilities related to arthritis, stroke, diabetes, heart disease, cancer, and dementia, among other, is expected to increase by more than 300% by the year 2050 (CDC 2011).

Disability can be thought of loosely as any medical or psychological condition that impairs a person's normal range of functioning, including participation in sexual activities. Disabilities can be the result of an acute event, such as a heart attack (that may or may not lead to a complete recovery), or they can evolve over time from either treated or untreated chronic illness, such as rheumatoid arthritis or

J. Hillman, *Sexuality and Aging: Clinical Perspectives*,
DOI 10.1007/978-1-4614-3399-6_5, © Springer Science+Business Media New York 2012

diabetes. Chronic illnesses and related disabilities can cause an individual to experience changes (typically declines) in relation to physical, psychological, sexual, interpersonal, and intrapersonal functioning. Older adults are often faced with the challenge of adapting to life with one or more chronic illnesses, or adapting to life as a caregiver for a partner or family member coping with such an illness.

Because an emphasis is often made, initially at least, upon the physical health of an individual after a chronic illness is diagnosed, patients, caregivers, and health care providers often neglect to acknowledge the impact that the illness may have upon the patient's sexual functioning. Even after an immediate medical crisis may be resolved, many health care professionals fail to provide patients with educational information about changes in their sexual functioning, or ways to effectively cope with any such changes. Reasons for this lapse in vitally important patient sexual education may be related to health care providers' embarrassment or anxiety (Zang et al. 2008) as well as limited training opportunities in sexuality (Nusbaum et al. 2003; Rosenbaum 2010). In other cases, the medications used to treat a patient's chronic illness can produce side effects that impair sexual functioning. To complicate matters further, the predominant societal stereotype that sexuality is a privilege reserved for the young and the healthy provides an additional barrier to successful adaptation.

Successful Coping

In order to cope successfully with any disability or chronic illness, an older adult must:

1. Acknowledge the presence of the disability or chronic illness.
2. Make a realistic appraisal of the illness via informed health care providers and become an active participant in a treatment plan designed to alleviate symptoms and maximize independence.
3. Establish "safe" personal and body boundaries.
4. Engage social supports, including family, friends, and partners.
5. Mourn the loss of any physical, cognitive, or emotional abilities.
6. Recognize that sexuality, defined well beyond the simple act of penetrative intercourse, can continue to be an important part of life.

Any individual must first acknowledge the presence of a disorder in order to cope effectively. Denial is a primitive defense that will allow an older adult to believe that things are operating on the status quo, but it requires a great expense of psychic energy and is not effective in the long run. Some degree of ego strength is required for an individual to acknowledge that his or her sense of self has been somehow violated or changed. Also, the very nature of certain chronic illnesses, such as schizophrenia and dementia, can limit an older person's ability to recognize the presence of an illness. In these situations, caregivers are often faced with the task of identifying the presence of a disorder and seeking help for their friend, relative, partner, or spouse.

Once patients recognize that they are faced with some potential, significant life changes, they are better equipped to mourn any losses and reestablish a sense of self based on their internal versus external attributes. A trusting therapeutic relationship can ease the pain of admission and allow for the articulation and venting of emotions in response to the presence of a chronic illness or disability. Life review (Serrano et al. 2004), interpersonal (IPT; Hinrichsen and Clougherty 2006), and cognitive-behavioral therapies (Laidlaw et al. 2009) can all be useful in allowing clients to experience and process anger, sadness, helplessness, curiosity, rage, and resignation in the mourning process, as well as to address interpersonal changes or challenges that may emerge. Both aging patients and their mental health care providers also should recognize that even though many older adults possess at least once chronic illness, advanced age itself is not an automatic "death sentence" from activities of daily living, including participation in one's sex life.

The Critical Need for Patient Education

Even though coping with changes in sexual functioning and body boundaries can be considered a psychological process, the presence of acquiring accurate information about a disorder, including specific information about its etiology and treatment, remains essential in order for an older person to adapt effectively. Providing information about the course of an illness can concretize an otherwise ambiguous life event and provide necessary structure and expectations for patient progress. Many older adults are not well informed about the nature of their illness, much less the impact of their disorder upon sexual functioning. Empirical studies suggest that the vast majority of patients fail to receive any information about sexuality in relation to their acute or chronic illness (e.g., Nausbaum et al. 2004; Yang and Jolly 2008).

Even though geropsychologists are not medical personnel, they have the responsibility to become generally familiar with the medical disorders of their clients, be available to discuss the impact of those disorders upon sexual functioning, and collaborate effectively with health care professionals from a variety of settings including nursing homes, hospitals, and outpatient clinics (e.g., Karel et al. 2010). In the words of one prominent geropsychologist, "we must wear many professional 'hats' in order to best serve the needs of our patients." In the following sections, a number of common chronic illnesses will be reviewed within the context of sexuality and aging. Various medical aspects of these disorders will be highlighted in order to provide some general background for mental health care practitioners. It also is important to note that mental health providers are not expected to become medical experts, nor should they act as one. One critical aspect of psychological care is to help guide older clients themselves to engage in effective communication with their health care providers (i.e., asking relevant questions; acting assertively) in order to obtain the best possible information and care.

Guidelines for Psychotherapy

A number of recommendations can be offered for mental health professionals working with a middle-aged or older adult who must come to terms with a chronic illness or disability. Because some older adults may be uncomfortable discussing sexual matters, the use of the PLISSIT model (Annon 1976) can be useful. In level one of the model, the client is given permission to openly discuss their sexual issues and concerns. In order to elicit client information, questions can typically be phrased in a way that normalizes potential problems and concerns. For example, a psychologist might ask an older client recently diagnosed with hypertension, "Many people with high blood pressure notice changes in their sexual response or in their sex lives. I'm wondering what kind of changes you have noticed" (Finger 2006). Another important aspect is to use terminology that the client is familiar with; do not assume that someone understands the medical terminology used by their physician or viewed in educational literature.

Another critically important part of the permission phase is to allow clients to feel comfortable defining and dictating their own boundaries in relation to sexual behavior. For example, some patients may need help feeling empowered or entitled enough to say "no" to a partner who may be pressuring them to "just act normal" and engage in a sexual activity that they find painful, embarrassing, or unfamiliar and new. Other partners may initially pressure a patient to engage in sexual activity too soon after a medical procedure or before a patient is psychologically ready to engage in sex. Although it is vital to include partners in discussions of diagnosis, treatment, and coping, patients must be assured that they themselves have the final say in what kind of sexual behaviors they feel comfortable with. Patients also can be reminded that sex is much more broadly defined than participation in penetrative intercourse in a specific position; touching, stroking, petting, and other sensual activities such as bathing and massaging are certainly important and included.

In the second step of the PLISSIT model, information is gathered and shared between client and therapist. Here, therapists can often guide clients to act as advocates for themselves within the medical system. Dispelling myths becomes vital for the patient to progress in counseling. Both parties work as a team to help the client gather as much accurate information as possible. Mental health professionals can also help clients evaluate information that they may receive through the internet or other community-based sources. In the third level of the PLISSIT model, the practitioner offers specific suggestions to the client to help her address an underlying sexual concern or problem. For example, a therapist could help a client with painful arthritis communicate his concerns about trying a new side-by-side sexual position that would place less pressure on a sore hip, or a client with a colonoscopy bag discuss having sex with her partner in a new sexual position which would allow him to enter her from behind, allowing her to focus less upon the medical device and more upon pleasure. Suggestions for a variety of techniques and alternatives can be discussed, which may range widely from the timing of pain medication, the scheduling of sexual activity to different times of day, trying new sexual positions, engaging in more sensual activities, to the use of different lubricants. Lastly, therapists must also acknowledge what information they do not know, and seek out help and consultation with the appropriate health care professionals.

Arthritis and Pain

Estimates suggest that up to 50% of older adults suffer from arthritis, including both rheumatoid and osteoarthritis (Speer and Schneider 2003). Writing, walking, and other activities of daily living become increasingly difficult and painful. Both men and women with arthritis may become concerned or ashamed about the appearance of their hands, which may become misshapen throughout the course of the disease. A vicious cycle often ensues in which an older patient experiences pain when walking or moving their extremities, which leads to a general physical inactivity, to muscle stiffness and reduced joint lubrication, to even more pain in response to movement. As a result, many individuals with arthritis may limit their activities of daily living. Participation in sex also is likely to decline, and this is likely to go unaddressed by health care providers (Rosenbaum 2010).

For someone with arthritis, sexual relations can become more painful than enjoyable. Many patients report experiencing the most pain and stiffness during the evening, the night, and the early morning, which unfortunately coincides with the times that most people tend to engage in sexual activity (Badeau 1995). Difficulties in flexion of the hips, knees, and back can cause significant pain during intercourse (Mayo Clinic 2011a, 2011b, 2011c, 2011d, 2011e) and require arthritis sufferers to seek out new, and potentially unfamiliar and threatening, sexual positions. Even if an older patient with severe arthritis cannot engage in intercourse, achieving satisfaction via masturbation may be challenging or difficult due to pain and lack of mobility in the arms, hands, and fingers.

To complicate matters further, the medications often prescribed to reduce the pain and swelling associated with arthritis may reduce interest in sex. The steroids prescribed to control swelling often provoke ED in male patients. Other side effects from such medications include hair loss, hair growth in undesirable places, and weight gain. It can be difficult to feel attractive and desirable when one is in pain, feeling unattractive, or unable to orgasm or achieve an erection. Also, unfortunately, throughout the last decade, significant bias in the medical community suggested that arthritis was a common and garden-variety illness that generally would affect an older adult's sexual functioning (e.g., Golan and Chong 1992).

Helpful suggestions can be offered to middle-aged and older adults with arthritis (or any individual coping with pain from muscular-skeletal disorders or fibromyalgia, for example) who wish to engage in satisfying sexual relations

(American College of Rheumatology 2010; Kautz 2007; Mayo Clinic 2010a, 2010b, 2010c; Nusbaum et al. 2003; Rosenbaum 2010). For example:

- Sexual activity can be pursued during the daytime hours when pain generally is less intense.
- Plan ahead for sex; patients may feel less pressured during sexual activity if they give themselves permission to turn off the phone and ignore the doorbell for a few hours.
- Warm baths and showers can loosen muscles and be incorporated as foreplay.
- Time pain medication to reach maximum effectiveness during sexual activity.
- Placing pillows or rolled sheets under the body, and sometimes the use of air or water beds, can help lessen pressure on sore joints.
- The use of electric blankets and heating pads can sooth joints and muscles.
- Masturbation, both individually and with a partner, can be encouraged as it can help normalize sexual response, increase pleasure, and lessen focus upon penetrative intercourse.
- Particularly with hip pain, avoid the missionary position for intercourse. For alternative positions, partners can lie down, side by side, with the receptive partner's knees drawn up. Sitting and kneeling also may be less painful on the hip joints.
- The use of vibrators and other sex toys during masturbation and partnered sex can help compensate for soreness in hands and fingers.

Some researchers even suggest that sexual activity can stimulate the natural production of endorphins and corticosteroids. These neurochemicals are natural pain relievers whose effect may last for hours after the conclusion of sexual activity (Badeau 1995). Sexual activity may also result in a more relaxed musculoskeletal system (Butler et al. 1994) and presumably a lesser experience of chronic pain overall.

From a psychological perspective, middle-aged and older patients coping with arthritis also must feel free to refrain from sexual activity if it is too painful or uncomfortable. When told about their options for sexual activity, some patients may feel guilty or like "a quitter" if they do not engage actively in intercourse with their partner. It is vital that patients with arthritis, like any chronic illness or disability, engage in sexual behavior that is comfortable for them, and not feel pressured to engage in sex to meet the expectations of a physician or therapist. As with individuals coping with any chronic illness or disability, couples therapy can improve communication between partners and establish new and enjoyable patterns of sexual relations that may incorporate sensual as well as sexual activities (c.f., Akkus et al. 2010). Consultation with physical therapists also can prove useful (Rosenbaum 2010).

Diabetes

Diabetes is a chronic illness that affects nearly 3% of the population worldwide, and is projected to affect nearly 5% of the population by 2030. Only 10% of individuals with diabetes have Type-1 diabetes (i.e., a lack of insulin production) and the remaining 90% suffer from adult onset or Type-2 diabetes in which there is decreased sensitivity to insulin. Associated with obesity and physical inactivity, the significant increase projected in diabetes worldwide is expected to occur primarily among older adults (Wild et al. 2004). Although most people who experience adult-onset diabetes can control its course with diet, medication, and exercise, effective treatment often requires strict adherence and life-style changes.

Diabetes also can result in significant anxiety and misunderstanding among its sufferers because it is commonly linked to sexual dysfunction. In men, the vascular and neurological changes associated with diabetes are significant risk factors for ED as well as difficulties with ejaculation and low sexual desire (Bhasin et al. 2007). The likelihood of experiencing ED with diabetes increases with the presence of advanced age, heart disease, obesity, tobacco use, high cholesterol, and diabetic complications such as neuropathy (Rosen et al. 2009). (Please refer to Chapter 8 for additional information related to male sexuality and the diagnosis and treatment of ED.) It also is important to note that for men who suffer from diabetes-related ED, underlying sexual urges, interest, and opportunities for sensual pleasure are generally not diminished in any physiological way.

The impact of diabetes upon female sexuality has typically been ignored, both clinically and in research trials (Nowosielski et al. 2009), and available empirically based evidence is limited (Nowosielski et al. 2009). Although myths persist in the medical community that diabetes has little impact on female sexuality (e.g., Golan and Chong 1992), recent estimates suggest that between 20–70% of women with Type-1 diabetes and 15–50% of women with Type-2 diabetes suffer from some type of sexual dysfunction. The most common types of sexual dysfunction reported include low sexual desire, decreased vaginal lubrication, and inability to orgasm (e.g., Bhasin et al. 2007; Ho and Rabi 2006). Among women, sexual dysfunction can result from a lack of blood flow to the clitoris and vagina during sexual arousal. Atrophy of the uterus and ovaries also has been observed (Golan and Chong 1992). Although women may continue to experience orgasm, an overall decrease in frequency of orgasm and an increase in the amount of manual stimulation required to achieve orgasm may result from diabetes. Depression in consort with diabetes can also increase the risk of difficulty with orgasm (Nowosielski et al. 2009). In sum, these aforementioned changes in sexual response among women with diabetes are real, and should not be ignored or overlooked by physical or mental health care providers.

Amputations as Complications

Another aspect of diabetes that is often not discussed in relation to sexuality and aging is that of limb amputation. Diabetes often causes vascular difficulties, and typically lower extremities such as toes, feet, and even lower legs must be amputated as a result of poor blood flow, infection, or gangrene. Despite general advance in health care, approximately 3–5% of individuals with diabetes will lose a lower extremity by amputation (van Houtum et al. 2004), and more than 85% of all amputations occur among patients over the age of 60 (Schulz et al. 1991). Coupled with ageist attitudes that only young, physically robust individuals are entitled to engage in sexual relations, health care providers themselves appear reluctant to discuss changes in sexual attitudes or functioning in response to the loss of a limb. Studies show that virtually no lower extremity amputees report having any physician initiated discussion of sexuality after their surgery (Geertzen et al. 2009).

Regarding sexual activity among amputees, barriers to sexuality include pain at the site of amputation, embarrassment about not being normal or "whole," and concerns that their partner finds them defective or otherwise unattractive. Concerns about loss of body integrity are often overlooked among older adult patients. Depression and anxiety also represent a common response to amputation, which can impact negatively upon sexual functioning (Bodenheimer et al. 2000; Horgan and MacLachlan 2004). Thus, it appears essential that mental health practitioners provide a source of information and support for middle-aged and older amputees who wish to return to their normal sex lives.

Heart Disease

As clinicians and researchers, we recognize that the brain is the seat of our emotions and consciousness, and that it alone oversees the activity and coordination of all organ systems. However, popular culture dictates that our heart is the most important part of our anatomy, and it has become imbued with unique spiritual and emotional characteristics. We speak from the bottom of our hearts, follow our hearts, and experience heartache and heartbreak. This muscular organ, only about the size of our fist, has come to represent love, caring, fidelity, courage, honor, and security. We seek out people who are kind hearted and lion hearted. It should come as no surprise that heart disease, experienced by nearly one-third of older adults and responsible for the number one cause of death in the USA (Xu et al. 2010) can be interpreted as a central assault to the self. If damage to the heart occurs, patients often wonder if they are still the same person, with the same beliefs, thoughts, and potential for love. The implications for sexuality in relation to heart disease, including heart attack, stroke, and high blood pressure, should not be underestimated from either a medical or psychological standpoint.

Recovering from a Heart Attack

Myocardial infarctions can range from mild incidents that are experienced as being slightly out of breath to severe incidents that lead to severe crushing pains and emergency hospitalizations. In either case, the heart muscle becomes deprived of oxygen and suffers from damage. Once stabilized, patients are allowed to return home to resume normal activities after a course of rehabilitation. For most patients, returning to normal life includes a desire to resume participation in one's sex life.

Despite excellent medical care and completion of physical rehabilitation programs, fear of another heart attack and performance anxiety can inhibit a heart patient's full return to, and enjoyment in, sexual activity (Mandras et al. 2007). Depression and marital conflict also appear to be associated with increased fears of resuming sexual activity (Kazemi-Saleh et al. 2007). In contrast, other survivors of heart attacks may respond with anger and frustration when they believe they are no longer able to perform sexually. Anxiety disorders, as well as depression, may develop after a heart attack. For older patients, these concerns may be heightened by fears of death and by submission to stereotypical beliefs that older adults, particularly those with "bad tickers," have no need for, or interest in, sexual activity. Studies also indicate that for both women and men, it is the patient's psychological response to the myocardial infarction (including perceptions of support from partners), and not the severity of the heart attack itself, that is the better predictor of sexual disability (Eyada and Atwa 2007).

Providing education about the energy expenditures required for sexual activity and the warning symptoms to be on the alert for during intercourse can be vital in allowing both male and female patients to begin to feel comfortable with themselves again. Although a wealth of information is available, medical personnel typically do not share this information with their patients unless they ask for it directly (Mandras et al. 2007). Older patients, adhering to their cohort's general prohibition against discussing sexual issues openly, may be less likely than their younger counterparts to approach their health care providers with questions about resuming sexual relations. Some very general facts and guidelines regarding sexual activity after a heart attack include (Mayo Clinic 2011a, 2011b, 2011c, 2011d, 2011e; Kautz 2007; Thompson and Ryan 2009):

- Speak to your physician, who may ask you to complete an exercise or stress test. If you are cleared for participation in exercise, you are typically cleared for participation in sex.
- The energy required to perform sexual intercourse is roughly equivalent to that of briskly climbing two flights of stairs or walking on a treadmill at 3–4 miles an hour.
- Fewer than 0.1% of all heart attacks are precipitated by sexual intercourse
- Recognize that the majority of hypertensive drugs can cause side effects that limit sexual function.
- The use of a couple's "usual" positions during sex is advisable in the early stages of recovery. Changing to novel positions have been shown to increase blood pressure and heart rate. Having the woman "on top" or lying side to side has not been shown to decrease cardiac workload (Hellerstein and Friedman 1970).
- Having sexual relations with a familiar partner in familiar surroundings with a comfortable room temperature is a consideration. Unusual extremes in heat or cold (from either a hot or cold shower) also can increase cardiac workload.

- Masturbation requires less energy output than sexual intercourse, and can provide stress reduction and sexual fulfillment, particularly in the early stages of recovery.
- Having sex in the morning is desirable because of the benefits of a full night of sleep.
- Taking prescription medications such as nitroglycerin before sex may prevent chest pain.
- Waiting to engage in sex 3 h after eating or drinking is helpful. Blood vessels dilate in order to digest food and in response to alcohol.
- Recovering patients should expect some discomfort during intercourse including shortness of breath, sweating, and fatigue.
- Knowing the warning signs for stopping sexual activity and consulting a physician is critical. Some of these more serious symptoms include angina (i.e., chest pain), heart palpitations, or shortness of breath during intercourse. A failure to return to resting heart rate 7–10 min after orgasm also is cause for concern.
- The use of foreplay is essential for both physical and emotional preparation for intercourse.
- Resuming sexual relations with an emphasis placed on mutual pleasure, without the specific goal of orgasm is helpful. Sensate exercises can be particularly helpful in alleviating performance anxiety.

Although mental health practitioners certainly cannot and *should not* attempt to provide advice and direction regarding their patient's physical care, a knowledge base of symptoms and treatment planning is essential. Clinicians often help heart patients communicate more effectively with their primary health care providers to gain the knowledge and support that they need. Therapists often find themselves freed up to help older patients differentiate between realistic concerns and nihilistic fantasies regarding their heart attack and recovery. Partners also can be included in such therapeutic discussions in order to allay their fears and anxieties. Many partners of middle-aged and older adults with heart attacks begin to fear that they are unattractive when their partners hesitate to resume sexual relations. Others have unrealistic expectations about the rate of recovery, the role of medications, or only a limited understanding of their partner's fears and anxieties.

One of the most important things a mental health provider can do when working with a heart attack survivor is to help their client speak directly with their doctor about their sexual concerns. Findings from an empirical study show that heart patients who do not discuss their sexual function at discharge from the hospital are significantly more likely to abstain from sex or engage in sex with lesser frequency 1 year after discharge, when compared to those patients who discussed sexuality with their physicians. Another important finding is that physicians were significantly less likely to speak to their female patients about resuming sexual activity (35%) when compared to their male patients (46%; Lindau et al. 2010). It becomes vital to encourage our clients to speak directly to their doctors about the potential for resuming sexual activity after a heart attack, particularly if the patient is a woman.

Stroke

Stokes or cardiovascular accidents (CVAs) represent another outcropping of heart disease. In this case, a decrease in blood flow to parts of the brain causes tissue damage and transient or permanent changes in functioning. The aftermath of a stroke not only includes physical changes (most commonly paralysis) but also psychological adjustment. In addition to coping with a major medical emergency and potentially drastic changes in functional movement and speech, the injury to the brain itself may induce depression, impaired memory, and other changes in mental status. Clinicians must be alert to the synergistic affects of trauma and neuropsychological injury itself. Brain injury resulting from the stroke itself can lead to a loss of interest in sex, although this phenomenon has been observed typically as short lived. As noted, depression often occurs after a stroke, and findings from a large-scale meta-analysis suggest that 33% of all stroke survivors will suffer from depression at some point in their convalescence or recovery (Hackett et al. 2005).

Poststroke depression is associated with overall lower quality of life, unwelcome changes in social roles, and limited participation in activities of daily living (Landerville et al. 2009), and most patients engage in sex with lesser frequency after their stroke whether or not they present with depressive symptoms (Tamam et al. 2008). Commonly reported poststroke challenges in relation to sexuality include fears of having another stroke, fears of being unattractive to one's partner, lack of interest in sex, fears of incontinence, muscle pain and stiffness, muscle rigidity and spasms, fatigue, vaginal dryness, ED, inability to orgasm, and concerns about being able to masturbate (e.g., Kautz 2007; Thompson and Ryan 2009). Unfortunately, physicians typically provide little or no information to their poststroke patients regarding sexuality (Schmitz and Finkelstein 2010).

To help combat such issues and concerns, patient education is vital (Nusbaum et al. 2003) and may be provided within the context of psychotherapy. For example, if a stroke survivor experiences muscle paralysis, typically occurring on one side of the body (i.e., hemiparesis), suggestions can be made to allow the partially paralyzed partner to participate more fully in sexual activity. Patients may be instructed to lie on pillows on their nonfunctional side, allowing for a greater range of motion with their free hand and leg to caress and hold their partner. Additional recommendations include an obvious focus upon mutual pleasure rather than penetration and orgasm per se, experimentation with various sexual positions, and the establishment of pleasurable sensual activities. (Most of the aforementioned recommendations for coping with chronic pain also may apply.) The ability to maintaining simple routines, such as sleeping in the same bed, when possible, can help build or reestablish intimacy. Receiving outside assistance with bathing, dressing, and applying makeup, for example, can also help some stroke survivors feel more positively about their appearance.

Stroke survivors who lose the ability to speak (i.e., aphasia) may find it particularly difficult to communicate with their partners, much less resume sexual activity (Lemieux et al. 2001). Although limited empirically based information is available to assist aphasic patients within the context of their sexual relationships, the need is great as approximately one-third of all stroke survivors experience significant speech impairment (e.g., Kautz 2007). Compounding the problem, many partners and some health care providers do not recognize that the areas governing the production of speech are different than those responsible for the comprehension of speech. A stroke patient who cannot talk does not necessarily lose the ability to understand what is going on around her, or to understand what is being said to her. Talking down to a patient with aphasia may further erode the patient's self-esteem and subsequent interest in resuming sexual relations. Sometimes hand signals or assistive technology devices (i.e., handheld computers) can be used to convey either messages of affection or assent to sexual activity to one's partner.

High Blood Pressure

Hypertension is a common chronic illness among older adults; more than 40% of adults over the age of 65 cope with elevated blood pressure (Speer and Schneider 2003). If untreated, symptoms can include headache, irritability, fatigue, dizziness, depression, anxiety, and ED. If left untreated, high blood pressure also can lead to stroke and heart attack. Although most cases of hypertension can be managed effectively with lifestyle changes and medication, the majority of drugs used to treat high blood pressure have side effects that negatively affect patients' sexual functioning. For example, diuretics, often referred to as "water pills" (e.g., hydrochlorothiazide), and beta-blockers (e.g., Atenolol) are consistently linked with ED and other sexual dysfunction. Many of these drugs also are used for patients recovering from heart attacks and stroke.

A listing of some of the more commonly prescribed diuretics and beta blockers that may cause sexual side effects (McVary 2007) is provided in Table 5.1. It is important to note that these side effects may occur among both men and women. Whereas older men commonly experience a loss of interest in sex, difficulty in reaching orgasm, and ED, older women are more likely to experience vaginal dryness and a lack of sexual interest.

Encouraging patients with high blood pressure to speak frankly with their physician or health care provider about the presence of side effects is essential, as some classes of medication for hypertension are unlikely to cause ED or other types of sexual dysfunction. For example, ACE inhibitors (e.g., Lotensin, Zestril, & Prinivil) actually widen blood vessels and are rarely linked with ED. Calcium channel blockers (e.g., Diltiazem, Verapamil, & Amlodipine), alpha-blockers (e.g., Cardura), and ARBs (angiotensin II receptor blockers; e.g., Losarten) also are less likely to cause ED and other types of sexual dysfunction (Barksdale 1999). In empirical studies, the use of drugs such as Losarten have been shown to decrease symptoms of sexual dysfunction and actually lead to an increase in sexual satisfaction and frequency among both men and women with high blood pressure (Caro et al. 2001). The goal is not for psychologists and other mental health providers to dispense medical advice, but to foster increased collaboration between patients and their physicians in order to acknowledge concerns about ED and to explore their options.

Table 5.1 Medications for hypertension with known sexual side effects	Drug	Trade name
	Atenolol	Tenormin
	Bumetanide	Bumex
	Captopril	Capoten
	Chlorothiazide	Diuril
	Chlorthalidone	Hygroton
	Clonidine	Catapres
	Enalapril	Vasotec
	Furosemide	Lasix
	Guanabenz	Wytensin
	Guanethidine	Ismelin
	Guanfacine	Tenex
	Hydralazine	Apresoline
	Hydrochlorothiazide	Esidrix
	Labetalol	Normodyne
	Methyldopa	Aldomet
	Metoprolol	Lopressor
	Minoxidil	Loniten
	Nifedipine	Adalat, Procardia
	Phenoxybenzamine	Dibenzyline
	Phentolamine	Regitine
	Prazosin	Minipress
	Propranolol	Inderal
	Reserpine	Serpasil
	Spironolactone	Aldactone
	Triamterene	Maxzide
	Verapamil	Calan

Note: Patients must be cautioned never to suddenly stop or switch any medication for high blood pressure without first consulting with their physician

Incontinence

Up to one-half of all older adult residents in nursing homes can be classified as incontinent. What may be more surprising is that estimates suggest that up to one-third of all *community-living* older adults in the United States will suffer from urinary incontinence at some point in their lives (McCormack et al. 1992). Although incontinence is one of the most common disorders among older adults, it is one of the least discussed and understood by middle-aged and older adults themselves. Many individuals who experience incontinence who are otherwise healthy and active become depressed, anxious, and socially isolated (Bogner et al. 2002; Oh et al. 2008) because they do not want to go out in public out of fears of having an accident or "giving themselves away" with some type of bulky clothing or odor. One can consider the corollary—that incontinence puts significant strain on anyone's sense of sexuality and can exacerbate any existing concerns about intimacy. The vast majority of women who experience incontinence, for example, show significant decline in frequency of intercourse or cessation in sexual activity entirely (Cohen et al. 2008).

Many older adults assume that incontinence is simply a by-product of aging, and that there is no treatment for this problem. To the contrary, estimates suggest that more than 80% community-living older adults who experience incontinence experience a complete cessation of, or at least a significant improvement in, their symptoms with appropriate treatment (Serati et al. 2009). To assist an older adult in treating incontinence, it becomes important to gain the appropriate medical consultation to identify the specific type of incontinence. Four primary types of loss of bladder and bowel control are: stress incontinence (loss of urine from sneezing, coughing, or lifting in response to pelvic muscle weakness), urge incontinence (preceded by a sudden overwhelming urge to urinate, the patient cannot get to the commode in time), overflow incontinence (in which urine leaks from a distended and overflowing bladder), and reflex incontinence (loss of bladder or bowel control related to a neurological disorder such as Parkinson's disease).

Although influenced by psychological and environmental factors, incontinence typically has its basis in underlying medical problems. Such medical problems are as diverse as stroke, diabetes, obesity, alcohol abuse, pneumonia, prostate gland enlargement, urinary tract infections (UTIs), Alzheimer's disease, Parkinson's disease, and multiple sclerosis. Sometimes conditions in the environment play a role in incontinence, particularly in institutional settings. Some of these situational barriers to bladder and bowel control include toilets without appropriate height supports or hand holds, poor lighting in bathrooms, binding clothing that is difficult to remove quickly, and cold room temperatures. Being constipated can contribute to the problem. Yet other older adults may drink multiple cups of caffeinated coffee during a busy morning, or drink too many martinis at one sitting and overflow their bladders inadvertently. It also is important to note that drinks with corn syrup, artificial sweeteners, citrus and tomato juices, and spices can be linked to bladder irritation and incontinence (Mayo Clinic 2011a, 2011b, 2011c, 2011d, 2011e).

A variety of drugs commonly prescribed to older adults also can induce changes in bowel and bladder habits. Because constipation can lead to bowel obstruction and loss of bowel control, a variety of medications can lead to difficulties with both urinary and fecal incontinence. See Table 5.2 for a brief review of these drugs and their potential side effects.

From an interpersonal perspective, in addition to causing discomfort and anxiety about circulating freely in public, many older adults with poor bladder or bowel control panic at the thought of being in bed or close to a partner and losing control. Sometimes simply being sure to empty one's bladder before intimate dinners, walks, or interludes is enough to prevent an accident. Older male patients are often relieved to learn that it is anatomically impossible for urine to "come out at the same time" as ejaculate when they are with a partner. Other older patients experience incontinence on a more global level. They feel infantilized when they have accidents, need to wear "diapers," and literally accept that this disorder represents a return to a helpless, sexless stage of life that resembles infancy. It is important for older patients to understand that having symptoms of incontinence in no way returns them to a

Table 5.2 Medications that can induce changes in bladder and bowel habits

Drug type	Potential side effects
Antidepressants	Constipation
Tranquilizers	Constipation
Anticholinergics	Urinary retention, constipation
Narcotic pain relievers	Constipation, confusion
Diuretics	Frequency and urgency in urination
Alpha-adrenergic blockers	Bladder neck relaxation
Caffeine	Frequency and urgency in urination
Antihypertensives	Unsteady posture
Analgesics	Constipation
Alcohol	Frequency and urgency in urination; unsteady posture

regressive state; they are adults struggling with a complicated medical problem. Education about the disorder, coupled with the ability to discuss overall fears of losing control and independence in a trusting therapeutic relationship, often allows clients with incontinence to regain their sense of vitality and agency. Discussion of urinary changes and fears with a partner also remains essential (Mattson-DiCecca et al. 2009).

For patients who wish to manage their incontinence with conservative measures, aside from medication or surgery, a number of options are available. These behavioral measures may include (e.g., Mayo Clinic 2011a, 2011b, 2011c, 2011d, 2011e):

- Pelvic floor or kegel exercises for men and women. There is some evidence that older women who engage in regular sexual activity perform natural kegel exercises that reduce urinary incontinence.
- Maintaining fluid intake between six and eight glasses of fluid a day, avoiding the use of caffeinated and alcoholic beverages.
- Undergoing behavior training that includes relaxation exercises when in the midst of an urge to urinate, biofeedback techniques that help promote pelvic muscle control, and timed visits to the bathroom (i.e., bladder training).
- Maintaining a normal body weight to avoid putting additional strain on bladder and bowel muscles.
- Scheduling bathroom times to avoid rushing and to make sure that the bladder and bowel can be completely emptied. Planning a specific time to have a bowel movement in one's morning routine, for example, can lead to increased regularity in bowel habits and thus decrease the likelihood of accidents.
- Engaging in some type of exercise to promote more appropriate movement of fecal material through the intestines and to improve muscle tone in the abdomen.
- Obtaining appropriate education about incontinence and fully exploring all avenues for treatment.

For individuals who experience chronic incontinence despite treatment (e.g., some poststroke and cancer survivors) additional recommendations can be offered to help them engage more fully in sensual and sexual activity. Discussion of such options appears vital, as incontinence typically affects participation in sexual activity very negatively (e.g., Serati et al. 2009). When incontinence or indwelling catheters can pose difficulty with intimacy, couples can shower or bathe together as part of foreplay (or as a sensual, intimate activity in itself). Plastic sheeting or pad can be placed on top of, or underneath regular bedding if a client is concerned about leakage during sex, and the affected partner can be sure to empty their bladder or bowels before initiating sexual activity.

If cleared by their physician, both men and women with indwelling catheters can have sex, if they so choose. Some patients may be able to remove their catheter before sex, for reinsertion later, but others may not be allowed to remove the device. For these women, the catheter can be taped to a thigh or abdomen, and crotchless underwear can be used to help hold the catheter in place. For men, an indwelling catheter can often be folded back over an erect penis, and a condom can then be put in place (Kautz 2007). Of course, discussion of these options between partners well ahead of time, in a typically nonsexual and nonthreatening environment, is essential. Individual and couples therapy can help provide assistance in having these sometimes challenging discussions.

In sum, we should not allow our middle-aged and older clients suffering from incontinence to accept naively that their situation is hopeless. With the benefit of a trusting therapeutic relationship, we can encourage them to discuss their anxiety and fears with both their health care providers and their partners. We can also encourage our older clients to seek medical and psychological consults for incontinence, as it can be treated in the majority of cases with up to an 80% success rate (e.g., Serati et al. 2009). Regarding different types of treatment, empirical evidence suggests that behavioral treatments such as biofeedback for pelvic muscle control, relaxation exercises during periods of increased urges to urinate, and voiding schedules are more effective in 60% of all cases than drug or alternative treatments (Glazer and Laine 2006; Wyman et al. 2009). Thus, mental health practitioners appear uniquely prepared to help assist these patients. At the very least we can inform our middle-aged and older adult patients that incontinence is *not* a normal part of aging.

Parkinson's Disease

Another chronic illness that affects up to 1 in 100 adults over the age of 60 is Parkinson's disease. In such individuals, the substantia niagra, a cortical structure in the midbrain, does not produce sufficient quantities of the neurotransmitter dopamine. The resulting lack of dopamine usually leads to classic symptom presentations of motor dysfunction. Some of these primary symptoms include a resting tremor of the extremities, difficulties in initiating movement, an overall slowing of movement, stiffness in the limbs, speaking in a low, raspy voice, and blank or mask-like expressions. These motor disturbances certainly can impact on an older adult's sexual expression; one is less likely to feel "sexy" with a visible, potentially embarrassing resting tremor or a perceived inability to communicate with others. Older adults with Parkinson's disease also are at greater risk for developing dementia and depression than members of the general population, which also are known to interfere with sexuality (e.g., Wiwanitkit 2008). Specific types of sexual dysfunction reported by women living with Parkinson's disease include decreased interest in sex and difficulty achieving orgasm. Sexual problems reported by men living with Parkinson's disease include ED, premature ejaculation, difficulty achieving orgasm, and sexual dissatisfaction (Bronner et al. 2004).

The drugs used to treat the symptoms of Parkinson's disease appear to have side effects that alter the sexual response cycle (see Cummings 1991, for a review). On a positive note, L-dopa and Sinemet, the primary drugs used to treat Parkinson's disease, have been reported to restore prior interest in sexual activity, improve erectile function, and increase rates of masturbation, as well as to improve overall motor skills. However, these drugs also have been documented to induce hypersexuality (i.e., excessive masturbation that interferes with normal daily functioning, initiation of extramarital affairs if one's spouse does not consent to increasingly frequent sexual activity) or sexually deviant behavior in up to 13% of patients. (Fortunately, these symptoms may be controlled in most cases by a reduction in medication or the addition of antipsychotic agents such as lithium.) Examples of observed deviant behaviors include self-injurious behavior, voyeurism, pedophilia, and exhibitionism, as well as other compulsive behaviors such as gambling or shopping (Cooper et al. 2009). It also may be assumed that our patients are unaware of the sexual side effects often associated with their medication. Older adults with Parkinson's disease would benefit from intensive, interdisciplinary treatment from both psychological and medical practitioners in order to manage their illness successfully and to navigate changes in sexual interest or function.

Depression

Differential Symptoms Among Older Adults

Depression is a relatively common disorder among older adults. Various estimates suggest that between 3 and 20% of community-living older adults suffer from major depression. Ten percent of chronically ill older adults and up to 30% of older adults in nursing homes may also meet criteria for clinical depression (Hybels and Blazer 2003). Even though a large-scale meta-analysis revealed that psychological and pharmacological treatments for depression are equally effective among older, as compared to younger, adults (Scogin and McElreath 1994), the differential symptom presentation of depression among older adults may make the initial diagnosis somewhat difficult. Specifically, older adults may manifest depression via somatic complaints versus general complaints about feeling sad, depressed, "down in the dumps," or "blue." Older adults also are more prone to display cognitive impairments and delusions when depressed. In any responsible discussion of depression among elderly adults, it must be noted that older adults are at greater risk for completed suicide than their younger counterparts, and that suicidal ideation must be explored thoroughly with any patient (U.S. Public Health Service, 1999.)

Depression among this cohort also can be caused by a variety of organic problems including medical disorders such as Parkinson's disease or even a urinary tract infection. Individuals with depression also are more likely to be simultaneously coping with cardiovascular, cerebrovascular, and musculoskeletal diseases (Charney et al. 2003). In making an appropriate diagnosis of depression, it is just as important to ask older patients about changes in their sexual habits and interests as it is their younger counterparts. It often is difficult to ascertain whether sexual difficulties among depressed older adults represent a symptom of the depression itself or whether they stem from other concomitant medical or psychological problems.

Issues in Treatment

Clinicians themselves are beginning to display age-appropriate, realistic attitudes toward older adults and their chances for success in response to appropriate treatment (e.g., Hillman 2008a, b). However, sometimes patients and their spouses may have unrealistic expectations for treatment. One patient's wife told her therapist, "Well, if I sleep with him more often…that should cheer him up and get him out of his funk!" Thus, including the partner in therapy and treatment planning, with appropriate permission from the patient, of course, may be vital. With this arrangement, both parties are encouraged to discuss current changes in their sex life, possible guilt about the etiology of the disorder, and realistic expectations for treatment.

Another aspect of treatment that can impact greatly on a middle-aged or older patient's sexual expression is the multitude of side effects that accompany most antidepressant medications. The antidepressant drugs listed in Table 5.3 have been shown to cause problems with the normal sexual response cycle, including diminished overall arousal and inability to orgasm among both men and women (McVary 2007). Popularly prescribed serotonin reuptake inhibitors (e.g., fluoxetine, sertraline, and paroxetine) which are used widely among depressed older adults because of their low incidence of anticholinergic effects (e.g., dry mouth and constipation) are responsible for many of these sexual side effects. The pervasiveness of these debilitating side effects has been documented extensively. Findings from clinical studies suggest that up to 70% of women (e.g., Clayton and Montejo 2006) and up to 80% of men taking SSRIs experience some type of sexual dysfunction including low desire, inability to orgasm, and ED (Balon 2006). The use of alternative or adjunctive medications such as bupropion (i.e., Wellbutrin), occasional drug holidays, or a decrease in dosage of the current medication may help alleviate patients' sexual dysfunction (Safarinejad 2010, 2011). In sum, psychotherapists should be well aware of any medications that their patients are taking for depression, and both ask and inform their patients about possible sexual side effects. There *are* ways to combat antidepressant medications' side effects.

Table 5.3 Antidepressant and other psychiatric medications with sexual side effects

Drug	Trade name
Amitriptyline	Elavil
Amoxapine	Asendin
Buspirone	Buspar
Chlordiazepoxide	Librium
Chlorpromazine	Thorazine
Clomipramine	Anafranil
Clorazepate	Tranxene
Desipramine	Norpramin
Diazepam	Valium
Doxepin	Sinequan
Fluoxetine	Prozac
Fluphenazine	Prolixin
Imipramine	Tofranil
Isocarboxazid	Marplan
Lorazepam	Ativan
Meprobamate	Equanil
Mesoridazine	Serentil
Nortriptyline	Pamelor
Oxazepam	Serax
Phenelzine	Nardil
Phenytoin	Dilantin
Sertraline	Zoloft
Thioridazine	Mellaril
Thiothixene	Navane
Tranylcypromine	Parnate
Trifluoperazine	Stelazine

Note: Patients must be cautioned not to stop taking any medication for depression without talking to their physician first

Other antidepressant medications, including selective serotonin reuptake inhibitors such as Prozac, may induce the opposite effect intended in its use among elderly patients. One such side effect is that of manic behavior. Although this manic response is generally rare, it may present itself in unusual ways and may reveal underlying sexual frustrations or conflicts.

Mario

Mario was a 68-year-old man diagnosed with severe depression. He took a daily dose of Prozac in conjunction with his weekly psychotherapy. After a few months of treatment, Mario became manic as a side effect of his medication. His speech became rapid and pressured, he became visibly agitated, he became grandiose, and his judgment became impaired. During a therapy session, he showed his psychologist a note written by his wife. Mario said that he promised his wife he would tell "his story," even though he believed it was just nothing. Apparently, Mario had begun to cross the street haphazardly, barely avoiding being hit by passing cars. When asked if his wife and his therapist had reason to be concerned about him getting hurt in traffic, he responded, "Well, I just dare them to hit me…I feel like they would just bounce off of me…look at me, I'm a strong guy for my age! See those muscles!"

Mario also said that his wife got very upset with him because he had begun to flirt with waitresses when they went out to dinner. "You know, I'm not hurting anybody. I'm just having fun!" His therapist calmly asked about the incident in detail, and asked Mario why he thought his wife got so upset when he did not stop; he admitted that the restaurant owner had come to their table to ask him to "keep it down" and "to keep his hands to himself." Mario also admitted that he had gotten up a few times during dinner to "make a speech" to celebrate the occasion. On consultation with his psychiatrist and input from Mario, his medication was adjusted.

Once his dosage had been adjusted and he was no longer manic, Mario recognized easily that he should not have touched the waitress without her permission. He also admitted that he wished he could continue to feel so "high" instead of feeling depressed. He also had enough ego strength to recognize that his desire to be desirable to women represented an underlying conflict in his own marriage and his difficulties in feeling "old, weak, and unmanly." With the help of his therapist, Mario was able to work on these issues underlying his depression, and develop more positive coping strategies.

Dementia

One of the most dreaded illnesses among older adults is that of dementia. Statistically, 14% of US adults over the age of 70 are estimated to have some type of dementia, with 10% or at least 5 million men and women suffering specifically from Alzheimer's disease (Plassman et al. 2007). Older adults may fear "losing their minds" and becoming a significant burden upon their partner or family members. Because a number of disorders can result in significant decline in mental status, including Parkinson's disease, vascular disease, Huntington's chorea, and Pick's disease as well as Alzheimer's disease, a differential diagnosis becomes vital in coping with these debilitating disorders. Regardless of the specific type of dementia affecting an individual, the life-altering course of the disease demands significant attention from geropsychological practitioners.

Caring for the Caregiver

A number of factors arise in consort with these dementing diseases in relation to sexual behavior. One of the most difficult problems often arises for caregivers who must redefine their role as both caregiver and spouse or lover. The daily demands of caregiving can leave one feeling tired, anxious, and depressed. Tasks may range from helping their loved one to climb the stairs to use the toilet, bathe, dress, and eat. Caregivers have described this perceived change in role as distressing. One older woman described her transformation from spouse and lover to "doting parent." She noted with disdain that after she "changed [her husband's] diapers and put on his bib to feed him," she felt too much like his mother to even consider kissing or cuddling him. It somehow felt safer for her to stroke his head and cheek, and literally pat him on the head as she often did with her sick children. It appears that ingrained incest taboos surface among caregivers who feel that they serve more of a parenting function, and thus could not possibly engage in an intimate, sexualized relationship with their spouse, who now effectively represents a childlike figure. It can be a significant challenge to separate caregiving and spousal or partner roles (ALS Association, 2011; Duffy 1995).

Many caregivers report that their loved one has two deaths—the death of the personality or "true self" followed by the slow death of the body. Many caregivers feel that their intimate relationship ends when their spouse can no longer recognize them or know their name. Other caregivers report that they are too tired after physically demanding daily routines to even consider sexual contact. Still others describe anger and resentment interfering with any thought of a love life. One older caregiver responded by saying that he missed his wife's cooking and cleaning. He was forced to take a new role at home and although he knew it was not his wife's fault that she was ill, he was angry that he had to learn how to cook, do the laundry, shop for groceries, and clean the house. He said that he could not imagine having sex with a woman who was clearly no longer his "real wife." Certainly, it is not a therapist's place to encourage a caregiver to seek out sexual contact with a spouse with dementia. Rather, it is more helpful to caregivers to vent their frustrations and to articulate their perceived change in roles to manage their frustration and anxiety.

Changes in Sexual Behavior

In contrast, some caregivers report enjoying sexual intimacy and closeness with a partner who has dementia. Some described feeling closer to their spouse because they were engaging in a familiar, comforting activity that did not require words. Others mentioned that they liked the increased sexual attention from their spouse; one symptom of dementia is diminished impulse control, which can often result in more frequent sexual advances in up to one quarter of men and women with dementia (Black et al. 2005; Gua 2008; Wallace and Safer 2009). Still others felt that sexual behavior had a calming effect on their partner, allowing both of them to relax and reduce overall levels of tension and stress in the household for a few hours. However, the majority of caregivers find such changes in sexual behavior puzzling and distressing. A number of questions and concerns have been reported among caregivers and these include:

- Anger and confusion when their, otherwise caring, sensitive partner wishes to engage in intercourse without any foreplay. This apparently insensitive behavior is more likely to be a reflection of difficulties in remembering the correct "order" or sequencing of events (Duffy 1995). Just as someone with Alzheimer's disease may attempt to put on his pants before putting on his underwear, damage to the temporal lobe may result in such a person's attempt to engage in intercourse before emotionally and physically readying their partner. Sometimes redirecting a confused spouse can be effective. For other caregivers, the simple knowledge that this abrupt behavior is motivated by neurological deficits and not a disregard for their emotional needs can allay anxiety [and upset].

- Fears that their partner may begin to expose him- or herself or engage in sexual behaviors in public. It is notable, however, that the majority of patients with Alzheimer's disease do not engage in public sexual displays (Duffy 1995). If a person with dementia does lift her skirt or unzip his pants in public, it is more likely to be a sign of uncomfortable clothing or a need to void rather than a sexual urge (Ballard 1995).
- Distress that after a sexual encounter, their partner attempts to engage in sex again, as though forgetting what had just taken place. Similarly, anxiety has been reported among caregivers whose spouses became hypersexual as a result of their Alzheimer's disease (Duffy 1995). These partners even described feeling frightened about the consequences (e.g., possible temper tantrums and angry outbursts) if they would not go along with their partner's wishes for sex. Sometimes distraction and redirection are useful in these cases, but it is important to note that the underlying emotional distress caused by these incidents does not fade as quickly as the temporarily thwarted behavior.
- A sense of duty to perform sexually for their partner, particularly among female spouses, even if they are emotionally unavailable or averse to the idea (Duffy 1995). Some women from the current older adult cohort subscribe to the notion that they are to perform for their partner both in the kitchen and in the bedroom; "a good wife never says 'no.'" A frank, objective discussion of this role within the context of a therapeutic relationship may help caregivers avoid the guilt associated with this belief and make a more appropriate assessment about whether they would like to engage in sexual relations with their disabled partner or not.
- Fears that others in their caregiver support group will not understand their sexual concerns. One elderly woman noted, "Well, I just assumed that people would think I was a dirty old lady or that I was just selfish if I wanted to talk about sex with Gerald…I love him and want to take care of him, but I just don't want to have sex with him now…what will the other ladies think of me?" Because group therapy and support groups can be so vital for older caregivers, group leaders can initiate discussions about such sexual concerns in order to promote supportive group norms and to validate existing concerns among members.

It also is important to note that most research regarding the impact of dementing illness on sexual functioning focuses almost exclusively on men and women in the advanced stages of the illness. It is notable that in most clinical contexts, patients suspected of suffering from dementia are often ignored during clinical interviews and therapy trials. Even if a patient has some intact mental capabilities, the caregiver is attended to almost exclusively in terms of decision making and needs for therapy. A patient's diagnosis of dementia may be so distressing for both clinicians and family members that the patient in question becomes ignored in a form of denial.

Although challenging, involving both partners in therapy if one is in the early stages of the illness can allow for a discussion of future plans for sexual relations as well as other future plans about long-term care, end-of-life issues, financial planning, and learning how to "say good-bye." One older man recently diagnosed with Alzheimer's disease said that he wanted that time to talk with his wife about his guilt for leaving her in the prime of their lives, to discuss ways to make the house safe for his grandchildren to come visit in case he forgot to turn off the stove, and to tell his wife about his love for her and the value he places on their marriage on both emotional and physical levels. Although this couple shed many tears in therapy, they resolved some important issues from their past and were better equipped to plan for the future. This man's wife reported that she was able to enjoy their sex life until he no longer recognized her. "We both discussed that when he didn't know me anymore, it was all right to think he was gone…I didn't have to feel guilty about saying 'no' to him after that…I can now honor his memory and take care of him out of duty and love."

Schizophrenia

Managing chronic mental disorders such as schizophrenia and bipolar disorder can be challenging at any age. However, advanced age can make the treatment of such disorders even more difficult because of increasing concerns about side effects from medications, medical complications from other chronic ailments, and toxicity effects over time. At times, the presence of a serious infection or cancer may prevent elderly patients from taking their psychotropic medications, and may lead to an emergence or exacerbation of symptoms. Caregivers of these patients often are not asked how these disorders impact on their relationship or their sex lives.

The use of antipsychotic and mood-stabilizing medications alone can result in altered sexual function. The prescription drugs in Table 5.4 have been shown to produce side effects that can drastically alter the normal sexual response cycle. Side effects can range from a decrease in sexual interest to difficulties achieving orgasm to ED and may affect nearly 50% of all men and 40% of all women taking this class of medication (Montejo et al. 2010).

If a patient with a chronic mental illness becomes *unable* to take his or her prescribed treatment medications, the resulting chance in functioning can be extremely distressing and confusing to a partner or spouse. Sexual acting out and sexualized delusions can be particularly difficult to cope with, for both a patient's partner and her health care provider.

Table 5.4 Antipsychotic
medications that can induce
sexual side effects

Drug name	Brand or trade name
Benztropine	Cogentin[a]
Chlorpromazine	Thorazine
Clozapine	Clozaril, FazaClo
Haloperidol	Haldol
Lithium	Eskalith, Lithonate
Mesoridazine	Serentil
Risperidone	Risperdal
Thioridazine	Mellaril
Trifluoperazine	Stelazine, Suprazine

Note: Patients should be advised that any change in their medication regimen should be discussed with their physician or other qualified health care provider first
[a]Often prescribed to treat side effects such as restless movement and tremors that may occur with some antipsychotic drugs

Paul

Paul was a 68-year-old man who came to a geropsychiatry outpatient clinic for treatment of depression. He was a quiet, religious man of the Catholic faith with an obsessive character. His history revealed that he married at 26, and at that time he was aware that "something was wrong" with his wife. She would sometimes be too frightened to leave the house, and she had to quit her job as a receptionist. A few years later, she was diagnosed with paranoid schizophrenia. Paul said that he enjoyed his role as protector and provider; it made him feel like he was fulfilling his marital obligations and he felt competent because he managed most of the household chores while working full time as a mechanic. With the development of more effective psychotropic medications, Paul's wife became able to enjoy being a housewife and mother. She could tolerate going to church, cooking light meals, and participating in small social gatherings. The couple had four children, and Paul was delighted to follow his religious beliefs and bear children. It was one of his most important roles in life, next to being a faithful husband. Unfortunately, two of the couple's children were institutionalized in their mid-30s after developing paranoid schizophrenia.

In the past year, Paul felt that his wife was becoming more and more emotionally distant. He had researched her illness thoroughly and felt that she was becoming increasingly paranoid. As had been the case early in their marriage, Paul's wife feared leaving their home and would sometimes scream out for no reason, or point to strangers and shudder in fear. Paul decided to seek help for himself for depression after his wife was diagnosed with bone cancer. Her chemotherapy and radiation treatments were difficult, and her physicians advised against her taking her psychotropic medication. A few months into her treatment for bone cancer, Paul's wife was diagnosed with inoperable brain cancer. Paul was overwhelmed with fear, anger, and grief.

Paul was most depressed about losing his life partner. They had forged a strong bond over the years, and Paul was content in his role as advisor, protector, husband, and lover. Although he gave limited details in therapy, Paul said that he always felt emotionally and physically close to his wife. He noted that when she was particularly upset (i.e., agitated or paranoid), sexual contact would calm both of them and "remind them of what life was really about." Although he was not able to articulate his position fully, Paul's feelings of inadequacy in his own family (he was seen as the "least capable" of 10 children in his family of origin) were negated when he took on the role of rescuer and potent husband in his own marriage to a woman with a chronic illness. Paul again began to feel inadequate as his wife's cancer metastasized. She began to lose control over her bladder and needed help feeding herself. When he could no longer take care of her daily needs at home, he placed her in a nursing home. He experienced this as a personal failure and was upset that he could not provide for all of her daily care needs.

Paul placed his wife's "needs" above his own, and his stoically obsessive character would not allow him to easily express his feelings of fear and loss. During one session, Paul's otherwise calm demeanor shattered when he balled his hands into fists and shouted angrily, "Damn her illness! Damn it! I can't take it, just not this!" When asked what had happened, Paul's eyes began to water and his voice softened. He said that his wife's delusions had begun to change. Even though he knew logically that they were likely to be influenced by the growing tumor invading her brain, he still could not accept that "some part of her really thinks that!" During his last visit to the nursing home (Paul drove over 1 h each way to visit her and did so almost every day), his wife recoiled from his touch. She screamed at Paul and said, "My husband is Jesus Christ! He is my husband! Get out of my bed. He is my lover, my true lover! I want him in me now! Now! Get away from me. God is my boyfriend…who the hell are you? Get your hands off me…You aren't holy!" Paul became suicidal when his wife said that God became her lover. His hands shook when he said plaintively, "I have given up my life to God and so, what, what is this, he, he takes my wife?! What did I do to deserve this?"

It took quite a bit of work in therapy for Paul to articulate his anger at God for allowing this tragedy to happen, and for his wife to forsake him when he felt he was fulfilling his religious duty to serve her as a true husband in both sickness and health. Paul's suicidal ideation ceased when he was able to openly discuss his more "carnal" needs and his anger about being rejected by his wife. On some level, Paul could accept that her delusions may be, in part, an unconscious reflection of his deep religious beliefs and also a means through which his wife could distance herself from the relationship in order to avoid feeling the pain of their impending separation. When he was able to articulate his anger, face his feelings of inadequacy, and acknowledge the bitter irony of his wife's choice of delusional lover, he became better able to reach out to others as well as to his wife. He rekindled friendships with men with whom he worked, his adult children who were living in other states, and a widowed male neighbor from across the street. Even though his wife could not understand what he was saying to her in her last days, Paul was able to tell her that he loved her and that he understood how she could "turn to God" so completely.

Sequela of Sexual Trauma

Although not a chronic illness per se, the effects of sexual abuse or trauma are believed to include long-term personality changes and stress responses such as post-traumatic stress disorder (Feiring et al. 2009; Schnatz et al. 2010), borderline personality disorder (Davies and Frawley 1994), and psychosomatic illnesses. A large-scale meta-analysis of empirical findings derived primarily from young adult samples indicates that sexual abuse such as rape and incest may be poorly integrated or processed to emerge later as somatic disorders (Paras et al. 2009). Young men and women with histories of sexual abuse have been shown to have a greater likelihood of somatic, dissociative, affective, and substance abuse disorders (Feiring et al. 2009; Schoedl et al. 2010), and to exhibit specific somatic symptoms such as vocal cord paralysis, vomiting and other gastrointestinal distress, fibromyalgia, nonspecific chronic pain, and chronic pelvic pain (e.g., Freedman et al. 1991; Paras et al. 2009).

Virtually no empirical evidence exists to suggest that early sexual abuse can lend itself to the development or continuation of such somatic disorders among older adults. However, anecdotal evidence (e.g., Feil 1995) is building to suggest that older adults simply do not outgrow or simply forget about past sexual traumas. Some clinicians may not even inquire about early sexual abuse during initial interviews, and may assume out of denial or optimistic ignorance that time heals all wounds. Unfortunately, developmental issues associated with aging itself can complicate past issues of sexual abuse. For an older adult who has lost mobility and independence as a result of some chronic illness, this loss of control may allow previously memories of being abused and out of control to surface. Previous memories of abuse also may be compromised when an older person's mental status becomes impaired through either dementia or delirium. Because of cohort effects, older adults who experienced sexual assaults as children may have been told by frightened parents that "it never happened" or simply to "forget about it." Most elderly people grew up in a time when societal prohibitions warned against discussing sexual issues in general, much less about discussing sexual abuse among family members, friends, or even their family physician.

Barbara

Barbara was an 89-year-old woman who lived in an assisted living complex. She had been living in the community, but could no longer care for herself after an extensive hip reconstruction. A few weeks after arriving at the center, she presented with a variety of somatic symptoms including nausea, headaches, dizziness, ringing in her ears, and "flashes of light" in her eyes. Extensive medical workups revealed no underlying strokes, high blood pressure, or any other physical basis for her ailments.

Barbara soon took to lying in bed for most of the day. She seemed to enjoy getting attention from the aides and physicians and began to ring for assistance more and more often during the day, while becoming more and more removed from her peers. Barbara agreed to begin work in psychotherapy after her fifth medical consult revealed no underlying physical problems.

In therapy, Barbara only reluctantly discussed her hip surgery. She often avoided her therapist's questions about her quality of relationships and engaged in little introspection. One afternoon, a large package arrived for Barbara at the center. Her therapist asked Barbara what arrived for her, and she replied that it was a selection of new clothes in response to her distaste for her current clothing. Barbara had become upset when she was forced to wear a "frumpy" hospital gown and was ushered into bed by young orderlies or "hoodlums" after her surgery. She currently was very upset about her declining appearance, and began to spend a significant amount of time talking about feeling trapped in her bed, her wheelchair, and her room. When asked if she ever felt trapped before, Barbara started to cry, but would not respond directly to her therapist's gentle inquiry.

After working together for a few months, Barbara began to engage in scheduled social events at her extended care facility. She told her therapist that she took pride in her appearance, and that a decent-looking man asked her to go for ice cream together later in the week. When asked what she thought about this man, Barbara's voice became high-pitched and loud, "The hell with him, honey. Take them for all they're worth, that's what I say…Who cares about him?" A discussion of her past relationships revealed that Barbara had been divorced twice, and that her husbands had always been furious with her constant flirting, batting of eyelashes, and wearing of dresses "cut up to there." Barbara told many stories about going out to fancy parties wearing scantily cut dresses, and getting men to take her to the finest restaurants and theaters. When asked if she wanted to settle down, she said, "That's all part of the game-to make them pay, you know? I could get their attention, that's for sure. A nice dinner, all the right gifts, but oh, no, you don't."

When asked what "oh, no, you don't" meant, Barbara said that that was "her power." When her therapist responded that she wanted to understand exactly what Barbara meant, she became frustrated and retorted, "Oh, my power, my power over all of them now…I got my way back at them." Further discussion allowed Barbara to recount an incident that occurred when she was 18 years old. She had graduated from high school and took a job as a file clerk for a large trucking company. One afternoon, she ventured into the large garage looking for some old files. She had become friendly with some of the workers there (who were all men except for Barbara and the boss's wife who only occasionally came to the office to help file things), and she felt comfortable asking them to move one of the large filing cabinets blocking some others in the corner. Barbara described that "out of nowhere," one of the guys grabbed her and said, "I'll help you, little lady." He pushed her into a back room of the trucking bay and raped her. He then called one of his buddies into the room and he raped Barbara, too. Barbara said that she had tried to scream, but they held their hands over her mouth and threatened to stab her with some sharp-looking tools.

After the rape, Barbara got dressed, and went back to her desk, dazed. One of the truckers who befriended her saw her disheveled appearance and asked her what happened. Barbara said that she just sat there and shook her head and began to cry; "I couldn't tell him anything, not that, it was so horrible and I was so ashamed." The trucker apparently put two and two together, and went out into the trucking bay. He gathered up some other employees from the office, and apparently they discovered the two rapists and beat them savagely, requiring them to make a trip to the hospital. After this happened, Barbara said that "everybody was sure nice to me," but that nothing was ever mentioned about the incident. Barbara said that she went to the hospital a few weeks later for "a nervous breakdown," and that her boss came to visit her and offered her a retirement package, even though she had only worked there for a few months. Barbara took the retirement package and left town shortly after to live with her aunt.

A few months later, Barbara married the first man who showed any romantic interest in her. He was a member of a minority group, and her family took the news of the marriage badly. She said that her husband became "understandably frustrated" with her because she found sex with him painful and anxiety provoking. (She made no connection between her rape and her difficult sexual relationship with her husband.) Her husband became even more frustrated when Barbara began to buy and wear sexy clothes and to flirt openly with other men, even when they were "out on the town" together. The marriage ended a year later when her husband asked for a divorce. Barbara moved into an apartment with a coworker from the local phone company, and began "moving up the social scene" by targeting successful men for flirtation. She always made sure that she had very limited time alone with her dates, and often ended a relationship "before it really got started." Her second marriage to a wealthy businessman ended when he had an affair with his secretary, who Barbara assumed was "more free in bed."

The breakup of her second marriage put Barbara into "second gear." She resumed her flirting and socializing with vigor, and recognized that her anger at men could best be engaged by employing her feminine wiles. As she became older, however, she became increasingly "sick." As her appearance deteriorated, related in part to her adult years of heavy smoking and drinking, Barbara found that she could not get the kind of male attention that she was used to. Her headaches, dizziness, and muscle aches worsened, and she became a frequent visitor to various physicians' offices and hospital clinics. By the time she arrived at the assisted-living community, she herself felt that she had little hope to "get anything from men anymore," and had begun to garner male attention exclusively through her psychosomatic symptoms. After she was able to discuss her rape in therapy, Barbara's symptoms subsided somewhat. Barbara noted that "even after all of these years," she did not realize how "fresh" those memories were when she talked about them. With each telling of her story, her anxiety level diminished, and she was able to accept that the rape was not her fault, and that she had nothing to be ashamed about. Barbara went on to develop some meaningful same- and opposite-sex friendships in the assisted-living community.

Summary

A number of challenges face many older adults who suffer from chronic mental and physical illnesses, as well as acute psychological and physical disorders. A variety of ailments themselves, such as arthritis and heart disease, can impinge upon an older adult's sexual functioning directly. Sometimes the treatment can be as detrimental as the disorder itself in terms of sexual functioning. For example, the medications commonly used to treat certain conditions including high blood pressure, diabetes, arthritis, and schizophrenia can alter an older adult's sexual response cycle. Other disorders, such as urinary incontinence, can be effectively treated with conservative measures, and most members of the public as well as members of the mental health community are unaware of these recent advances. Other aspects of a person's life, such as the experience of sexual trauma or disfiguring surgeries in later life, have the potential to impact negatively on older adults' sex lives. In many cases, the caregivers of these patients require as much attention as the older patients themselves. It has become increasingly vital for clinicians to become knowledgeable about various illnesses that become more common with age, and about how both their course and treatment can impact upon their older patients' sexuality. Psychoeducation for older patients with chronic illnesses appears to be a key feature of more successful coping.

Chapter 6
HIV/AIDS and Other STDs among Adults over 50

Men and women over the age of 50 now account for more than 15% of all new HIV/AIDS cases and are becoming infected at rates four times higher than young adults. More than half of all people living with HIV/AIDS will be aged 50 and older within the next decade.

Health care providers can no longer afford to regard HIV and AIDS as a disease of youth and young adulthood. According to the Centers for Disease Control (CDC, 2008a), approximately 15% of all new HIV/AIDS cases in the USA occur among men and women over the age of 50. The rise in documented cases of HIV/AIDS and the increasing incidence of other STDs among adults over the age of 50 is unprecedented. Two distinct populations emerge upon examination of older adults with HIV: middle-aged and older adults who contract HIV/AIDS and other STDs for the first time at that point in their lives and middle-aged and older adults previously infected with HIV who are now living into midlife and beyond. Each of these groups must cope with unique age-related risk factors for infection and treatment, as well as underlying medical, economic, social, and psychological stressors including stigma and discrimination.

Relevant Statistics

New HIV/AIDS Diagnoses

According to the Centers for Disease Control and other experts, approximately 15% of all new HIV/AIDS diagnoses are made among adults over the age of 50 (CDC 2008b; Linley et al. 2007). Rates of HIV/AIDS infection are increasing nearly four times faster among older than young adults, and the primary means of infection among older adults is through sexual contact (Chiao et al. 1999). Notable disparities also exist in terms of ethnic differences in diagnoses. Older adults of color are at

J. Hillman, *Sexuality and Aging: Clinical Perspectives*,
DOI 10.1007/978-1-4614-3399-6_6, © Springer Science+Business Media New York 2012

increased risk of infection. Specifically, older African-Americans are 12 times more likely than older Whites to be diagnosed with HIV infection, and older Latinos are 5 times more likely to receive an HIV diagnosis than older Whites. Put another way, the incidence rate of HIV/AIDS for White US citizens over the age of 50 is 4.2/100,000 people. For older Latino citizens it is 21.4/100,000, and for older Blacks it is 51.7/100,000 (CDC 2008b; Linley et al. 2007).

Specific areas of the country that draw retirees in greater numbers, such as Florida, California, and Arizona, also report substantial increases in numbers of HIV and AIDS cases. In Florida, approximately 17% of all new AIDS cases were among adults over the age of 50. Of those newly diagnosed, 74% were male, and 26% were female. Nearly two-thirds of the infected women and nearly one-half of the infected men were Black. For both men and women, the primary mode of transmission was sexual contact (including 25% of infections among older men via heterosexual contact), followed by IV drug use (Florida Department of Health 2007). In California, 16% of all new HIV/AIDS cases and 40% of the current population of people living with HIV and AIDS (PLWHA) are over 50 (Foster 2011). In New York City, 17% of all new HIV infections and 37% of PLWHA are over the age of 50. In addition, increasing numbers of older women in New York City are being diagnosed with HIV/AIDS each year (NYC Department of Health and Mental Hygiene, 2010).

Despite the wealth of information we now have about HIV among older adults, it is unclear how many adults currently living with HIV remain undiagnosed and untreated. For example, symptoms of HIV infection may be different between older and younger adults. Early symptoms of AIDS among older adults often include memory loss, poor attention, and confusion, which can be misdiagnosed as Alzheimer's disease or simply dismissed as "normal" signs of aging. Although the symptom presentation can vary significantly, the cognitive changes that may accompany HIV infection are often referred to as HIV-associated dementia, AIDS dementia complex, or HIV/AIDS encephalopathy. CDC statistics also are likely to underestimate the actual numbers of HIV-infected adults over the age of 50 due to underreporting (e.g., family members often ask that their physician identify a non-AIDS related cause of death on their relative's death certificate), misdiagnosis, and the exclusion of older adults who may be asymptomatic except for the cognitive changes associated with HIV-induced dementia.

Older People Living with HIV and AIDS

Current estimates suggest that 1 in 3 PLWHA are over the age of 50 (CDC 2008a), with the numbers of older PLWHA increased by more than 75% between the years of 2001 and 2005 (CDC 2008a). With the improved effectiveness of antiviral medications, notably highly active antiretroviral therapy (HAART), infection with HIV is no longer viewed as an imminent death sentence. Many PLWHA and their health care providers now consider life with HIV much like living with a chronic illness.

In response to both the increased number of new HIV diagnoses, as well as the effectiveness of HAART, estimates suggest that that within the next decade more than 50% of every two PLWHA will be over the age of 50 (CDC 2008b).

This significant increase in HIV infection among older adults is often referred to as the graying of the AIDS epidemic (Gorman 2006), and it poses significant implications for not just the USA but global health (Emlet 2006a, 2006b, 2006c). AIDS represents the 15th leading cause of death among men and women over the age of 65 (Kaye and Markus 1997). More older US adults have now died of AIDS than the number of American GIs killed in the Vietnam War, and more than one-third of all people who currently die from AIDS are over the age of 50 (CDC 2007). Clear ethnic disparities also appear in which nearly half of all older PLWHA in the USA are Black (CDC 2008b). The health disparities involved in treatment of older adults with HIV, particularly those who are from a low SES or minority group status, cannot be ignored.

Increased Incidence of Additional STDs

Although the highest incidence of STDs among older adults is that of HIV/AIDS, significant increases in other STD diagnoses have emerged in relation to Chlamydia, syphilis, and gonorrhea (Jena et al. 2010), as well as the sometimes sexually transmitted Hepatitis C (Institute of Medicine 2010). For older men, the incidence of Chlamydia is 11 per 100,000 and the incidence of syphilis is 3 per 100,000, with significant increases in the last 5 years. Among older men and women combined, the incidence of syphilis is 1 out of 100,000 and approximately 10 out of 100,000 for gonorrhea (CDC 2001). Although overall prevalence rates for STD infection among adults over 50 remain lower than those of young adults, the incidence rates of STDs among this older age group are on the rise (Calvert 2003).

Some unique symptom presentations and diagnostic issues appear in relation to STDs in people over the age of 50. Diagnosis and treatment of Chlamydia is vital, as untreated it can cause significant infection of the cervix and urethra (Calvert 2003). Syphilis is more likely to be diagnosed in the secondary and tertiary stages of the disease among adults over 50, with symptoms often emerging 5–20 years after the initial infection. Symptoms commonly reported at the time of a syphilis diagnosis include stroke, dementia, deafness, and reduced vision (Calvert 2003) as well as cardiovascular-related symptoms including aortic aneurysms (Swartz et al. 1999). Many older adults receive their initial diagnosis when a routine screening for syphilis is initiated upon hospitalization for stroke (Calvert 2003). Symptoms of gonorrhea among older adults may include a rash on the extremities (present in nearly 2/3 of all infected) as well as arthritis (Calvert 2003).

Up to 25% of adults over 50 may be infected with the sexually transmitted herpes simplex virus (HSV), with increased age serving as a significant risk factor. Incidence rates for herpes infection also are higher among Mexican (40%) and African-Americans (60%; Fleming et al. 1997). The incidence and prevalence of genital warts, otherwise known as human papillomavirus (HPV), among older adults is generally unknown. Among older women, greater concerns generally exist regarding HPV-related cervical cancer. Although the frequency at which older women should have a pap test is greatly contested (Calvert 2003), findings indicate that approximately 25% of all new cases of cervical cancer occur among women over the age of 65 (Cornelison et al. 2002).

Up to two-thirds of individuals diagnosed with Hepatitis C are over the age of 50. Many individuals in this age cohort contracted the disease decades ago, when using or experimenting with IV drug use. Up to 75% of those infected with Hepatitis C do not know they have the virus, as its symptoms may remain dormant for up to 30 years. The virus is transmitted through blood, which can occur during high-risk or unprotected sexual activity. In addition, older adults with HIV infection or other STDs face increased risk for contracting Hepatitis C due to already compromised immune systems. Although detection of the Hepatitis C virus involves a simple blood test, few physicians order such tests for their older adult patients or ask about related risk factors (Institute of Medicine 2010).

Age-Related Risk Factors

At the beginning of the AIDS epidemic in 1980, older adults were at greater risk than younger adults to contract HIV through infected blood products. Older adults were more likely to undergo lengthy operations that required transfusions (e.g., knee and hip replacements), open heart surgeries, and exploratory surgeries for internal bleeding after car accidents than their younger peers. Fortunately, as a result of increased safeguards and routine testing introduced to the US blood supply in 1985, the number of new HIV/AIDS cases due to infection via tainted blood products has essentially dropped to zero (Goodnough et al. 2003).

Unfortunately, many older adults and health care providers themselves are unfamiliar with the age-related factors that *currently* place older adults at increased risk for HIV infection. For example:

- Older adults are less likely to use condoms than younger adults for a variety of reasons, including diminished concerns about pregnancy, lack of familiarity with condoms (including how to put them on and use them correctly), and discomfort with condom negotiation
- Due to decreases in estrogen levels and related thinning of the vaginal wall and a decrease in naturally produced lubrication, post-menopausal women are more likely to experience micro- and macroscopic tears of the vagina during intercourse, allowing the HIV virus easier access to the bloodstream. Like most younger adult women, the majority of older adult women are not even aware that sexual intercourse is often accompanied by such trauma to the vaginal wall because these tears are often painless and generally undetectable after intercourse.
- Decreased levels of immune function, typically associated with increased age, make older adults more susceptible to infection, per exposure, when compared to younger adults.

- Older adults have never been the target of any national campaigns about STD prevention, and many assume that HIV/AIDS only affects young, gay, or IV drug users.
- No national programs exist that are specifically tailored for HIV prevention among minority group elders. As noted, older Black and Latinos are at increased risk for HIV infection compared to their White counterparts.
- Health care providers are significantly less likely to discuss issues related to HIV/AIDS with their older patients than their younger patients and least likely to discuss sexual issues with older female patients (Lindau et al. 2006).
- Older adults are unlikely to know that oil-based lubricants are not safe for use with condoms (Hillman 2008a, b). For example, older adults may believe incorrectly that baby oil and hand lotion are safe to use with condoms, and do not realize the need to use only water-based lubricants (e.g., KY-jelly; Astroglide).
- With older age cohorts displaying rising rates of separation and divorce, increasing numbers of older adults are finding themselves dating for the first time in years and may engage in related unprotected sex.
- Viagra and other drugs for the treatment of erectile dysfunction (ED) allow more older men to engage in intercourse than ever before. Physicians are not required to discuss STDs with their patients who take such performance-enhancing drugs, leading to increased risk of high-risk behavior. Older men who use drugs for ED, including widows, are 50% more likely to have HIV than those who do not use such medication (Jena et al. 2010; Smith and Christakis 2009).
- Despite common societal views that older adults are asexual, older men and women can and do engage in high-risk behaviors such as IV drug use and sex with multiple partners and prostitutes. The oldest person to have a documented case of HIV infection was an 88-year-old White widow. She is believed to have contracted HIV through sexual activity with her husband, a recreational IV drug user (Rosenzweig and Fillit 1992). The use of illicit substances also has been increasing among older adults, including the use of intravenous drugs such as heroin.
- The use of alcohol and drugs, including crack cocaine, is known to impede effective decision making for participation in safer sex practices. For older adults, who typically process alcohol and other drugs less effectively than younger adults due to age-related declines in liver function, impaired judgment after even one drink can increase risk.
- Older women face increased risk for HIV infection. Coupled with the general decrease in immune system functioning with age, an increased risk of vaginal trauma, and lower rates of condom use, an older adult woman has a higher risk of contracting HIV through each act of heterosexual intercourse than her younger adult, female counterpart.
- Men are more likely than women to influence whether a condom will be used during sexual intercourse, and older adult males are the least likely age group to use a condom,
- Many older adult clients do not even consider heterosexual contact as a risk factor for HIV, particularly if they are widow and widower who assume that their respective partners were involved in a long-term monogamous relationship.

Women and men over the age of 50 are significantly less likely to use condoms than their younger counterparts, particularly when a female partner is postmenopausal. In fact, only 25% of both men and women over the age of 50 have reported using a condom during casual sex. In contrast, teenagers between the ages of 14 and 17 report using condoms 80% of the time during casual sex (Schick et al. 2010). When asked about their use of condoms during sexual relations, one 69-year-old female client laughed and responded, "Oh, my. You know that I don't have to worry about a baby on the way! But that is flattering that you would think about me being so young to even ask. You are so sweet."

Other patients have responded to questions about condom use with some variant of, "Oh, that's just for people who are dirty, you know what I mean? My Reggie and I have been together for over a year. I don't have to worry about anything. He's squeaky clean." When asked to elaborate upon what it means to be "squeaky clean," it was revealed that this statement was based upon her visual examination of her boyfriend's genitals, and not any kind of medical examination or result. This client had also never discussed condom use with her boyfriend, out of fears that he would reject her for "being a hussy, because only dirty girls need to use condoms in the first place." When asked about condom use with her boyfriend, another 52-year-old female patient responded simply, "I just don't like to think about it. I'm over 50 now. What are the odds of me catching something, anyway?" Significant psychoeducation and therapeutic work need to take place in order for older adults to acknowledge and realistically assess their risk factors.

Cultural Factors

Various cultural factors can pose unique risks for HIV/AIDS infection among adults over 50. In some Black cultures, for example, men having sex with men (MSM) is not labeled as gay or homosexual, and those MSM may not believe that HIV/AIDS prevention messages are relevant to them. In traditional Latino culture, the concepts of machismo and marianismo, gender roles that devalue education for women, and subscription to certain Catholic beliefs that prohibit condom use may contribute significantly to increased rates of HIV/AIDS infection. As noted, older Latinos are five times more likely to be diagnosed with HIV/AIDS than older Whites (CDC 2008b.)

Machismo dictates that Latino men possess a vigorous sex drive that they are typically unable to satisfy or control. In traditional culture, Latino men are virtually expected to engage in promiscuous sex with a variety of partners. Within the context of STD infection, it is important to note that it is acceptable for Latino men to have both male and female sex partners. Male partners are those who are discussed as being "on the down low." Latino men are likely to have more than one partner at a time, even if they are married (Beaulaurier et al. 2009; Gonzalez et al. 2009).

Machismo also dictates that Latinos serve as a sole provider and that they are dominant in their household. Conversely, Latino culture also has a female role of marianismo, defined by submission, dependence, and lack of freedom (Rios-Ellis et al. 2005). A Latina is expected to be submissive to her father and then later to her husband. For a Latino woman to even ask her husband or boyfriend about his sexual activity is viewed as brazen and shocking, and Latino women who ask their significant other to use a condom are typically viewed as promiscuous themselves (Gonzalez et al. 2009). If a man is macho enough, the expectation is that his partner will be too fearful to even ask him to wear a condom (Rios-Ellis et al. 2005).

Marianismo also views education for women as superfluous and even selfish, as a woman's primary role is to remain at home to care for her husband and family. Limited formal education is certainly associated with lesser knowledge about physical and sexual health. Findings from a recent community-based study revealed that half of the older Latino participants were unaware that HIV could be passed through vaginal, anal, and oral sex. Results from the study also revealed that the majority of older Latinas were unaware of their risk for contracting HIV/AIDS, and that more than half feared that asking a macho partner to use a condom would damage their relationship (Hillman 2008a, b).

The Need for a Correct and Timely Diagnosis

One of the most intriguing aspects of the case of the oldest woman to be diagnosed with HIV/AIDS (Rosen and Fillit 1992), who also happened to be White and middle-class, is that she was a widow. This woman was first diagnosed with AIDS 7 years after the death of her husband. It took careful risk assessment and sensitive questioning by open-minded clinicians to arrive at the appropriate conclusion and provide appropriate treatment, as older women typically tend to be overlooked by health care professionals in relation to potential HIV infection (Zablotksy 1998).

Current studies show that only 25% of adults between 55 and 64 have been tested for HIV (Beaulaurier et al. 2009). For adults aged 65 and older, only 12% have ever received testing (CDC 2008c). Although Medicare pays for HIV tests, the CDC only recommends routine HIV testing for individuals only up to the age of 64 (CDC 2006). Unfortunately, most older adults who receive a diagnosis of HIV do so while they are hospitalized and not as part of routine medical care. Older adults also are more likely to receive their initial diagnosis after they have already developed AIDS (Mugavero et al. 2007).

The implications for an incorrect or a delayed diagnosis of HIV among older adults are insidious. Such errors postpone the delivery of appropriate medical, psychological, and psychosocial interventions and result in significant decreases in quality of life and ultimately in increased mortality (Linsk 1994). Unfortunately, due in part to a natural decline in immune system functioning with age, adults over 50 who contract HIV are likely to die sooner than their younger counterparts, even if both people are diagnosed and treated at the same time. Current CDC reports indicate that more than half of individuals who are first diagnosed with HIV when they are 50 or older develop full blown AIDS or die from AIDS-related illnesses within 1 year of their initial diagnosis (Linley et al. 2007). It also is important to note that of all racial groups, older Black women and men have the shortest length of time reported between their initial diagnosis and death from AIDS (CDC 2001). Significant health disparities persist in terms of both HIV detection and treatment for older adults of color.

Individuals with an HIV infection in mid and late life also are more likely to suffer from complications from chronic diseases commonly associated with advanced age such as heart disease and diabetes. The drugs associated with HAART can introduce negative side effects such as high cholesterol, high blood sugar, and insulin resistance (Silverberg et al. 2007; Wigfall et al. 2010). Unfortunately, it remains unclear to what extent individuals who were being treated for HIV when they were young or middle-age adults (when drug regimens were less sophisticated and caused greater side effects) are likely to age well into older adulthood. It also remains unclear to what extent older adults being diagnosed and treated initially with HAART will be able to live into advanced age without significant side effects and increased mortality from other chronic illnesses.

HIV-Associated Dementia

Like syphilis, HIV can cause significant neuropsychological dysfunction that often present in the form of cognitive impairment or dementia. Autopsy studies suggest that up to 80% of individuals who contract AIDS will develop HIV-associated dementia complex (HADC; American Academy of Neurology AIDS Task Force 1991). Clinical studies also suggest that more than one-third of all people currently infected with HIV will meet diagnostic criteria for dementia (Buckingham and Van Gorp 1988). Recognizing the causal relationship between such a sexually transmitted disease and a debilitating dementia is particularly important among older adults, whose symptom presentations of HADC may mimic, and subsequently be mistaken for, Alzheimer's disease. Another concern is that older individuals with HIV-associated cognitive deficits are significantly less likely to remain compliant with their HAART medication (Barclay et al. 2007). Older adults with HIV infection also are likely to show initial cognitive declines in attention and processing, which may impair their ability to drive a car or pilot an airplane safely (Hardy and Vance 2009).

Although many reviews of the cognitive deficits associated with HIV dementia among young adults are available, significantly fewer (e.g., Hardy and Vance 2009; Morgan et al. 2011) provide comprehensive summaries that differentiate between the symptoms associated with HADC in older as compared to younger adults. Such symptoms among older adults include but are not limited to cognitive changes such as impaired attention, impaired concentration, poor short-term memory, confusion, and impaired abstract thinking; affective symptoms such as apathy, indifference, and social withdrawal; and behavioral symptoms such as psychomotor slowing, diminished coordination, unsteady gait, difficulty with writing, impaired occupational functioning, and a significantly decreased ability to engage in activities of daily living. It

Table 6.1 Differences between HIV-associated dementia and Alzheimer's dementia

Symptoms	HIV-associated dementia	Alzheimer's dementia
Onset	Acute	Gradual
Progression	Rapid	Gradual
Time until acute stage	6 months–1 year	More than 1 year
Affect	Appropriate	Labile
Mood	Apathy, mania, depression	Depression
Use of language	Intact	Impaired (aphasia)
Short-term memory	Impaired	Impaired
Encoding	Intact	Impaired
Psychomotor speed	Significant slowing	Minor impairment
Tremor	Common	Rare
Gait	Impaired	Intact
Opportunistic infections	Common	Rare
Cerebrospinal fluid	Elevated protein levels	Average
T-cell count	Below average	Average

Note: These characteristics represent a summary of typical symptoms. Each client's symptom presentation is unique, and deviations from typical patterns must be expected

is important to note that unlike Alzheimer's dementia, HADC typically includes both cognitive and psychomotor impairment. The American Academy of Neurology AIDS Task Force (1991) also identifies HIV-Associated Minor Cognitive Motor Disorder (MCMD) as a neurological syndrome in which an individual does not yet meet criteria for dementia, but exhibits mildly or moderately impaired functioning along with psychomotor slowing and tremor.

Practitioners must be aware that the symptoms of HADC often mimic various symptoms of Alzheimer's disease. Table 6.1 summarizes the similarities and differences typically observed in their respective symptom presentations. Both disorders present with a debilitating picture, in which patients display significantly diminished social and occupational functioning, and short and long-term memory loss. However, some vital distinctions can be made between the typical presentations of HADC and Alzheimer's dementia. Regarding onset, HADC tends to be sudden, whereas recognizable onset of Alzheimer's dementia typically is gradual. HIV-induced dementia often has a rapid, aggressive progression over a period of 6 months to 1 year, whereas Alzheimer's dementia has a gradual progression of symptoms, often over many years. Language impairment, such as aphasia (an inability to speak), is not a part of the typical HADC presentation, unlike the typical presentation of Alzheimer's dementia. No pronounced language deficits are observed in HADC, until its very end stages.

Although short-term memory is significantly impaired in both HADC and Alzheimer's dementia, patients with HIV-induced dementia, unlike their counterparts with Alzheimer's, appear to maintain their ability to encode and learn new information. For example, although most patients with HADC and Alzheimer's dementia display impairment in their recall of a list of words, even after numerous repetitions, the patient with HADC would be more likely to recognize certain words as part of the original list (indicating that she or he was able to encode the new information, but was unable to retrieve it without cuing).

Patients with HADC, compared to those with Alzheimer's disease, also appear to have significant impairment in psychomotor speed. When asked to connect thought with action, a patient with HADC is expected to display significant difficulty. Additionally, patients with HADC often display other symptoms of AIDS, such as low-grade fevers, depressed mood, skin lesions, opportunistic infections (e.g., pneumonia, shingles), diarrhea, headache, night sweats, sudden weight loss, and incontinence. Neuropsychological testing can also show a relatively large "scatter" or dispersion of scores on test batteries for older PLWHA when compared to older adults with Alzheimer's disease (Morgan et al. 2011). Clinicians, clients, and their significant others and family members should recognize that reduced sex drive, depressed mood, memory impairment, and incontinence are *not* a normal part of healthy aging, and should be explored as a sign of underlying pathology or illness, whatever form that might take.

Adults Over 50 Living with HIV/AIDS

Stigma and Isolation

For older PLWHA, the potential stigma and discrimination related to HIV/AIDS from peers, health care providers, and society at large poses significant challenges. As a result, recent studies based upon HIV-positive adults over 50 in the New York area suggest that more than one-third of older PLWHA suffer from symptoms of clinical depression. Loneliness, social stigma, and cognitive impairment (e.g., early symptoms of HIV-associated dementia) also accounted for a significant role in explaining the older participants' depressive symptoms (Grov et al. 2010). Because the New York City metropolitan area serves as the base for a number of national HIV/AIDS, LGBT, and older adult HIV/AIDS service programs, and participants in the study were already linked with various social service agencies, the numbers of older PLWA with depressive symptoms are likely to be significantly higher for those who live outside such a large urban area.

Another contributing factor in depression among older PLWHA is that of the side effects often associated with HAART. Although HAART has significantly improved the quality of life for HIV-infected individuals, older adults are more likely than younger adults to encounter side effects and are more likely to be taking multiple drugs with various interactive effects. One side effect not commonly discussed among health care providers and their older patients living with HIV/AIDS is that of diarrhea. Studies suggest that nearly one-third of middle-aged and older adults being treated with HAART experience periods of debilitating diarrhea and fecal incontinence (Siegel et al. 2010). Due to fears of having an accident in public, many older PWLA in the study reported that they became homebound, which significantly limited their social interactions with others. Still other study participants expressed fears that the presence of diarrhea meant that they would lose weight quickly (i.e., wasting), indicating that they were approaching the end stage of the disease. Still other older PLWHA with diarrhea felt consistently ashamed, embarrassed, helpless, humiliated, and "dirty," contributing to further depressive symptoms and social withdrawal.

A consistent theme that emerges from all studies of older PLWHA is that of social isolation, brought on primarily through social stigma and fears of triggering such stigma (e.g., Emlet 2006b; Grov et al. 2010; Siegel et al. 2010). Anxiety, chronic stress, and depression appear to be closely linked. Qualitative analysis of responses from older HIV-positive men and women suggests that these middle-aged and older adults face a "double threat" of stigma associated with aging and HIV status if White (e.g., Emlet 2006c), and if they are of color, a "triple threat" of stigma associated with aging, HIV status, and minority group status (e.g., Haile et al. 2011). Older PLWHA in these studies also report that they experienced stigma and discrimination from both peers and health care professionals. Examples of such discrimination by professionals ranged from hostile and unkind remarks, to lapses in care, to the unethical and illegal breach of confidentiality regarding the older person's HIV-positive status (Emlet 2006c). One can only image the devastation brought upon an older PLWHA when their own health care providers single them out for hostility and rejection.

Secondary Prevention

Studies of self-reports by older PLWA indicate that they continue to engage in sexual activity with relatively high frequency (Illa et al. 2008). Many lay people do not realize that safe sex practices are recommended for HIV-positive individuals, even when both partners are HIV positive, in order to help prevent them from reinfecting each other and to avoid passing different strains of HIV to each another. Unfortunately, estimates suggest that significantly more than one-third of older PLWA engage in unprotected vaginal and anal sex, with gay men having lower rates of condom use than both heterosexual men and lesbian and heterosexual women (Golub et al. 2010). The use of drugs and alcohol before sex, as well as limited knowledge of HIV/AIDS, is also associated with unprotected sex and other high-risk activities (Cooperman et al. 2007; Lovejoy et al. 2008). Greater fear of stigma is also associated with a decreased likelihood of even sharing their HIV-positive status with current or potential sex partners.

In the first study of its kind, Golub and colleagues (2011) sought to discover what individual factors were associated with older HIV-infected adults' proactive use of condoms during sexual activity—and focus upon risk reduction. This group of researchers discovered that various aspects of psychological well-being, including one's sense of personal growth and potential, of having quality relationships with others, and of having control over one's life, were positively associated with condom use during sexual activity. Even when the older HIV-positive participants reported using substances, the potentially negative effect of drugs and alcohol upon condom use appeared to be overridden when psychological well-being was high.

Golub et al.'s (2011) study is important because their findings highlight the value of personal psychological resources among older HIV-infected individuals, and it suggests that therapeutic and social support interventions may serve as preventive health measures. Empirical studies have demonstrated consistently that both group therapy and interpersonally based support groups are very effective in reducing depressive symptoms among older PLWHA (Heckman et al. 2011), as well as increasing condom use when combined with condom negotiation skills (Illa et al. 2010). Because such studies suggest that group interventions are just as effective as individual interventions for older HIV-positive individuals, organizations and communities with limited resources can feel more confident about providing more typically cost-effective group interventions. Consistent with the value of social support for reducing feelings of social isolation and symptoms of depression, the contact information for various organizations available to assist both adults over 50 and their practitioners in coping with HIV/AIDS are provided in the appendix at the end of the chapter.

Knowledge Among Practitioners

Limited attempts have been made to explore health care providers' knowledge and attitudes about HIV among older adults. One such study employed a sample of physician assistants, psychologists, and nurses at a large mental health facility with specialized geriatric units (Hillman 1998). The results showed that these health care providers correctly identified gay men and IV drug users as among the top risk groups for HIV transmission among adults over the age of 50. However, although a significant number of the health care workers in the study were aware of the increase in HIV transmission among heterosexual older adults, the vast majority remained unaware of the increased risk of HIV transmission via intercourse in post menopausal women. They incorrectly estimated that an older adult woman who was date-raped by an HIV-infected rapist had the same chance of contracting HIV as a young adult woman raped by the same assailant.

Contrary to expectation, exposure to various patient populations including adults over the ages of 50 and 70 and work with patients who were HIV positive were not associated with more accurate knowledge of HIV or its transmission among adults over 50. Also contrary to expectation, specialized training in HIV, geriatrics, neuropsychology, and human sexuality was not associated with greater knowledge of HIV/AIDS or HIV-associated dementia among aging adults. It is unclear whether any of the training that these health care providers received broached the subject of HIV and aging specifically. It also is interesting to note that in debriefing after the study, various participants spontaneously remarked that they had never ever considered that one of their older patients might have HIV, and that they never received any formal or informal education about HIV and aging. It is clear that as practitioners, we must correctly educate ourselves and our clients.

Issues in Assessment

Questions at Intake

Agism is evident when health care providers fail to ask, or even consider, whether their older adult clients are at risk for contracting HIV/AIDS or other STDs. To make an accurate diagnosis, it is vital that clinicians ask their middle-aged and older clients candidly and directly about a variety of issues including (see Hillman and Stricker 1998, for a review):

- Sexual history, including extramarital affairs and multiple partners.
- Current sexual behaviors, including vaginal, oral, and anal intercourse.
- Partner selection, including heterosexual and LGBT.
- Recreational drug use, particularly IV drug use.
- Sharing of needles for insulin or other prescription medications.
- Major operations and blood transfusions that occurred before 1985.
- Use of blood products for hemophilia, before 1985.
- Caregiving for HIV/AIDS patients.
- Sexual abuse or assault.
- Changes in mental status such as apathy and confusion.
- Physical symptoms such as swollen glands, loss of appetite, night sweats, weight loss, and a nagging cough or cold.

Health care providers should never assume that an older client is free from the risk factors associated with HIV transmission. The potential emotional discomfort associated with asking older adults about their sexual activities and sexual history is well worth the effort if it reveals a possible link to an HIV/AIDS or other STD diagnosis and to timely, appropriate treatment. It also is vital that clinicians remember that a 4- to 7-year window may pass between infection with the HIV virus and overt symptoms. Thus, sexual history remains a vital part of any intake assessment, regardless of a patient's age, marital status, or current sexual abstinence. Recall that the oldest patient to have a documented case of HIV infection was an 88-year-old woman who had been widowed and sexually abstinent for more than 7 years.

We know that middle-aged and older adults, as well as younger adults, use IV drugs, have affairs, employ prostitutes, and are survivors of sexual assault. It is essential to ask older adults about their numbers of sex partners, and if they have engaged in unprotected sex with both male and female partners. Another salient item for investigation regards caregiving activities for individuals with HIV/AIDS, particularly for adult children who may be living at home and receiving nursing care from an older parent, or older adults who may be caring for grandchildren with HIV/AIDS. Although these topics may be difficult subjects to broach, the information gained is invaluable.

The Presence of Family Members

Due to the sensitive nature of these questions, particularly for some older adults who may not be socialized to openly discuss their sexual activities, it is advisable to speak to an older client without their spouse, significant other, or children being present for the entire interview. For better or worse, many older adults come to an initial interview with a spouse, a caregiver, or children. Additional information from family members can provide important information that might otherwise be lost, particularly if the older person has some cognitive issues, and their presence allows the clinician to observe family dynamics firsthand. However, it is advisable to discuss at the beginning of the intake that some time will be allotted to speak to the designated patient alone. Many clients, as well as their family members, spouses, partners, and concerned friends, become noticeably more relaxed about the procedure when they are simply told that it is standard procedure.

It also can be helpful to preface questions about a client's sexual history and high-risk behaviors by telling him or her that these are standard questions asked of every client at every interview. It is important to inform patients and clients that whatever information they reveal will be confidential, and that this information will not be disclosed to other family members without their expressed, written, informed consent, with the exception of mandatory reporting for danger to self or other, and elder abuse as mandated in some states. (One should not assume that all older adults are familiar or fully understand all HIPPA rules and regulations.) Normalizing the discussion of such typically private matters as sexual behavior and drug use often allows clients to speak freely and to express additional concerns (e.g., Hillman and Stricker 1998). It also is important that clinicians use the same sexual terms as the ones used by their patients. For example, if older clients use the term rubbers instead of condoms, one should follow their lead.

Queries about Caregiving

Many middle-aged and older adults find themselves assuming the role of caregiver for their adult children (or grandchildren) who are infected with HIV or AIDS. With the increasing number of AIDS patients in this country, difficulties with medical insurance coverage, an inability to acquire hospital or hospice care, and a lack of other available caregivers, many older parents, particularly older mothers, are providing daily nursing care for adult children with AIDS. Some of these parents describe having a positive experience, in which they are able to show their children that they love them, even to the bitter end. Other older clients in this caregiving role have spoken about acknowledging and accepting their child's LGBT lifestyle for the first time and about forgiving their child for turning to drugs as a way to deal with their problems. Other older patients have been more reluctant caregivers and have voiced concerns about who will care for them when they are sick and in need of care. Still others have discussed how preoccupied they are with keeping the nature of their child's illness a secret from friends and family members. They fear the stigma significantly, predicting that no one will visit them or want to socialize with them once their child's HIV-positive status is known.

The Benefits of a Team Approach

Regardless of the clinician's degree status, professional title, or proscribed role on a treatment team, each mental health professional should take personal responsibility to ensure that a potential HIV diagnosis is examined and, ideally, ruled out. The use of a team approach can provide a professional cohort to discuss and work through any feelings of discomfort regarding the discussion of high-risk behaviors and potential HIV status with a patient. Just as we often inform patients that we often have little control over how we feel, but that we are responsible for how we manage those feelings and our behavior, it is vital that team members remain respectful of other professionals' feelings, even if they appear negative, ambivalent, or anxious. However, once such feelings have been aired and processed within the group, team members typically are better able to manage those feelings so that they do not interfere with the assessment and treatment of patients. Often the team member who has the greatest rapport with the patient, or the team member who has the greatest comfort with discussions about sexuality can play the role of mediator and patient educator, with favorable results.

The Importance of Follow-Up

Even after assessment takes place, it remains vital that clinicians remember that adults over 50 are the least likely group to be targeted for educational interventions regarding HIV/AIDS and its transmission. In most cases, client education is as important as proper assessment. Despite agist stereotypes, client of all ages should be educated about the risks associated with unprotected sex and other high-risk activities. Among older adults, particularly those who may have postmenopausal female or who have same-sex partners, the fear of conception is absent. Thus, this older age group is least likely to engage in safer sex because their associations with condoms are typically for prevention of pregnancy. Older adults may not even know how to use a condom properly. Still other older IV drug users may not be aware of the use of bleach and the avoidance of needle sharing to minimize viral transmission. Older adults also are less likely to be aware of universal health precautions. Such information is critical, particularly when an older adult may provide nursing care for an adult child or grandchild with HIV/AIDS.

Case Examples

Marsha

Marsha was a 58-year-old woman enrolled in a geriatric day hospital program. Marsha was admitted on the recommendation of her adult daughter who was concerned that no one was available to help her mother manage her obsessive-compulsive disorder and keep up with her activities of daily living. Marsha's daughter is her only child from her first marriage, which ended more than 20 years ago in divorce. Marsha arrived on time each day for the program and was always dressed and groomed neatly for someone of her size. Marsha reports that she gained over 75 pounds in the last 5 years; she is obese. She also appeared to demonstrate some difficulties with attention and concentration.

Although she agreed readily to participate in group psychotherapy, Marsha had significant difficulty avoiding her compulsions, even under the influence of appropriate medication. During group sessions, she would only be able to sit for a few minutes at a time before she would feel compelled to walk seven steps forward, three steps backward, turn 180°, and repeat the procedure in varying degrees. She also whispered to herself as she counted numbers in a complicated pattern. It initially was unclear how much Marsha would benefit from the day program's group therapy and milieu therapy because of the severity of her symptoms. Her daughter reported that they intensified about 5 years ago, when she also changed her eating habits and began to gain a significant amount of weight. Neither Marsha nor her daughter could identify a precipitating incident.

Three months into the program, Marsha experienced a breakthrough. She had begun to socialize with two elderly single women who had also been secretaries who enjoyed big band music. Marsha began to eat lunch with these women and to be able to stave off some of her compulsions long enough to talk with them in brief conversation. During a group therapy session, one of these women discussed how vulnerable she often feels in her own home, because her late husband is no longer there to look after her. Marsha, who had been whispering to herself, gave her new companion a sideways glance and began to rock back and forth and hum loudly. The group leader paused to ask Marsha how she was doing, and Marsha began to cry. She admitted that about 5 years ago, a maintenance man she had hired to do some chores around the house suddenly became violent and raped her. She swore that she would never tell anyone about the attack because she felt so ashamed and afraid. Marsha received overwhelming support from other group members, and was able to discuss the event more openly over the next few weeks. She learned that it was not her fault about what happened, and that it was normal for her to feel angry, embarrassed, and upset. Marsha also was able to explore the relationship between her desire to control her environment via her compulsions and her attempts to feel comforted and reassured with the consumption of food.

Despite the positive response that Marsha received in group therapy among her elderly peers, the treatment team did not demonstrate such cohesion when Marsha's group therapist suggested that Marsha be informed about the risks of contracting HIV and other sexually transmitted diseases and to assist her in scheduling an AIDS test. The initial response among team members was that it was just "water under the bridge." Marsha's individual therapist asserted that she did not want to broach the subject of HIV testing because it would only retraumatize her patient and lead to a resurgence of binging and compulsive behavior. After all, this happened more than 5 years ago!

A lengthy discussion ensued in which many team members were able to discuss their anger and discomfort about the thought of a gentle and innocent older woman contracting HIV through a senseless and violent acquaintance rape. When the team leader asked how they would proceed if Marsha were a 25-year-old woman who had been raped, they quickly recognized their conscious and unconscious desires to deny the entire unfortunate incident. Many of the team members also acknowledged their implicitly held stereotype that "well, elderly women just don't get AIDS."

In a parallel process like that observed in the patients' group therapy session, in which group members were able to rally around Marsha after she aired her secret about the rape, the treatment team was then able to support Marsha's group therapist in her need to discuss the issue openly. The team agreed that it was vital to speak sensitively but frankly with Marsha about the rape and about her subsequent risks for contracting HIV. Because one of the nurses on staff had a particularly good relationship with Marsha, she volunteered to meet with her to address specific medical questions and the hospital's policy regarding HIV test results.

Gerald

Gerald was an 83-year-old man who was brought to the geriatric assessment unit of a hospital after having a serious fall at home. Concerned neighbors in the apartment complex called the police when they had not seen Gerald for a few days. Gerald appeared malnourished and unkempt, and he presented with a low-grade fever and swollen glands. On admission he was confused and apathetic when staff members tried to communicate with him. He preferred to spend his time on the unit alone, curled up on his bed. After being contacted by the staff social worker, Gerald's estranged son agreed to go to his father's apartment to gather his pajamas and toiletries. While at his father's apartment, Gerald's son discovered a large collection of empty liquor bottle, and vials of white and brown powder. The treatment team concluded that Gerald was suffering from major depression and substance abuse, and suggested that he enroll in an inpatient treatment program for drug abuse. The team physician ordered a series of chest X-rays and considered using antibiotics to treat a probable case of pneumonia. The treatment team did not even consider HIV as a possible diagnosis, even when Gerald admitted to engaging in intravenous drug use, a clear high-risk behavior.

Florence

Florence was a 75-year-old widow who exhibited symptoms of depression and dementia. Her niece insisted that she enroll in a day hospital program in order to help lift her spirits, and Florence reluctantly agreed. Florence arrives at the day program each day, dressed appropriately and neatly. She is eager to make friends with other patients in the program and quickly acquires a close knit circle of five female companions for breakfast, lunch, and conversation between group sessions. During group therapy, Florence is eager to participate. She responds supportively to patients dealing with a variety of problems, yet can also gently challenge their denials and cognitive distortions. Before long, most patients in the day program clamor to sit with Florence, to hear her tell a humorous story or to offer a kind word. Although Florence is reluctant to discuss her own problems in group therapy, her depression appears to lift, and her mood brightens. She tells the group that she has been doing more cleaning and baking at home. She even brings a loaf of her special apple bread to share with her group of friends.

About 1 month into the program, it became obvious that Florence was having trouble remembering things. She often forgot to take her hat home, or forgot to bring an umbrella when it was raining. She sometimes lost track of what people were saying in group therapy, but could use her sense of humor to smooth over any rough spots. Her primary therapist asked for a neuropsychology consultation to rule out dementia, and Florence agreed to participate. The neuropsychology intern said that Florence was superficially cooperative, but that she became very frustrated when she couldn't trace a series of lines, remember a series of words, or recite parts of a story. The official report suggested that Florence had mild to moderate dementia, with unknown etiology; the pattern was not entirely consistent with Alzheimer's or vascular dementia. The test report also contained suggestions that Florence use a notepad to help remind her about daily medications and physicians' appointments, and that she retested within 6 months to see if she would benefit from instrumental assistance around the house. Florence did not take the news well, and insisted that, of course, she forgot things once in a while, but that she was "as fit as a fiddle."

About 2 months into the program, Florence failed to arrive on time for the morning session. Because she had not called in sick, her primary therapist called her home and initially thought she had dialed a wrong number because a younger woman answered the phone. The therapist was surprised because Florence had told her that she was a widow of more than 15 years who lived alone. The woman on the phone said that she was Florence's daughter, and that Florence was taking a little longer than usual to help her with her morning routine. Florence herself refused to discuss the issue over the telephone, and said that she would come to the program at the regular time on the following day only if her therapist promised that they would discuss the issue "in private."

When asked about her situation at home, Florence admitted that her daughter was dying of AIDS. She implored her therapist to keep it a secret from the other patients, and only reluctantly accepted that this information would be shared (in confidence) among the treatment team members, as per the usual agreement for acceptance into the program. Florence said that it was her duty as a mother to take care of her daughter, but that things were very difficult for her. When asked why she never talked about this serious problem behavior, she said that she didn't want to burden others with her problems or "waste time crying over spilled milk." Despite her therapist's urging, Florence felt strongly that there was no need to discuss this problem with others in group therapy, because it would probably just make others upset.

Although she never had formal nursing training, Florence was proud that she learned how to help her daughter with her catheter, change her dressings (she apparently had open wounds that were not properly healed), and help her eat, bathe, and go to the bathroom. She said that she was faring much better with this daughter than with her son, who died of AIDS 5 years previously, also under her care. When asked how her two children contracted the virus, Florence tersely replied that they had turned to "dirty street drugs." Although Florence never provided a wealth of details, it also appears likely that her daughter turned to prostitution to finance her drug habit, and that she also could have contracted HIV through unprotected heterosexual sex. Florence failed to mention whether or not she wore latex gloves and goggles when taking care of her daughter and son, or that she even knew how to properly dispose of potentially contaminated medical waste.

Within the treatment team, concerns were raised that Florence was unaware of the basic universal precautions required for caring for someone with AIDS, and that even if universal precautions were explained to her, her dementia would prevent her from employing them accurately and consistently. However, the treatment team was reluctant to discuss universal precautions with Florence, much less ask her to consider an HIV test for herself. Some staff members admitted openly that the topic of HIV and AIDS made them uncomfortable, and that they simply did not want to "put Florence through that with everything else she has been through."

Staff members also argued about whether Florence should be encouraged to discuss her situation with her peers. Many staff members expressed fear that the older adults in the program would ostracize her and even engage in a mass exodus from the program in order to avoid contracting the disease. Heated discussions also arose about whether staff members and patients alike should take universal precautions when dealing with Florence on a daily basis. Ultimately, Florence agreed to have a visiting nurse help with some of her daughter's caregiving. She refused to take an HIV test and never disclosed to her peers that her daughter had AIDS.

A Call for Prevention and Advocacy

The development of primary prevention programs for HIV/AIDS among older adults represents one of the most proactive steps we can take to guard against agist decision making. An essential weapon against the inaccurate diagnosis of HIV and HIV-associated dementia is available through both practitioner and patient education. To complicate matters, of all age groups older adults are less likely to be targeted for HIV prevention strategies, and are less likely to be knowledgeable about the nature, transmission, and progress of the disease (Orel et al. 2004). Studies suggest that even healthy, independent, community-living older adults are generally unaware of the age-related risk factors associated with HIV/AIDS (e.g., postmenopausal women experience thinning of the vaginal walls; older adults die sooner of AIDS than their younger counterparts), and some even maintain incorrectly that HIV can be passed via public toilets, mosquito bites, and blood transfusions (Hillman 2007, 2008a).

Older adults also are less likely to be cognizant of the significant time that may pass between initial infection with HIV and the emergence of symptoms, as well as the overall risk of infection in their age group (e.g., CDC 2000). Thus, adults over 50 may be more likely to unknowingly pass the virus to their partners and loved ones through sexual contact and IV drug use. Few educational programs about HIV and its transmission exist for older adults (e.g., Orel et al. 2010). Among health care providers, only a handful of training programs in academic and on-the-job settings incorporate geriatric issues, much less specific issues about HIV among older adults, into their required curriculum. No national educational programs exist regarding HIV/AIDS education among adults over 50.

Fortunately, some new statewide initiatives for HIV/AIDS prevention among older adults have appeared in New York and Florida. New York City's most recent educational program, sponsored by the New York State Health Department, features posters with the slogan, "Age won't protect you from AIDS," along with various facts about HIV prevention (e.g., HIV prevention is a lifelong job) and a phone number for additional information and questions. Variations on a theme are available as these posters offer various options, including close-ups of women and men from various ethnic groups, as well as one of a large birthday cake with multiple lit candles. It also is notable that all of these materials are available in English and Spanish (New York Department of Health 2011). The Senior HIV Intervention Project in Florida's West Palm Beach and Broward Counties trains peer educators, often referred to as "safe sexperts," and "condom grandmas," to educate older adults about condom use and to convince them to seek out HIV testing (Gearon 2008). As noted, the CDC only mandates HIV testing in medical settings for adults up to the age of 64 (CDC 2006). Asking health care providers to advocate for changes to this policy could have life-saving impact.

Summary

The rise in documented HIV/AIDS cases among older adults is unprecedented. The rates of infection are increasing nearly four times faster among older adults than among young adults. The primary means of infection among older adults include homosexual contact, heterosexual contact, and intravenous drug use. Despite the wealth of information that we now have about HIV among older adults, it is unclear how many older adults currently suffering from HIV and HADC remain undiagnosed and untreated. It also remains unknown how well equipped clinicians are to make differential diagnoses regarding HIV-associated dementia and Alzheimer's dementia. Practitioners must guard against falling victim to their own agist beliefs and stereotypes; dealing with older adults and HIV can be a daunting but necessary task.

The development of educational primary prevention programs, and of group interventions designed to help combat the stigma, social isolation, and depression often faced by older PLWHA are some of the most proactive steps we can take. It is vital to guard against agist and inappropriate diagnostic decision making. We also need to highlight the needs of the rapidly growing population of men and women over the age of 50 living with HIV/AIDS in terms of minimizing stigma, reaching minority group members, reducing social isolation, and celebrating psychological well-being and resilience. An essential weapon against the inaccurate diagnoses of HIV/AIDS and HIV-associated dementia, and the discrimination sometimes unfortunately associated with such a diagnosis, is available through both practitioner and patient education.

Appendix

Resources for older adults and professionals regarding HIV/AIDS and aging

American Association of Retired Persons (AARP)
601 East Street NW
Washington, DC 20049
(202) 434-2260
www.aarp.com

AARP has a Social Outreach and Support *(SOS)* division that provides links to various referral services.

Health Watch Information & Promotion Services
589 Eight Avenue, Sixth Floor
New York, NY 10018
voice: (212) 564-7199
fax: (212) 564-7189
www.hwatch.org

Offers Community HIV Information/Education for Seniors (CHIEFS): A training for providers to assist them in training seniors to become peer educators, and Seniors & HIV/AIDS: Serving African-Americans Over 50, a training to assist healthcare and social service providers in understanding the needs of African-Americans 50 and over with HIV/AIDS and to integrate service models that promote access to and retention in care.

HIV/AIDS in Aging Task Force
425 East 25th Street
New York, NY 10010
(212) 481-7670
http://www.thetaskforce.org/issues/health_and_hiv_aids

Arranges educational seminars and conferences for health care providers.

HIV Wisdom for Older Women.
Founded by Jane Fowler
http://hivwisdom.org/index.html

Provides educational presentations, typically by an older woman living with HIV.

National AIDS Clearinghouse
P.O. Box 6003
Rockville, MD 20850
1-800-458-5231
http://www.cdcnpin.org/

Provides information about local resources and access to free government publications.

National AIDS Hotline
1-800-342-AIDS
1-800-344-SIDA
for Spanish 1-800-AIDS-889 (TTY)

This hotline is manned 24 h a day, 7 days a week. It can provide referrals to local programs and general information about the disease.

NAHOF: New England Association of HIV over Fifty
23 Miner St, Ground floor
Boston MA 02215
http://hivoverfifty.org/en/

Offers training and annual conferences

New York Association on HIV over Fifty, Inc.
J. Edward Shaw, Chairperson
119 West 24th Street
New York, NY 10011-1913
(212) 367-1009
http://www.nyahof.org/mission.htm
e-mail: info@nyahof.org or ednys2003@yahoo.com

Seniors in a Gay Environment (SAGE)
305 7th Avenue, 16th Floor
New York, NY 10001
(212) 741-2247
http://www.sageusa.org/index.cfm

SAGE offers referral services and HIV/AIDS information primarily to older lesbian, gay, bisexual, and transgender adults

Social Security Administration
1-800-SSA-1213
Social Security provides two different disability programs for eligible AIDS patients.

Chapter 7
Women's Issues in Sexuality and Aging

Nearly one-half of postmenopausal women report vaginal dryness and pain or discomfort during sexual intercourse. This is not, and should not be considered, normal.

Older women are faced with conflicting messages from society about their role as women, sexual beings, and contributors to society. These social demands placed upon aging women, coupled with their own internalized stereotypes and expectations, can be enormous and may be inappropriately internalized and manifested in the form of sexual dysfunction or satisfaction, body image distortions, depression, eating disorders, or poor participation in breast and cervical cancer screenings among others. Because of a general lack of knowledge among the health care community as well as the general population, older women remain unaware of the *normal*, expected physiological changes that take place with aging. Most stem primarily from a decrease in production of estrogen. Many older women are delighted to learn that treatments are available to assist them in coping with changes in their sexual response cycle in order to restore their enjoyment in sexual activity. Clinicians also need to acknowledge the sensual as well as sexual needs of middle-aged and older female clients, who may or may not have a partner. Informed practitioners can help their patients enjoy and relish their status as a woman throughout the life span.

Middle-aged and older women in the USA are faced with conflicting messages about their sexuality. Many segments of society, including those within the mass media, practically dictate that aging women must consistently appear beautiful, healthy, wrinkle free, physically fit, and sexually active and exciting. As noted in Chap. 1, significant changes in the media's portrayal of sexuality among aging women, including the concept of "cougar" (i.e., a sexually aggressive older woman who actively seeks out younger male partners) and "milf" (i.e., attractive middle-aged and older women with children; the acronym for "mothers I'd like [to] fuck") have emerged within the last decade. Sustaining such high standards in the pursuit of eternal youth is unrealistic for a variety of reasons, including the lack of available male partners in older age groups. In contrast, other segments of the population espouse that older women should accept their role as passive, virtually invisible, unattractive, and undesirable characters who would not even feign interest in sexual activity. Both of these directives conflict greatly with many older women's individual realities. In order to assist women in managing issues regarding their own sexual identity within the context of aging, it often is necessary to provide them with concrete information about the underlying physical changes that accompany normal aging as well as to explore related psychological and psychosocial issues.

Physiological Changes

Empirical studies of the oldest-old segment of our population (aged 80 and older) suggest that men are approximately two times more likely than women to engage in sexual activity such as intercourse (Fisher 2010; Lindau et al. 2007). Arguments have been made that a significant amount of variance in this relationship reflects the lack of availability of a partner for older women and the persisting gender differences in patterns of sexual behavior that first appear in adolescence. However, middle-aged and older women have been found to engage in a variety of sexual behaviors, both with and without a male partner, that include more than coitus (Fisher 2010; Lindau et al. 2007). Because an individual woman's level of participation in sexual activity appears to be relatively constant throughout her life span, on average (e.g., Janus and Janus 1993), it may be more beneficial for clinicians to focus on a woman's expectations and desires for sexual activity than any observed statistical differences between her rate of behavior and that of her male counterparts.

Unfortunately, more positive attitudes toward sexuality and aging do not automatically translate directly into greater sexual expression among older women. If a post menopausal woman wants to engage in more frequent sexual activity with her partner but experiences dyspareunia (i.e., pain or discomfort during intercourse), she must be made aware of the underlying physical changes that accompany typical aging, as well as ways to mitigate such changes, in order for her to better fulfill her sexual desires. Other women may be concerned that certain physical changes they are experiencing are abnormal and a sign of mental weakness, illness, or that they represent a general loss of femininity. Due to an overall lack of information about the expected, physiological changes associated with aging in the sexual response cycle (e.g., Donaldson and Meana 2011; Pancholy et al. 2011), these changes are frequently misinterpreted, misunderstood, overlooked, or dismissed by both aging women and their health care providers alike.

Menopause

Most physiological changes related to sexual function and aging in women begin somewhat abruptly with the onset of menopause, at age 51 on average for US women, and often include symptoms such as hot flashes, insomnia, increased abdominal fat, and vaginal dryness (Mayo Clinic 2011a, 2011b, 2011c, 2011d, 2011e). Although certain symptoms such as hot flashes and headache have been suggested as culture or individual specific, but as equally valid and distressing (Chen et al. 2010), other symptoms of menopause such as vaginal atrophy (e.g., Nappi and Kokot-kierepa 2010) appear to be biologically universal. During menopause, the production of the hormones estrogen and progesterone decreases, while the production of follicle-stimulating and luteinizing hormones (FSH and LH, respectively) increase in an attempt to promote estrogen production. What this increase in FSH and LH actually tends to produce is an increase in testosterone. Some estrogen also is produced in the body through the conversion of remaining adrenal androgens.

During this alteration in hormone production, the size of the uterus, cervix, and ovaries is reduced in response to the body's shifting of resources away from needs for reproduction; conception is no longer biologically possible (Zeiss et al. 1991). The uterus itself may be reduced in size, via changes in collagen and elastic content, by up to 50%. Related to other changes that may be more directly apparent in sexual function, decreases in estrogen levels can result in a thinning of the vaginal lining, a loss of vaginal elasticity, and a significant decrease in natural vaginal lubrication (Barbach 1996; Mayo Clinic 2011a, 2011b, 2011c, 2011d, 2011e).

Sometimes, clients have described situations in which health care providers have told them that menopause is "all in their heads," and that their symptoms are exacerbated primarily by negative or anxious emotions. In contrast, other women have complained that even though they have positive feelings about aging and their bodies, they feel like failures because they have allowed themselves to "fall victim" to menopausal symptoms of one variety or another. They assume that a positive mental attitude will prevent them from experiencing any biological changes or physical symptoms. Although some symptoms of menopause appear to be mediated by individual, societal, and cultural expectations (e.g., Chen et al. 2010; Nappi and Kokot-Kierepa 2010), a number of clear physical changes do occur at this developmental milestone. In response, every woman's unique experience should be validated. Accordingly, it is vital for both patients and practitioners to recognize that despite a postmenopausal woman's positive feelings about sexuality and high levels of emotional readiness and arousal, sexual intercourse may be experienced as painful and unpleasant without the use of appropriate interventions.

Sexual Response Cycle

Some researchers suggest that once concerns about conception are alleviated by menopause, postmenopausal women may find greater enjoyment in their sex lives. They are no longer saddled with beliefs that they must produce something in relation to sex, and sexual activity becomes pleasurable for its own sake. Others assert that our society views menopause as a medical disorder or deficiency, with negative expectations for social status, rather than as a normal developmental phase signifying greater freedom and an opportunity to reaffirm or redefine roles (e.g., Nosek et al. 2010). Many American women may miss out on the opportunity to view menopause

Table 7.1 The postmenopausal female sexual response cycle

Stage	Action	Age-related changes
Excitement	Vaginal lubrication	Delayed; May take up to 5 min compared to 10–15 s among young adults; Reduction in quantity
	Genital vasocongestion	Reduced
Plateau	Uterine elevation	Reduced
	Labia majora	Reduced elevation and swelling
	Breast changes	Less vasocongestion; Diminished nipple erection
	Clitoral stimulation	Maintained or heightened sensitivity; May produce irritation
Orgasm	Vaginal contractions	Two to three contractions versus five to ten contractions among young adults
	Uterine contractions	Weaker; Shorter in duration
	Subjective experience	Pleasurable sensations maintained
Resolution	Capacity for orgasm	Potential maintained for multiple orgasms
	Genital vasocongestion	Rapid loss and return to prearousal state

as a positive step in their lives, with accompanying sexual pleasure and expression. Nevertheless, many middle-aged and older women can expect that physiological changes related to aging will alter in their sexual response cycle, if not their actual enjoyment of sex. Understanding the typical changes that take place in sexual arousal may help aging women and their partners adapt and enjoy sexual activity in spite of these changes.

The sexual response cycle can be categorized in four stages: (1) the excitement, (2) plateau, (3) orgasm, and (4) resolution stages. A variety of estrogen- and nonestrogen-related changes can impact on a postmenopausal woman's response to sexual arousal. Please refer to Table 7.1 for an overview.

One of the most obvious changes in sexual response with advanced age (in response to a decrease in estrogen production) is the reduction in quantity of vaginal lubrication. The majority of postmenopausal woman find that it takes significantly longer for them to become lubricated. While it once took only a few seconds to become "wet" or aroused by her partner, producing less than adequate levels of lubrication may now take well over a few minutes. In addition to experiencing painful intercourse as a result, an older woman may mistakenly assume that she is no longer feminine, or may misinterpret her lack of lubrication as an emotional cue that she is somehow no longer interested in her partner. In a parallel fashion, her partner may feel inadequate because more time is required for her to get "ready" for sex, and her partner may feel upset, annoyed, or even angry that she requires more foreplay.

Postmenopausal women's partners also may feel that they are no longer sexy or appealing to their partner, particularly if they feel anxious or ambivalent about their own advanced age. In contrast, sometimes increased skin sensitivity makes breast, nipple, and clitoral stimulation irritating instead of arousing (Galindo and Kaiser 1995). In a worst-case scenario, an older woman does not inform her partner that she simply needs more time and additional lubrication in order to mutually enjoy sexual relations. Such painful intercourse or dyspareunia may be accompanied by vaginal bleeding, itching and burning sensations, increased urinary frequency or urgency, and swelling and soreness both during and after sex (Bachmann 1995; Mayo Clinic 2010a, 2010b, 2010c). These women may suffer in silence out of shame and misunderstanding as well as out of a poor sense of entitlement.

An older woman and her partner also may be pleased to learn that her subjective experience of arousal may be just as strong and pleasurable as that experienced by her younger female counterparts. Pleasure from stimulation of the breasts remains intact, despite a lesser likelihood of vasocongestion and nipple erection (Kaiser 1996; Roughan et al. 1993). And, despite changes in the objective experience of orgasm (e.g., an older woman is likely to experience more shallow, less frequent vaginal and uterine contractions; Leiblum and Rosen 1989; Zeiss et al. 1991), the overall subjective experience of orgasm as pleasurable remains virtually unaffected by age. Postmenopausal women also remain capable of experiencing multiple orgasms within the context of one sexual encounter (Leiblum and Rosen 1989).

The most detrimental impact of advanced age appears to take place during the excitement or arousal stage of the sexual response cycle. However, once an older woman becomes adequately aroused and lubricated (either naturally or with assistance), she can expect to experience subjective sexual pleasure and satisfaction consistent with that experienced by women many years her junior. The other physiological caveat is that once an older woman experiences orgasm, her body refracts rapidly to her prearousal state. Care and time must be taken to reestablish proper arousal and lubrication before additional acts are attempted.

Prevalence of Sexual Dysfunction

Findings from large-scale epidemiological studies provide insight into the numbers of postmenopausal women who experience sexual dysfunction. The most commonly reported types of sexual dysfunction reported among this cohort of middle-aged and older women include dyspareunia (i.e., pain or discomfort during intercourse), vaginal dryness (which typically leads or contributes to dyspareunia), and low sexual desire. As noted in Chap. 2, estimates indicate that at least one-third, and up to one-half, of middle-aged and older women report some type of sexual dryness or discomfort during sexual activity (Santoro and Komi 2009; Waite et al. 2009). What also is interesting to note is that self-reported rates of vaginal dryness and dyspareunia do not climb significantly among older as compared to middle-aged women. Population-based data suggest that approximately 18% of both middle-aged and older women report pain during intercourse (Waite et al. 2009).

It also becomes important for mental health providers to help aging women work with their health care providers to discriminate between dyspareunia and vaginismus. Dyspareunia represents pain during intercourse, that may be accounted for by a variety of issues including psychological factors (e.g., depression; anxiety), interpersonal issues, cultural issues, and physiological factors such as vaginal dryness. Vaginismus, which may appear in consort with dyspareunia, represents a specific type of sexual dysfunction in which the vaginal walls involuntarily clench and spasm. In many cases a woman with vaginismus cannot achieve penetration with any object—whether it be a penis, dildo, finger, or tampon. Typically, vaginismus is reported more often by young adult women, with dyspareunia reported more often by middle-aged and older adult women. Although the DSM-IV-TR (APA 2000) maintains that these two sexual disorders are distinctly different, other experts argue that this distinction is not clinically relevant, and that it detracts from the overall assessment of vaginal pain, fear of vaginal pain, pelvic-floor muscle dysfunction, and related medical problems. It may be more useful to think of these disorders as part of a singular genital pain and penetration disorder (Blinik 2010).

In terms of low sexual desire, population-based studies indicate that significant numbers of middle-aged and older women, including those with available partners, report having a sustained lack of interest in sex. For example, approximately 35% of middle-aged and 50% of older women in the US population representative samples reported that they have low sexual desire (Waite et al. 2009; West et al. 2008). (Interestingly, these rates are virtually identical to the proportion of women who report vaginal dryness; it remains unclear if there is a causal link between the two factors.) What also is interesting to note is that the proportion of postmeno-pausal women who report that their low level of sexual desire is actually *distressing* to them (or perhaps their partner) is smaller. Approximately 8–10% of women in large US samples report symptoms that meet the criteria for Hypoactive Sexual Desire Disorder (HSDD; Simon 2010; West et al. 2008), in which "the deficiency or absence of sexual fantasies and desire…[cause] marked distress or interpersonal distress" (APA 2000, p. 496).

Although virtually no data is available regarding lesbian, female bisexual, trans-gender, or minority group members' sexual dysfunction per se, one empirical study regarding a middle-aged, postmenopausal population that was composed primarily of low income, Hispanic women revealed that more than 75% of those women reported having some type of sexual dysfunction when they arrived at their local health clinic. Of this unique sample, the women with sexual dysfunction also were significantly more likely to present with depression, anxiety, and insomnia (Schnatz et al. 2010). It is unclear if these concomitant factors are related to socioeconomic challenges such as possible concerns about personal safety, lack of education, or decreased availability of quality health care, or other factors. Additional research must be conducted to determine if the base rates, as well as the subjective experience of sexual dysfunction, are different among various minority group populations.

Clinical Assessment

The proper assessment of a postmenopausal woman's sexuality and sexual func-tioning is vital. Mental health care providers can serve as indirect or direct liaisons between older female patients and their physicians or other physical health care providers (e.g., Sacerdoti et al. 2010). This role appears essential, as only 22% of middle-aged and older women report discussing sex with any physician since their

50th birthday (Lindau et al. 2007). Sometimes an older woman may feel more comfortable discussing sexual issues with a therapist with whom she has developed a long-term relationship than with a gynecologist whom she visits once a year. Alternatively, physicians may refer older patients to a psychologist or other mental health practitioner in response to specific concerns over sexual dysfunctions that were revealed in an introductory interview. (Of course, out of respect for a patient and in deference to ethical and legal issues, the patient's consent must be obtained in order to share even cursory information among professionals.) Psychologists must be attuned to appropriate modes of assessment, and also to general medical knowledge that may illuminate aspects of sexual dysfunction.

Dangers of Incomplete Assessment

Estimates suggest that nearly half of middle-aged and older postmenopausal women experience vaginal dryness and pain or bothersome irritation during intercourse (Santoro and Komi 2009). Although such pain can certainly be associated with medical disorders such as skin lesions or cysts, such dyspareunia often is caused by atrophy of the vagina coupled with inadequate lubrication during intercourse. Mistakes in making such a determination have led to serious consequences for elderly women. Anecdotal reports suggest that older women who reported having vaginal bleeding after intercourse underwent cystoscopies and D&C's under general anesthesia simply because their practitioner assumed that these women were not sexually active, and because they did not even ask their elderly female patients about participation in sexual activity (Butler et al. 1994). Many older women themselves, particularly because their cohort has not been well educated about sexuality, do not recognize that vaginal bleeding and inflammation may be their body's natural response to painful sexual intercourse without appropriate lubrication.

Specific Areas of Inquiry

A variety of pointed questions may be asked when working with an older female patient who has concerns about her sexual health and activity. As with any client, it is important that practitioners tell their patients that being open and honest will allow for the greatest benefit to them. Older patients also often respond well to being *asked* if they would mind answering questions about their sexuality. This asking of permission provides an older client with some sense of control over a potentially anxiety-provoking situation, and also provides for an established sense of commitment regarding her cooperation during the interview. If an older woman defers from discussing her sexuality, it is best to respect her decision, but to inform her that you are open to a discussion of such important issues with her at any time in the future (Galindo and Kaiser 1995). It also can be helpful to identify the source of resistance by noting and inquiring, "I respect your decision not to discuss matters of sex with you, and will not ask you any questions about it as you wish. However, I do wonder if you *would* be willing to share with me *why* you decided not to talk about it at this time. How do you think I would respond if you told me about your sex life?" Confidentiality and rules about sharing (or the lack of sharing) of information with team members of other professionals also are essential in order to provide a safe environment and to delineate professional boundaries for the client.

Depending on your familiarity with the client, use terminology consistent with her experience. If you are meeting for the first time, it often is helpful to use anatomically correct terminology (e.g., vagina, penis, vaginal lubrication) but then shift to the client's vernacular as required. Some questions can be adapted to a paper-and-pencil format, in which clients are instructed to check off concerns and items for future discussion. In individual interviews, it often is helpful to preface sensitive questions with an initial assessment of the older woman's potential religious or cultural prohibitions and marital or partner status. Certain subsequent questions could be tailored to accommodate the client's specific situations or concerns.

Examples of questions that address specific issues related to sexuality and aging among women (e.g., Galindo and Kaiser 1995; Kennedy et al. 1997) include:

- What kinds of sexual activity do you engage in? (Offer a variety of options including finger penetration, vaginal intercourse, oral sex, anal intercourse, penetration with dildos or vibrators, masturbation, caressing, kissing, massage, petting, or making out.)
- How important is sexual activity to you in your life?
- Would you like to engage in more or less sexual activity?
- If you would like to engage in more sexual activity, what do you think is keeping you from doing so? Are you tired, feeling sick or ill, without a partner, is your partner ill or tired, have you lost interest, are there relationship problems, are you worried about what you think, do you have religious or cultural concerns?
- Do you experience pain or discomfort during penetration or while having sex (i.e., dyspareunia)? When did such pain or discomfort begin?
- Do you have vaginal pain or dryness, either before or during sex?
- Do you experience bleeding during or after sexual intercourse?
- Have you considered using a lubricant during sex such as KY jelly, Vaseline, or hand lotion? (Vaseline and hand lotions are not necessarily an appropriate choice for vaginal lubrication because these lubricants are oil based and inappropriate for use with latex condoms, but many older adults use them because they are readily available.)
- Have you ever experienced orgasm with your partner? Has that changed at any time in the past few years?
- How often do you masturbate or touch yourself to feel good?
- How often do you have the urge to have sex? Is this different than from before? In what way?
- Have you or any of your partners had sex with anyone who used intravenous drugs, visited prostitutes, or engaged in sex with other men or multiple partners? (Allows for an assessment of high-risk behaviors.)

- Has anyone ever forced you to have sexual relations when you did not want to have? Has anyone ever touched parts of your body when you did not want them to? (Be sure that the client knows that this includes historical as well as recent events.)
- Do you feel safe in your own home or residence? Is there anyone you live with who makes you feel afraid or unsafe? (Allows for an assessment of potential domestic or partner abuse.)
- How do you feel about your appearance? About your body? Have you had any surgeries that have changed either the appearance of your body or how it works, such as a mastectomy or colostomy?
- How are your eating habits? Are you concerned about your weight? Do you ever binge, purge, or use laxatives to control your weight?
- When is the last time you were seen by a gynecologist or had a pelvic exam? What is it like for you when you go to the gynecologist?
- When is the last time you had a pap smear? Have you ever had cervical cancer or an abnormal pap smear? Do you have a family history of cervical or ovarian cancer?
- When is the last time you had a breast exam or mammogram? Have you ever had breast cancer or cysts? Do you have a family history of these illnesses?
- What questions, if any, do you have for me regarding sexual function or behavior?
- Do you have any chronic illnesses such as diabetes, arthritis, anxiety, or depression? How does that affect how you feel about yourself and your body?
- Are you currently taking any prescription or over-the-counter medication?
- Do you exercise or meditate?

It is often challenging to ask any client, much less an older woman, if she has engaged in high-risk sexual behavior. It is our obligation as clinicians to search for the truth, and not make any age-biased assumptions about our aging female patients. Straightforward but diplomatic wording of such questions is vital. For example, when trying to assess whether a woman's partner visited prostitutes, one might ask matter of factly, "Do you think that your partner was always faithful?" The less anxious the interviewer appears about potentially sensitive topics, the less anxious the older adult patient will be in response.

With older women in particular, it is essential to assess issues regarding subjective body image (Lapid et al. 2010; Sacerdoti et al. 2010) as well as issues regarding objective frequencies of sexual behavior and breast cancer screenings. Sometimes asking an older woman about whether she has difficulty finding clothing that she likes may bring up issues of shame regarding a mastectomy or difficulties in feelings attractive regarding a dowager's hump.

Side Effects of Medication

Although discussed less frequently in the literature, older women are just as likely as older men to experience negative, sexual side effects from various prescription and over-the-counter medications. Some discrepancies in reporting may reflect

Table 7.2 Medications that can induce sexual side effects older in women

Medication class	Drug	Generic or brand name
Antidepressants		
MAOI	Phenelzine	Nardil
SNRI	Duloxitine	Cymbalta
	Venlafaxine	Effexor
SSRI	Citalopram	Celexa
	Fluoxetine	Prozac; Prozac Weekly
	Paroxetine	Paxil; Paxil CR
	Sertraline	Zoloft
Tricyclic	Clomipramine	Anafranil
	Imipramine	Tofranil
Antihypertensives	Clonidine	Catapres
	Chlorothiazide	Diuril
	Digoxin	Lanoxin
	Prazosin	Minipress
	Reserpine	Diupress
Mood stabilizers	Lithium	Lithonate; Eskalith
Antipsychotics	Chlorpromazine	Thorazine
	Haloperidol	Haldol
	Thioridazine	Mellaril
	Trifluoperazine	Stelazine
Raloxifine	Fulvestrant	Faslodex
	Tamoxifin	Nolvadex

Note. This listing is meant to be illustrative—not comprehensive. A variety of medications may elicit similar, negative sexual side effects

the more objective, easily observable criteria related to male sexual function (i.e., maintenance of an erection). The female sexual response is more difficult to measure quantitatively and reports typically rely upon older women's subjective experiences, which typically were overlooked by researchers in the past. Table 7.2 presents a variety of prescription medications that are believed to disrupt the female sexual response cycle, fostering inadequate vaginal lubrication, a decrease in sexual interest or libido, and the inability to achieve orgasm.

It also is important to note that antidepressant selective serotonin reuptake inhibitors (SSRIs), including Prozac and Zoloft, are commonly associated with negative sexual side effects such as diminished libido and inability to orgasm. In contrast, some antidepressants that are less likely to cause negative sexual side effects include the atypical antidepressants Bupropion (brand name Wellbutrin) and Mirtazapine (brand name Remeron). Inquiring about a client's use of such medication is vital, as findings from population-representative studies suggest that 13% and 43% US middle-aged and older adults take prescription drugs for depression and high blood pressure, respectively (Barrett 2005). In sum, women who take any prescription medication should be advised to consult with their health care provider if they are experiencing sexual discomfort or problems.

Evidence-Based Interventions

Although many sexual problems among middle-aged women stem from a combination of psychological, relational, and psychosocial causes (Butler et al. 1994; Chrisler and Ghiz 1993; van Lankveld et al. 2010), physical changes with age can underlie and compound many existing problems. Fortunately, a variety of interventions are available to assist postmenopausal women and their partners in mitigating many of the hormonally induced changes that occur with aging, particularly in relation to vaginal dryness and pain during intercourse. Many remedies employed are designed specifically to combat the loss of estrogen, and have been employed successfully with community-based samples.

Other empirically based treatments include cognitive-behavioral therapy and physical therapy to help restore pelvic floor function (e.g., Bergeron et al. 2010; Ter Kuile et al. 2010). Interestingly, a number of these interventions have limited empirical support via large-scale, randomized clinical trials. This lack of large-scale trials likely speaks more to the previous lack of interest (and funding) in evidence-based approaches to treatment of sexual dysfunction among older women, however, (e.g., van Lankveld et al. 2010) rather than the actual clinical utility of the interventions themselves. It also is important to note that treatment for dyspareunia appears more effective when health care providers become involved; attempts to self-administer cognitive or behavioral treatments often meet with only limited success (Donaldson and Meana 2011).

Topical Lubricants

Prescription lubricants. One of the most easily employed interventions designed to combat the vaginal dryness that often accompanies aging is via the use of supplemental lubrication. One way to increase lubrication is to introduce additional estrogen to vaginal tissues in the form of topically applied prescriptions. Although a physician must prescribe these treatments, their effectiveness has been well documented and only limited amounts of estrogen are expected to leech into the body's bloodstream (Krychman 2011). Vaginal estrogen can be delivered via topical creams (e.g., Premarin; Estrace) that are typically used daily for a few weeks, and then approximately two to three times a week. Vaginal estrogen also can be applied via a small, flexible, plastic vaginal ring (e.g., Estring) that is put into place by the cervix by either the patient or her health care provider once every 3 months. The third approach is to use a small, disposable applicator to insert an estrogen tablet into the vagina (i.e., a vaginal suppository) once a week for approximately 2 weeks and then twice a week. The tablet dissolves in the vagina and its use is painless. As noted, all topical medications that introduce estrogen can be obtained only by prescription, and may carry the risk of side effects. (It also is vital that all older women with concerns about their sexual function to visit the appropriate health care provider to rule out any potential, underlying disorders or diseases).

Table 7.3 Characteristics of non-prescription vaginal lubricants

Petroleum-based[a]	
Generic names	Mineral Oil, hand moisturizers, most skin creams
Brand names	Vaseline Petroleum Jelly, Stroke 29, Jack Off
Positives	Inexpensive, easily accessible
Negatives	Irritating to vagina, not safe with latex condoms, stains fabrics, bad taste and odor
Natural oil-based[a]	
Generic names	Vegetable, corn, olive, and peanut oil; butter
Brand names	Crisco
Positives	Inexpensive, easily accessible, does not irritate vagina
Negatives	not safe with latex condoms, stains fabrics, some smell and odor
Water-based with glycerin	
Generic names	–
Brand names	KY Jelly/Liquid, Astroglide, Wet, Embrace, Good Head, KY Warming Jelly, Astroglide Warming Liquid, Replens[b]
Positives	Inexpensive, easily accessible, commonly recommended, safe with latex condoms, does not stain fabrics, may taste sweet
Negatives	Dries out quickly, can become sticky, can foster yeast infections, some odor
Water-based without glycerin	
Generic names	–
Brand names	Liquid Silk, Maximus, Saliva, Slippery Stuff, Oh My, Probe, Sensual Organics
Positives	Relatively inexpensive, safe with latex condoms, does not stain fabrics, lasts longer and is thicker than water-based lubricants with glycerin
Negatives	Found primarily online or in adult stores, bitter taste
Silicone-based	
Generic names	–
Brand names	Eros, Wet Platinum, Id, Millennium, Pink, Gun Oil
Positives	Safe with latex condoms, lasts three times longer than water-based lubricants, does not irritate vagina, odorless, tasteless
Negatives	Expensive, found only online or in adult stores, may need to be removed with soap and water, cannot be used with silicone sex toys

[a]As noted, not safe with latex condoms
[b]A vaginal suppository

Non-prescription lubricants. Estrogen-free topical lubricants also present viable options for postmenopausal women who have difficulty with vaginal dryness. These kinds of lubricants are accessible without a prescription. Many women have at least heard of KY Jelly or Astroglide, which are topical, water-based lubricants available at most drug and grocery stores. These types of products are easily obtained without a prescription, and they do not carry a risk of hormonal side effects. It also is important to note that there are five distinct types of non-prescription lubricants available, and each type presents with a unique set of pros and cons (Andelloux 2010). Please refer to Table 7.3 for additional information.

Based upon the relative merits of the five different types of over-the-counter topical lubricants available, it becomes apparent that different types of sexual activities, as well as personal sensitivities and preferences, will help determine the appropriate selection. For example, various case reports indicate that many older women and their partners assume that Vaseline petroleum jelly or a moisturizing hand or skin cream will serve an effective, inexpensive, and readily available lubricant for sex. However, as noted in the table, such petroleum-based lubricants can cause vaginal irritation and infection and are not safe for use with latex condoms. (Petroleum-based lubricants may work well for men who wish to masturbate, but their use with vaginal and other mucus membranes does not). The petroleum in these lubricants actually eats away at the latex in condoms, causing microscopic holes that are large enough for the HIV virus to pass through. Even though postmenopausal women may not be concerned about unwanted pregnancies, they should be aware that sexually transmitted diseases including HIV are not reserved for the young. Similarly, natural oil-based products like vegetable oil and butter can irritate the vagina and are not safe with latex condoms.

Although KY Jelly, Astroglide, and other water-based lubricants are the most commonly recommended type of topical lubricant to women, they may not be the ideal choice if they lead to vaginal yeast infections (due to the relatively high glycerin or sugar content of the lubricant), or if they dry out quickly and need to reapplied often. Many women using such water-based lubricants may not realize that multiple applications may be needed to avoid increased friction, irritation, and discomfort, and that the use of these lubricants may actually lead to yeast infections in some individuals. However, in comparison to oil-based lubricants, the use of water-based lubricants, both with and without glycerin, are safe for use with latex condoms.

Clinicians should help middle-aged and older women become aware of these five different types of lubricants, and that they have additional choices beyond what is offered at the local grocery or drug store. For example, if a woman wants to avoid yeast infections, needs longer-lasting lubrication, and does not want to compromise on the safe use of a latex condom, a silicone-based lubricant may be the best choice. The lack of taste and odor with silicone lubricants also bodes well for women who are interested in using them with oral sex. Of course, silicone-based lubricants typically are the most expensive on the market.

It also is important that clinicians help normalize the discussion and purchase of such lubricants. In clinical practice, the information on Table 7.3 could be copied as a handout to be given to patients for their review. Because an older woman may be too embarrassed to seek out and buy water-based non-glycerin and silicone-based lubricants at the local adult store, she may be relieved to learn that she (or her partner) can order them discretely online. Older women also can be encouraged to discuss their selections with their health care providers and their partners. Fostering a frank discussion about the merits of the different kinds of lubricants, as well as the interpersonal challenges that may arise with their selection or use (e.g., What if my partner thinks I'm a hussy because I got some kind of fancy lubricant instead of just using hand cream?; How can I make myself realize that I'm worth the $20 that this costs?; How can I ever bring this up with my gynecologist?), would be quite therapeutic for any woman and particularly for those who are feeling anxious, ambivalent, or ashamed about their sexuality.

It also is important to recognize the psychological implications of using any "aid" for sex at any age. Although it appears a simple and elegant solution to employ topical lubricants during intercourse to address vaginal dryness, many women may be uncomfortable applying them in front of their partner, or they may feel embarrassed or inadequate because of their need to use them at all. Effective communication between an older woman and her partner becomes essential in order for both parties to feel comfortable. The use of lubricants can sometimes be incorporated into intercourse as foreplay, which also benefits a postmenopausal woman with typical, slower arousal phase. For women who opt not to use lubricants immediately before intercourse, other non-prescription alternatives have been offered. These additional options, which also should be discussed with one's health care provider, include over-the-counter products such as Replens, a water-based vaginal suppository that can be used in advance, or vitamin E oil from capsules applied vaginally on an alternate day schedule (Barbach 1996).

Masturbation

Masturbation or self-stimulation is another effective, nonhormonal treatment for vaginal dryness and atrophy that may come as a surprise to both aging women and their practitioners. Women who remain sexually active, either through sexual intercourse with a partner or through mutual or solo masturbation, have shown lesser declines in vaginal lubrication with age (Roughan et al. 1993). Although the exact, underlying mechanism is unknown, it is posited that masturbation increases blood flow to the surrounding vaginal tissue, which promotes increased lubrication on a daily basis (Galindo and Kaiser 1995). Because of numerous religious and societal prohibitions against masturbation, however, suggestions to engage in self-stimulation must be discussed sensitively, particularly with older patients. Respect for a woman's religious prohibitions against masturbation must take highest priority, and may potentially circumvent any further discussion of the issue.

In contrast, other women who have ascribed to social taboos about masturbation may feel liberated when the practice is "prescribed" by their psychologist or other health care provider (Galindo and Kaiser 1995). Anecdotal evidence suggests that a number of female patients have been cured of insomnia, vaginal dryness, and symptoms of anxiety after being instructed by their clinicians to experiment with masturbation before bedtime. Older women also may be encouraged to learn that masturbation does not have to involve penetration into the vagina with a finger, dildo, or other foreign object. Manual stimulation of the clitoris and labia is often enough to induce sensations of pleasure, orgasm, and the desired increase in blood flow.

Hormone Replacement Therapy and Supplements

Within the last decade, a number of controversial findings have emerged in relation to hormone replacement therapy. Previously, physicians often prescribed estrogen replacement therapy to assist older women in dealing with the physiological changes induced by a lack of hormonal estrogen. Estrogen replacement therapy was touted as effective in combating hot flashes, vaginal atrophy and dryness, and in reducing the risk of cardiovascular disease and osteoporosis. However, findings from the Women's Health Initiative (WHI), a large-scale randomized, double-blind clinical study of hormone replacement therapy compared to placebo among postmenopausal women funded by the National Institutes of Health revealed that use of such estrogen rich medications actually increased a woman's risk for cardiovascular events (e.g., heart attacks, strokes, and blood clots; Toh et al. 2010), invasive breast cancer (Rossouw et al. 2002), and even dementia (Coker et al. 2010). Alternative treatments under investigation currently include supplements such as soy-based isoflavoids, vitamin D, black cohash, and DHEA (Mayo Clinic 2010a, 2010b, 2010c). What remains central for most mental health practitioners, however, is not what type of medication or supplement our postmenopausal female clients are prescribed or taking What is vital, however, is our ability to help our patients and clients communicate openly and effectively with their physicians in order to make the best choice about their medications or supplements.

Cognitive-Behavioral therapy

Cognitive-behavioral therapy, couples' therapy, sex therapy, and physical therapy designed to restore pelvic floor function represent additional, empirically based treatments for dyspareunia and vaginismus (Bergeron et al. 2010; ter Kuile et al. 2010). Before such treatments even begin, however, it is important to assess what factors underlie a middle-aged or older woman's low sexual desire or pain during intercourse. It is vital to understand if a woman has religious prohibitions, fears of contracting HIV or another STD, a history of trauma or abuse, depression or anxiety, or if there is lack of interest or sensitivity in the partner (e.g., Knoepp et al. 2010). Taking a thorough medical, sexual, and psychosocial history is essential (van Lankveld et al. 2010). It also is important to track when the pain or lack of interest began. The interpersonal factors involved cannot be underestimated, even within the context of potential muscle or nerve damage to the pelvic floor. Much of the cognitive-behavioral treatment involves cognitive interventions to minimize catastrophizing, sensate focus exercises, and relaxation protocols including guided imagery. An integrated treatment including psychotherapy and physical therapy appears to promote the best outcomes (Bergeron et al. 2010).

Changes in Sexual Self-Concept

In American society, women typically are judged on their physical appearance. Despite unrealistic expectations from the media and other sources to remain in a state of perpetual youth, women can naturally expect to experience changes in their body's shape and appearance as they age. As well as inciting changes in internal organ structure and vaginal lubrication, menopause is believed to initiate changes in a woman's distribution of body fat, and thus induce changes in her overall appearance. Specifically, women tend to gain weight at the onset of menopause (Mayo Clinic 2010a, 2010b, 2010c). Additionally, locations of body fat storage may be redistributed to the waist, breasts, and upper back (Lovejoy et al. 2008; Sowers et al. 2007), which may change some women's physiques and even require them to find different clothing to accommodate their body's changing shape. Hence, some post-menopausal women bemoan the development of what some refer to as their "men-o-pots" (i.e., menopausal pot bellies).

Other, more general physical changes associated with normal aging impact on an elderly woman's appearance and her experience of sexuality. As noted, our society equates youth with sexuality; an older woman in particular must navigate difficult issues involving both her own attitudes toward her body's changes and society's general disdain for the aging female body, which should be swaddled in discrete clothing and cloaked in restorative creams and makeup (Chrisler and Ghiz 1993). Empirical studies suggest that at all ages, women, as compared to men, manifest greater concern and dissatisfaction in relation to their body shape and weight (Mellor et al. 2010). At an extreme, some older women have expressed concerns that young children may find their aging faces and bodies "scary". How women interpret and internalize these changes greatly influence their sense of self and subsequent expressions of sexuality and sensuality.

Sexual Identity for Women with and Without Partners

For some women, body image represents a central construct, a core feature of their identity that gives them pause for consideration on a daily (or in some cases, an hourly) basis. These women's daily experience of their aging body may serve as a narcissistic injury to their previous sense of self, which relied heavily on an outwardly youthful countenance. Accordingly, when a sample of older women were asked about their body image in relation to their sexuality and sensuality, the vast majority cited that they were concerned about their energy level, physical health, and nutritional status. Most flatly denied that they had any concerns about their sexual identity; in fact, most responded that it simply was not really a consideration "at my age." However, the vast majority of these women also expressed significant interest in wearing fashionable makeup, clothing, and wigs, belying their interest and apparent pleasure, concern, and pride that they took in their appearance (Campbell and Huff 1995). In another sample of older adults, women aged 79–85 were found to maintain a more positive body image than women aged 65–71. The women in the oldest age group also reported that they spent more time grooming and focusing upon their appearance (Baker and Gringart 2009).

In contrast, other older women view their body's appearance and function as less central to their identity. These women view their femininity and inherent worth as less of a reflection of their external appearance, but rather as a function of their satisfaction in personal relationships, their personal accomplishments, and sense of spirituality. Empirical studies also suggest that women who engaged in compensatory activities such as exercise (Baker and Gringart 2009) who felt that they had greater control over their lives, who placed less importance on arbitrary standards of physical attractiveness, and who felt more positively about their worth as human beings also had a more positive body image (Rackley et al. 1988). Thus, an internal locus of control appears beneficial for an older woman, particularly regarding her sense of acceptance and mastery over her health, appearance, and body image.

For many middle-aged and older women, even for those who reported satisfaction with their appearance, issues of body image may become particularly salient after the death of a spouse or long-term partner. After mourning the partner's death, a renewed sense of urgency may occupy her thoughts as she prepares to begin dating and faces the pervasive double standard that submits that women, but not men, gain access to new romantic partners primarily as a function of their appearance. Other women find that once their lifelong partner is no longer available as a sexual partner (e.g., either through death or because of personality and physical changes as a result of advanced dementia), their interest in maintaining a sexual identity is equally diminished. For the vast majority of women, however, concerns about one's appearance persist throughout life, and these concerns originate from both internal (e.g., desires to feel feminine and attractive) and external sources (e.g., social pressures to be attractive to gain acceptance). Clinicians can often provide aging women with the permission that society does not grant them to address these sexually related pleasures, fears, and anxieties, whether they are with or without a romantic partner.

Margaret

A number of older women participated regularly in group therapy for caregivers. Each woman's husband was seriously ill, and required significant nursing demands. Nearly all of the patients concurred that the greatest difficulty they had in their new role of caregiver was that they felt that they had little time or energy for themselves. For many of these group members, the hour spent in group therapy was one of the only activities they engaged in that focused on their own needs and interests. During one session, Margaret mentioned that while she accepted with some measure of pride that the marriage vows she took included loving and caring for her husband "in sickness and in health," she missed his arms about her at night and their intimate moments together. (Her husband was in the end stages of Alzheimer's disease and he remained virtually motionless in a hospital bed). The other group members then collectively discussed their loss of interest in sex and sensuality. One group member quipped, "Who has time for that anymore, anyway? I'm lucky if I can get out to come here, or to run to the store to buy some groceries." Others debated about whether their sex lives really were that important anymore, "since [we] are older and we don't have husbands to be with that way anymore, anyway."

In response, the therapist made a direct request for each woman to buy and use scented body lotion when bathing and getting dressed in the morning. When some of the women balked at this homework assignment, stating that they did not have time for something so "silly," the therapist reminded them that they could easily purchase such products at the local drug or grocery store, or even by mail order in a cosmetics catalog. She encouraged them to humor her and "give the activity a try," and then make their individual decision about how they felt about it. She also encouraged the group members to sample a variety of products and to select the fragrance that they liked best. She also directed them to allow themselves an extra few minutes in the morning to rub the lotion on their legs, stomach, chest, and arms as they got dressed.

During the next week's session, Margaret and a few other women in the group actually brought their scented bath oils and lotions with them to the group to show each other. Margaret said that she initially felt guilty about taking the time to go shopping and buy herself something, but that she began to feel better once she got home and used her lotion. "It is as though it follows me throughout the day…it reminds me about the group, too, when I start to feel down." Other women said that they had not paid much attention to different parts of their bodies in a long time, and that "it wasn't as bad as I thought; my stomach isn't that big after all." Another woman said, "I hadn't been in the make-up aisle in a long time. I bought myself some new blush, too." She was able to acknowledge that she still received pleasure from tending to her appearance and "feeling feminine," even though she had not done so in a long time because of her involvement in caregiving. Still other women reported that the use of the lotions allowed them to acknowledge and mourn their personal losses, "I had forgotten how Sam used to help me with my lotion on my back. I miss his touch on my skin…It makes me sad, yes, but it also makes me remember…And now I can do some of that for myself…I'm still a woman, too, even if my husband can't hold me anymore."

In sum, these women renewed their interest in their bodies and in basic sensual pleasures. The therapist's "prescription" for the use of perfume and scented oils allowed these women to explore and begin to focus on their own bodies. Many were pleasantly surprised to renew their own sexual identities, without direct input or feedback from their husbands. As caregivers, they had been granted permission to tend to themselves.

Concerns About Body Image

Although it is obvious that older women have concerns about their body image as do their younger counterparts, the specific concerns and anxieties that older women have are likely to be somewhat different from those held by their juniors.

The following have been identified as common concerns related to body image among older women:

- Facial changes (wrinkles, age spots, sagging chins)
- Graying and thinning hair
- Changes in body fat distribution
- Loss of vigor and strength
- Incontinence (including fears of odor and dislike for the baggy, loose clothing sometimes required to accommodate large pads)
- Poor posture and gait resulting from arthritis, hip replacements, or other ailments
- Impaired vision (may reduce ability to ambulate and to groom oneself)
- Hearing loss (making it difficult to communicate and socialize)
- Gnarled hands caused by arthritis (both for appearance sake and for the resulting lack of mobility to get dressed, groom oneself, and engage in tactile stimulation of self or partner)
- Denture appearance, fit, and odor; also concerns about how to manage them during intimate moments
- Inability to exercise and to ambulate independently (including embarrassment over the use of a walker or wheelchair)
- Dowager's hump caused by osteoporosis (postural concerns and difficulties with finding attractive clothing)

It is notable that many of the body image-related concerns espoused by older women do not focus on appearance per se, but on sensory deficits and changes in body function (e.g., incontinence and arthritis stiffness) that can impair a woman's autonomy. Because many sensory deficits make it more difficult for older women to groom and dress themselves, many of these women cite additional concerns that their loss of independence makes them feel less feminine and sensual. For others, a loss in hearing, vision, or taste means that they are not able to gain as much enjoyment from previously enjoyable sensual activities such as listening to a favored type of music, admiring beautiful art, working out hard at the gym, going on long walks, or savoring a gourmet meal.

Practical Issues

To complicate practical matters further, the fashion industry has not yet met increasing demands for older women's sometimes special needs for clothing (Chrisler and Ghiz 1993; Lee and Sontag 2010). Older women who have arthritis or other ailments that result in difficulties with manual dexterity may need clothing with large fasteners, buttons, or Velcro patches to allow them to dress independently. A large dowager's hump may make it difficult to find blouses or dresses that have enough fullness through the upper back to fit properly and comfortably. Because of certain medical conditions and poor circulation, some older women must find clothing that provides additional warmth indoors as well as comfort and good looks.

Only a few specialty stores and catalog houses offer such clothing, as the clothing industry is slow to recognize this growing market in the population. In response, many older women either stop shopping altogether or are forced to buy extra large sizes and perhaps have them altered. Others lament that while shopping used to be an enjoyable, sociable activity that allowed them to affirm their femininity, it has evolved into an unpleasant chore that serves as an unpleasant reminder of their bodies' undesirable changes in appearance and function. The use of online specialty merchants (e.g., Silvert's Adaptive Clothing; Buck & Buck) can provide a viable alternative, as can the development of alternative means of expressing one's femininity via other types of social outings (e.g., going to the movies, the gym, for walks) or activities (e.g., massages; manicures).

Eating Disorders

A dangerous assumption made by some health care providers is that eating disorders are manifested primarily among adolescent and young adult women. In the past few decades, the numbers of patients diagnosed with eating disorders including anorexia nervosa and bulimia have increased dramatically. Because some proportion of these individuals tend to develop chronic symptoms, one would expect that these younger women develop into middle-aged and older women with eating disorders. Recent findings suggest that approximately 40% of older adults with some type of eating disorder displayed symptoms of an eating disorder in their adolescence or young adulthood (Lapid et al. 2010). Although the specific etiology for eating disorders is unknown, women with a history of comorbid depression or personality disorder may be at greater risk for maintaining these disorders into late life (Hsu and Zimmer 1988). In sum, evidence suggests that up to 3% of middle-aged and older women may suffer from eating disorders (Gadalla 2008).

What may be more insidious about the prevalence of eating disorders in late life is that within the past decade, evidence is mounting that the majority (60%) of cases in older women are of late versus early onset, and that more than 80% of these cases represent anorexia nervosa, with a very high mortality rate in which more than 1 in 5 of those women will die (Lapid et al. 2010). The increasingly intense pressure on women from all age groups to maintain a youthful, slim appearance certainly may contribute to this emergence of symptoms in later life. Additional triggers for onset of an eating disorder in later life can include concerns about weight gain during menopause, the loss of a long-term partner through death or divorce, or the overall loss of control felt during the occurrence of a chronic illness or hospitalization. Although few studies have been conducted on the topic, the incidence of binge eating disorders also can be expected to increase among middle-aged and older women (e.g., Marcus et al. 2007).

Unfortunately, health care providers typically overlook the presence of eating disorders in middle-aged and older women. Some practitioners may simply believe that eating disorders occur primarily among adolescent populations. Clinicians should be particularly alert to any older adult who is described as a "problem eater" by staff members in an institutional setting (Ronch 1985), or to any rapid fluctuations or change in weight among community-living patients (Lapid et al. 2010). Most importantly, one must be careful not to assume that rapid weight loss among an older female patient represents a stressful response to aging or a response to a chronic illness. An interdisciplinary team approach that includes therapists, geriatricians, and nutritionists, coupled with a thorough intake interview about binging, purging, and body image, can help reveal a more complete diagnostic picture and specialized treatment.

Ruth

Ruth was a 68-year-old divorced woman who moved to a nursing home because her rheumatoid arthritis and advanced glaucoma prevented her from caring for herself independently. She used a walker to ambulate and was not able to drive because of her diminished vision. Instead of moving in with relatives, Ruth decided to move into a nursing home. She had no children and had only a tense relationship with a niece who lived in the area. While at the nursing home, Ruth presented with depression and symptoms of an eating disorder. She would eat her meals in the dining room, but would often pick at her food and rearrange it on the plate rather than eat it. One nursing assistant observed Ruth placing spoonfuls of vegetables and meat into an empty milk container, ostensibly to make it look as though she had eaten more of her meal. It also was unclear whether Ruth purged by vomiting after meals. She had a private bathroom in her room, and staff members were unclear as to how closely they should (or could, for ethical reasons) monitor her. When asked by staff members why she was not eating or if she was making herself vomit after she did eat, Ruth responded with a flat denial. However, she was overheard to tell other patients that she was on a diet because she was "so big and fat." In reality, Ruth was within the normal weight for a woman her size and age.

After one outing with her niece, Ruth returned to the nursing home in tears. When approached by the staff psychologist, Ruth admitted that she was so upset because her niece always rushed her when they were shopping. When asked about it further, Ruth described how frustrated she was about her inability to find pants that fit over her thighs and her dowager's hump. "I can't stand it!…Everything just fits wrong and I look so big and fat and ugly!…And, she just rushes and rushes me…she doesn't care, but I do…I can't stand it." Ruth then admitted that she was trying to diet because she felt that if she lost enough weight, she might be able to fit into some of her older clothes that she liked, and not have to go shopping anymore. When pressed, she said that she believed if she lost enough weight, her thighs would shrink and her dowager's hump might even get smaller, despite her physician's information to the contrary. She also admitted to starving herself at times, and using her finger to make herself vomit in the bathroom after she ate a dessert or a "starchy food" like pasta or a baked potato. Because of the professional concern and willingness of the staff members to put aside age-related stereotypes that eating disorders were exhibited exclusively among younger adults, Ruth was interviewed and diagnosed with bulimia. With the support of the staff psychologist, she agreed to begin a program of treatment.

Exacerbation of Existing Psychopathology

Women with longstanding psychological disorders also are at increased risk for body image disturbances. Depression and psychotic depression certainly foster inaccurate perceptions of a changing, aging body. Because psychotic depression is often undiagnosed among older adults, as it often is assumed to be a symptom of dementia, many older women may develop delusions about their appearance or body function. For women with narcissistic or histrionic personality disorder, aging itself poses many egregious developmental tasks. For a woman with narcissistic traits, reliance upon a cane for increasingly severe arthritis can be viewed as an assault against the self. Even being around other older adults who are naturally showing signs of their age can be taken as a narcissistic assault as the personality-disordered older woman is literally forced to see what does not mirror her own needs and expectations. An older woman with histrionic traits who may have relied heavily on her abilities to relate to members of the opposite sex in primarily sexualized ways may feel at a loss if she perceives that her "looks are gone" and her prior means of attracting attention is no longer effective.

Fortunately, both dynamic and cognitive-behavioral approaches in therapy can be used to assist older women in coping with some of the bodily changes that they experience with age. From the dynamic perspective, an exploration of internalized object relations can help aging women acknowledge that a previous sense of self that was based primarily on outward, physical characteristics is bound to suffer. Allowing older women to acknowledge, accept, and grieve their perceptions of lost youth and independence will allow for more adaptive coping behaviors to emerge. A more global, integrated view of the self will emerge that will allow older women to view their aging bodies as "outward signs of inner wisdom" (Chrisler and Ghiz 1993). Cognitive-behavioral measures can also foster adaptive coping with physical aging. Techniques such as occupational therapy, structured exercise programs, and art, dance, and music therapy have been shown to increase older women's self-esteem and improve their body image (e.g., Pratt 2004). Even though a frail older woman may have limited strength and mobility, exercising, dancing, or practicing yoga while sitting in a chair or in a bed from a prone position can also generate significant benefits.

Yolonda

Yolonda was a 71-year-old woman who attended a partial day hospital program for depression. She had never married, had been a successful stenographer, and had maintained her own apartment for the last 45 years. During the past year, however,

Yolonda developed glaucoma. She had always enjoyed needlepoint, and said that she became depressed when she could no longer pursue her hobby and when she had difficulty crossing the street and go shopping by herself. It had become almost impossible for her to see oncoming cars or to read the prices on items in the store. She was admitted to the day hospital program when the woman who helped her with meals and laundry each day noticed that Yolonda was no longer eating and that she would not get out of bed.

After being in the program for a few weeks, Yolonda began to make friendships with some of the other women. She began to explore knitting and listening to music as other possible hobbies. (She could "feel" the needles and stitches for knitting.) As the therapeutic relationship grew between Yolonda and her therapist, Yolonda began to address other issues that lent themselves to her depression. She noted that she had been in love with a young man in her early 20s, but that he had died tragically in a car accident. She had other offers for dating and even for marriage, but Yolonda felt that she could only have loved this one man. Now that she was approaching the last decade of her life, however, she sometimes regretted her decision not to seek a companion.

When asked how she felt about trying to date or find a companion at this point in her life, she responded flatly, "How can I with this hair?" Her therapist was curious, as Yolonda was an attractive older woman with shoulder-length gray hair. She inquired as to what was wrong with her hair, and Yolonda acted very surprised. She expressed concern that her therapist couldn't understand that with "no hair," no one would possibly think she was attractive. In fact, Yolonda literally meant that she had no hair; she had a circumscribed delusion that her hair had been falling out in clumps, and that she was almost bald. She said that she tried to always "dress nice," and sometimes even wear makeup, but that she was too anxious about her hair to even think about dating. When her therapist attempted to explore the delusion carefully, and asked what Yolonda saw in the mirror or when she brushed her hair, Yolonda simply shrugged and said, "I suppose sometimes it is worse than others...I guess a part of me does think I have hair...but that doesn't last for very long."

After intensive therapy in which Yolonda explored her ambivalence about closeness and intimacy with men, her major depression began to subside. Through work with an interdisciplinary team, Yolonda also received some psychotropic medication to help alleviate the psychotic features of her depression. After another few weeks, Yolonda's observing ego became strengthened. She was able to recognize that she did, in fact, have thick hair, but that it was her fears about becoming close to others that allowed her to think that she had no hair. Yolonda's therapist also was accepting and patient in that Yolonda's delusions were very real to her, and that they must be taken seriously. Ultimately, this older woman recognized that she did not have to have male companionship to feel loved and attractive. With intensive therapy and medication management, she was able to develop caring friendships, mourn the loss of her eyesight, and adopt more suitable hobbies. Although it remained unclear whether Yolonda had never resolved or mourned the loss of her first love or whether she had an unexplored lesbian orientation, this older client ultimately was able to discount her body image distortion and feel pride and acceptance in her appearance. Her distortions in body image were taken seriously, and were thus treated appropriately by all members of the treatment team.

Nancy

Nancy was an 89-year-old woman who resided in a nursing home. She was known by staff members as one of the most challenging patients in the facility. Nancy was diagnosed with narcissistic personality disorder, and often made inappropriate demands on both staff members and other residents. Throughout her life, Nancy had relied on sexuality and her appearance for her sense of self-esteem, approval, and femininity. From a dynamic perspective, due to poor internal object relations developed through poor modeling by her parents and significant others at a young age, Nancy literally sought out others to mirror and appreciate her own external image. Her only sense of self was what she saw reflected in the eyes of others. In the nursing home, Nancy consistently demanded that others acknowledge her presence, and even would go so far as to say that she would only wear pink, blue, and purple as they were "the colors of royalty and princesses." Nancy also would only associate with other residents who were at least as healthy as she was. She said that she could not tolerate "having to look at old, sick people," and sometimes would refuse to eat in the dining room if certain patients in wheelchairs were present.

Although Nancy sometimes caused management problems, staff members could often effectively use limit setting and behavioral (e.g., token economy) techniques with her. However, Nancy's behavior soon escalated after she fell in the bathroom and broke her hip, requiring immediate surgery for a hip replacement. After the surgery, Nancy was forced to rely on a wheelchair to ambulate and was told that she needed to attend physical therapy daily in order to guarantee the success of the procedure. The staff psychologist was asked to see Nancy after she had began to refuse her physical therapy treatments, and her physician, physical therapist, and occupational therapist could not persuade her to continue.

During therapy, Nancy's therapist recognized that making any long-term personality changes was highly unlikely in the short term, and she focused instead on encouraging Nancy to articulate her feelings in order to encourage her to attend her physical therapy sessions. Nancy became a regular participant in twice-weekly psychotherapy and began to tell her therapist about her desires to be wealthy, independent, and beautiful. "You know, I was quite a woman in my day…I turned heads…I wore all the right clothes in the right way, if you know what I mean." When asked gently how she felt now that she was older, and had an unfortunate fall, Nancy was able to vent her anger and articulate some of her narcissistic injuries, "This is a disgrace…I can't even get myself to the bathroom…what's the point now, anyway?" Once she vented some of her anger and frustration (most of the staff members were so frustrated with Nancy that it was difficult for them to tolerate, much less entertain her constant tirades and complaints), Nancy was willing to go to some of her appointments. However, the physical therapists and nursing aides often spent more than 30 min yelling, screaming, talking, and begging in order for her to attend. The situation was spiraling out of control, and Nancy's narcissistic needs were being met in a maladaptive way. She was gaining significant attention, but only because she was being passive, argumentative, and demanding.

Nancy's therapist then educated the nursing home staff members about the fear underlying Nancy's refusal to attend physical therapy. (Confidentiality issues were not broached in any of these discussions.) With guidance, the physical therapists were willing to set limits for Nancy and to only give her positive reinforcement when she did attend her sessions. For example, staff members were instructed to ask Nancy to attend her session once in the morning, and to give her only one reminder a few minutes before her appointment. They also were told not to cajole or beg Nancy to attend her appointment, but to stress to her that it was her personal choice and responsibility to attend therapy; they would respect *her* wishes. This tactic provided Nancy with some sense of dignity and control, and forced her to move from a passive-dependent position to a more assertive one in which she gained attention, mirroring, and reinforcement by engaging in adaptive behaviors. The physical therapists also provided a spontaneous therapeutic intervention for Nancy in which they crafted a homemade birthday card for her that included the phrases "sexy lady," "beautiful," and "fun and feisty" that was presented to her after 10 sessions of work.

Through the combination of appropriate direction and support from her therapist, physical therapists, and other support staff, Nancy soon regained her mobility. In the short run, her personality structure became less rigid, and a more behavioral approach allowed her to complete her sometimes painful trial of physical therapy. Nancy also went on in therapy to work on changing some established patterns in her interpersonal relationships and was able to develop more positive coping skills in response to her changing body image.

Positive Changes

Although the majority of older women report some level of discomfort or disdain regarding their body image, some women maintain positive attitudes toward their bodies and appearance throughout life. In fact, others actually grow to enjoy a level of comfort with their body for the first time in later life (e.g., Baker and Gringart 2009). It thus is important that practitioners ask their middle-aged and older female patients about their body image, rather than simply assume that their emerging perceptions are negative and disturbing.

Amy

One older female outpatient in psychotherapy reported that she had been considered very beautiful when she was younger; everywhere she went she heard cat calls, whistles, and shouts. Amy described that she was constantly approached by a variety of men, whether she was out shopping at the grocery store or at a nightclub with a date. She found it sometimes difficult to make female friends; they often felt jealous and threatened by her appearance. Some female acquaintances even told her that they wouldn't introduce her to their husbands. What upset Amy even more was that she felt that no one ever took her seriously. "After all," she said, "back then, beautiful women were automatically assumed to be dimwitted…maybe they thought you could be a model or a dancer…people didn't even think I could possibly be a good secretary…no one would even ask me who I voted for in the elections."

Once Amy reached her mid-50s and 60s, however, she found that it was a huge relief to walk on the street without being "a target" any more. She described having a refreshing sense of freedom to go where she wanted, when she wanted without living in fear or notoriety. Amy also found it was easier to begin friendships with women, and that both men and women seemed to be more interested in what she had to say than what she looked like. Near the completion of her therapy, Amy enrolled in a college course in politics at the local community college and felt validated by her scholastic performance. Her therapist allowed Amy to recognize both the positive and negative aspects in her assumption of the role as "blond bombshell," and how she may have played a part in perpetuating this dynamic in the form of a self-fulfilling prophesy. But for now, age had leveled the playing field for Amy, and it allowed both herself and others to focus on her inner qualities instead of exclusively on her outward appearance.

Breast and Cervical Cancer

In any responsible discussion of female sexuality, the detection and treatment of breast, cervical, and other gynecological cancers must be addressed. In the USA, breast cancer is the second leading cause of death for US women and is the most commonly diagnosed cancer in women next to skin cancer (Centers for Disease Control and Prevention; CDC 2010). Nearly one in eight women will receive a diagnosis of breast cancer in their life time (Horner et al. 2006). This aspect of sexuality becomes increasingly important for the growing population of older adult women. As women age, their risk for both breast cancer and cervical cancer, caused primarily by certain strains of the HPV virus, increases. The median age of diagnosis for breast cancer in the USA is 61 years of age, and the median age of diagnosis for cervical cancer is 48 years of age (Howlader et al. 2011).

Detection

Despite this age-related increase in pathology, epidemiological studies show consistently that older women, compared to their younger counterparts, are significantly less likely to have mammograms or pap smears on a regular basis, despite their greater use of primary health care services. More than one-third of all US women over the age of 55 do not adhere to recommended guidelines for mammography screenings (Rakowski et al. 2004).

Screening measures such as mammography and pap tests are vital in order to provide early detection and treatment. Although some controversy exists regarding the most recent screening guidelines from the US Preventive Services Task Force of the Department of Health and Human Services (DHHS; 2010), their recommendations for mammography indicate that routine mammography for women with an "average" risk for breast cancer should begin at age 50, that women should receive a mammogram every other year, and that routine screening should end at age 75.

The American College of Obstetricians and Gynecologists screening guidelines for pap tests (2009) recommend that women aged 30 and older can be screened every 3 years if they receive negative test results for 3 consecutive years. Physicians are told that women between the ages of 65 and 70 can stop having the pap test done entirely if they have had three consecutive normal test results or no abnormal results for the last 10 years. These guidelines also indicate that if a woman over 30 has increased risk factors, including those who may be HIV positive or have a compromised immune system, may require more frequent screening.

Significant controversy remains regarding these guidelines, however, because many women over the age of 65 may take on new sex partners, increasing their risk for HPV exposure, and they may certainly have long-term sex partners who engage in affairs or unprotected sex with other female or male partners. It is as though these professional health care organizations themselves engage in ageism, and simply assume that older adults either are, or become, asexual.

To compound the problems with recommended screening guidelines, studies suggest that many older adults are unaware that pap smear tests detect the presence of cervical cancer in its early stages and that the virus responsible is transmitted through sexual contact (Ives et al. 1996). Physicians also have shown less interest in recommending mammograms for women with dementia and for women living in nursing homes (Marwill et al. 1996). Disparities in health care seeking and availability also mean that more minority group women fail to receive routine screening, as well. Fortunately, older women who have been educated about cancer risks and who have been invited to participate in screening programs have shown similar rates of compliance as their younger counterparts. Physician recommendation appears to represent the best predictor of compliance (Schonberg et al. 2007), and studies suggest that nearly 90% of adults over the age of 70 would like to discuss cancer screenings with their doctors (Lewis et al. 2006).

Treatment Issues

In addition to difficulties in obtaining appropriate physician referrals, patient interest, and compliance with cancer screening measures, treatment issues regarding breast and cervical cancer among postmenopausal women typically pose significant medical, ethical, and psychological problems. Although a psychotherapist certainly is not qualified to make medical diagnoses and treatment plans regarding breast lumps, abnormal pap smears, and breast and cervical cancers, mental health practitioners can assist their clients in coping emotionally with individual barriers to screening, and with their potentially abnormal test results.

Changes in body image are certainly central to breast, cervical, and other gynecological cancer survivors. For example, women survivors of such cancers have described themselves as "maimed, damaged, asymmetrical, [and] incomplete," and reported low levels of self-esteem, poor body image, and decreased levels of sexual desire (Sacerdoti et al. 2010, p. 538). What often escapes discussion, however, is the aftermath of treatment in relation to sexual functioning. Studies indicate that oncologists typically do not address issues of sexuality with female cancer patients before, during, or after treatment (e.g., Juraskova et al. 2003) out of embarrassment, a lack of knowledge and experience, and expectations that patients were responsible to bring up those issues (Stead et al. 2003). However, female cancer patients themselves typically do not bring up concerns about sexual function, including fears of pain during intercourse, out of embarrassment or because they believe their oncologist will view such discussions as limited in importance in relation to their primary treatment (Molassiotis et al. 2002).

Fortunately, mental health providers can engage in a variety of activities to help alleviate the concerns of female cancer survivors. Because few oncologists make referrals to mental health professionals or sex therapists, clinicians can seek out collaboration with hospitals and individual oncologists to help provide women survivors with needed, related psychological treatment, psychoeducation, or at least adequate referrals (Sacerdoti et al. 2010). The role of psychologists and other health care providers is essential, as primary predictors of quality of sexual life and functioning for female cancer survivors are typically based upon personality traits and other psychological factors, rather than clinical outcomes themselves (Den Oudsten et al. 2010).

The role of psychoeducation appears central for both female cancer patients and their health care treatment providers. Many women are unaware that various chemotherapy drugs, including particularly those that bind to estrogen receptors on tumor cells (e.g., the drugs Fareston, Faslodex, and Nolvadex) can lead to significant vaginal dryness and dyspareunia. Sometimes, oncologists may wish to prescribe medications to combat vaginal dryness even before chemotherapy commences, in order to help alleviate this effect (Sacerdoti et al. 2010). In addition to individual psychotherapy, and couple's therapy, even brief psychoeducational interventions have demonstrated effectiveness in combating female cancer survivor's low levels of sexual desire and arousal (Brotto et al. 2008). Because anecdotal reports indicate that cancer survivors often use KY Jelly other another water-based lubricant recommended to them by their health care provider or that is readily available in the local grocery store, which often fail to provide longer-lasting lubrication, educational about possible options for alternative lubricants (e.g., silicone-based lubricants) also can be helpful.

Additional factors may complicate treatment issues for women with breast and gynecological cancers in the oldest-old age cohort for women with cognitive impairments. Women in these groups may be intimidated by their physicians, particularly if their physicians themselves are not comfortable discussing issues of sexuality with their older patients. Psychologists and other mental health care practitioners may be able to assist family members in identifying and articulating their views about screening and treatment for an older mother or aunt who may be unable to make appropriate decisions for herself due to dementia or other chronic mental illness. Questions also may be raised about whether treatment for breast or cervical cancer will provide an increase in quality of life for a woman who is in her 80s or 90s, or whether these treatments will only induce pain and suffering that exceeds any benefits in decreased mortality. A variety of survival benefit curves and predictions based on general population statistics are available to assist patients and physicians in determining treatment course. However, care must be taken to incorporate quality of life issues (e.g., pain relief and the ability of family members to provide care outside of an institutional setting) and religious preferences into any otherwise sterile discussion. Such tasks can be made even more difficult if an elderly patient diagnosed with cancer is impaired mentally. For example, is it appropriate to order a mastectomy without obtaining the older woman's permission if she has some type of cognitive impairment?

Some health care providers may feel that for an older woman, particularly for one who is widowed or single, breast loss is not as emotionally difficult or as central to their identity as it would be for younger women. For a number of older women, many feel that their breast represents one of their only remaining, concrete reminders of their femininity in the face of chronic illness and age. It becomes vital to provide open lines of inquiry regarding invasive medical procedures that may alter a woman's physical appearance. On discussion, some physicians are able to consider reconstructive surgery after mastectomy if their patients request it.

What is most important for mental health providers is to assist their patients in communicating effectively with their physicians, caregivers, friends, and family members in order to provide the most effective treatment plan. Empirical research suggests that older women would benefit from an advocate and supporter during treatment planning; older women with breast cancer have been shown to seek less information about their treatment options and make hastier decisions about their treatment than younger women (Meyer et al. 1995). It also is important to recognize and respect the fact that for some older women, aggressive treatment for breast or cervical cancer is their preferred choice, whereas the withholding of treatment becomes the preferred choice for others. As effective clinicians, we must inform and support our older female clients in their difficult decision making, without imposing our own moral and religious views.

Summary

Older women are faced with conflicting messages from society about their role as women, sexual beings, and potential contributors to society. These societal demands placed on middle-aged and older women, coupled with their own internalized stereotypes and expectations, can be significant, and may be inappropriately internalized and manifested in the form of sexual dysfunction or dissatisfaction, body image distortions, depression, eating disorders, or poor participation in breast and cervical cancer screenings, among others effects. Due to a general lack of knowledge among the health care community as well as the population at large, most aging women are not aware of the *normal,* basic physiological changes that take place with aging. Many older women are delighted to learn that treatments are available to assist them in coping with such changes in order to restore their enjoyment in sexual activity. Clinicians also need to acknowledge the sensual as well as sexual needs of their female clients, who may be with or without a partner. There is no reason why a woman cannot enjoy and relish her status as a woman throughout her entire life span.

Chapter 8
Men's Issues in Sexuality and Aging

*Although population-based studies indicate that men typically
engage in a variety of sexual activities well into their 80s and
90s, additional findings suggest that nearly two out of three men
will experience erectile dysfunction at some point in their lives.*

The expression of male sexuality encompasses significantly more than the number
of times per month one has intercourse or a man's ability to have an erection. At the
same time, clinicians must recognize that many older men do place significant
emphasis upon their ability to have penetrative intercourse and that any complaints
or concerns about erectile dysfunction (ED) must be taken seriously. ED represents
one of the most common sexual problems that older men experience, and clinicians
can play a vital role in providing patients with the ability to seek appropriate medical consultations, to allay anxiety about diagnostic procedures, to manage psychological problems related to sexual dysfunction, to involve their partners in assessment
and treatment, and to help resolve issues related to sexual identity and prowess.

Historical and Societal Context

One of the cornerstones of Freud's theory was his articulation of the differences
between male and female sexuality and sexuality's overall relation to personality
functioning and development. The penis, as reflected in its importance as an external
sexual organ (as compared to the internalized, female uterus), was paramount. From
early childhood, women were thought to envy the penis both symbolically and literally. The penis represented something that a woman was "missing" as well as more
general opportunities to advance and enact change in a male-dominated society.
Although it can be argued that Freud was a champion of women in that he took their
views seriously, promoted nonsexualized physician–patient relationships, and believed
that psychic trauma rather than an inflamed uterus was responsible for neuroses and
psychosomatic symptoms, he remained steeped in his culture and championed the

J. Hillman, *Sexuality and Aging: Clinical Perspectives*,   199
DOI 10.1007/978-1-4614-3399-6_8, © Springer Science+Business Media New York 2012

inherent position and power of men. The psychological price to be paid for such an emphasis on male anatomy is that men could be expected to experience castration anxiety in the presence of aggressive women and mothers, competing fathers, and personal failures at home or at work. Perhaps such fear of decline in sexual prowess has helped to drive the recent change in the popular lexicon from a discussion of impotence to the less emotionally charged term erectile dysfunction or ED.

Popular culture reflects this general emphasis on the male sexual organ and its imbued abilities to wield power and foster competition. Consider the pervasive use of slang and curse words to illustrate the power of the penis: fuck you; bite me; suck it; blow me; piss off; dick head; jerk off; prick; tool; and he's got balls. There are no such parallel expressions featuring female anatomy. Something is inexorably tied between the penis and competition, aggression, and physical satisfaction in the public eye. Men over the age of 50 including Pierce Brosnan, Denzel Washington, Denis Quaid, Richard Gere, Liam Neesan, and Sean Connery command large audiences as action heroes and sex symbols. Despite the emergence of the aging female "cougar," it remains much more common and socially acceptable for older men to marry and date younger women. Although some changes are taking place, sex and power remained intertwined in Western culture, particularly for men. As a corollary, any sexual dysfunction among men (caused by physiological or psychological difficulties) must be taken seriously as a significant psychological assault.

Normal Changes in the Sexual Response Cycle

Even though aging is not a pathological process, normal aging typically produces some changes in male sexual functioning. Many middle-aged and older male patients are unfamiliar with these changes, and may experience significant anxiety when they occur. Sometimes the simple dissemination of knowledge of these age-related changes is enough to ease the mind of older male patients and to encourage them to take proactive measures and resume their enjoyment of sexual activity. A mental health provider can help provide such psychoeducation and can also serve as a conduit for more effective communication between a male client and his family practitioner, internist, cardiologist, oncologist, urologist, or geriatrician when it appears that true physiological or medical problems may underlie such sexual changes.

There is some debate about whether men experience age-related hormonal changes as women do in menopause. These hormonally related changes in males are believed to be associated with a very gradual decline in the production of testicular testosterone sometime after the age of 50. More specifically, some researchers attribute a general loss in sexual interest or libido and a decrease in the ability to obtain and maintain an erection to this decline in androgen production (Kaiser and Morley 1994; Morley 1996; Patnaik and Barik 2005). Alternatively, other researchers assert that testosterone production has little to do with an ability to maintain an erection (Segraves and Segraves 1995), and that other factors such as vascular disease appear to play a major role (Kaiser et al. 1988).

Whether or not a decline in testosterone, specifically, is responsible for changes in the sexual response cycle among middle-aged and older men, a number of normative age-related changes are commonly observed (e.g., Kuzmarov & Bain, 2009):

- Decline in sexual desire or libido.
- Decrease in vasocongestion (blood flow) to the scrotum.
- Decline in the number of both spontaneous and morning erections.
- Reduced tension in the scrotal sac both before and during intercourse.
- Decrease in penile sensitivity.
- Increase in time required to achieve erection.
- Tactile stimulation may be required to achieve an erection.
- Decline in the rigidity of erection.
- Increase in time required for orgasm and ejaculation.
- Decline in the force and volume of ejaculate.
- Longer refractory period between erections.
- Although it is not an age-related change, most men (and their partners) are unaware that the ability to experience or feel an orgasm is separate from the act of ejaculation. In other words, men can orgasm without ejaculating.

Some men find that while they were once able to achieve an erection in seconds, they may now require a few minutes to achieve arousal sufficient for penetration. This may not necessarily indicate underlying pathology; a decrease in the number of adrenergic and cholinergic receptors may interfere with smooth muscle relaxation and the rapid, autonomic flow of blood to the penis to produce an erection (e.g., Schiavi & Rehman, 1995). Unfortunately, feelings of panic and concern may be common for men who are unaware that this time delay in achieving erection is normal in many regards. Many older men immediately jump to the conclusion that they have ED and unnecessarily avoid or cease involvement in sexual activity.

Because both middle-aged and older men also may experience a general decline in sexual desire, their partners may or may not question whether their attractiveness or desirability is also in decline. Some partners and spouses may become confused, frustrated, or angry when they find that they are asked, or simply need, to physically stimulate their partner's penis in order for him to achieve an erection. One long-time spouse remarked, "I always thought it was just good enough for him to just see me naked…Now he wants me to touch him with my hands and my mouth…We haven't done that sort of thing in years…Why aren't I good enough the way I am?…Is this some sick fantasy of his or is he having an affair and maybe *she* does those things?" Effective communication between partners can become critical. Many older men are distressed deeply that they "can't get things going" as quickly as they used to, and feel powerless and embarrassed. Learning that their partner or spouse is willing to discuss the issue can alleviate further performance anxiety in such men. Some partners are happy to place less emphasis on the actual quality of a spouse's erection than on the quality of their foreplay or on their relationship in general, and both parties can move forward to simply enjoy each other without heightened performance-related demands or expectations.

Not all age-related changes in the male sexual response cycle are inherently negative, however. For example, many middle-aged and older men who previously experienced difficulties with premature ejaculation, an increased need for physical stimulation, and a slower buildup to orgasm can prolong the sexual act, which can be associated with greater enjoyment for both partners. Some older men have described feeling free to enjoy themselves and their sexual partners now that they no longer have to worry about "holding back" or "controlling themselves" during intercourse. Other partners have reported that they enjoy the slower pace of sex now that their husband "takes his time" and seems to enjoy himself instead of rushing ahead to climax. Again, open communication and basic education among both partners (often assisted by a clinician) can be critical.

Clinical Interviewing and Assessment

Proper interviewing and assessment is vital to assist older men in discussing previously undisclosed issues in order to foster patient education, interpersonal exploration, and consultations with appropriate medical specialists when necessary. Open-ended questions and questions that assume the presence of some difficulties may make it easier for initially reluctant patients to respond. Sometimes the presence of a younger, female clinician can be cause for concern among older male patients who may not be accustomed to women as professionals or who were raised in a cohort in which it was considered inappropriate to discuss sexual matters with women. If a client appears reluctant to discuss sexual concerns with a female professional, a frank discussion of that concern is in order. Similar issues can arise when an older male patient is confronted with a younger male clinician. Most patients respond very positively when their clinician is willing to address the "process" as well as the "content" of their interview. Sometimes simply asking, "What is it like for me to ask you these questions about sex?…Some people do find it a little unsettling at first" is enough to engender a meaningful discussion of the patient's underlying anxieties, fears, and social mores. Once these issues are addressed, a more open discussion of a patient's sexual concerns and symptoms is likely to follow.

A number of standard questions can be addressed during a patient interview, or when a patient in longer-term therapy suddenly announces concerns about sexual function. These include (e.g., Galindo & Kaiser, 1995; Sbrocco et al. 1995):

- Most people have some difficulties with sex at some point in their lives. What concerns do you have about your sex life or sexual functioning?
- How do you think your partner feels about your sex life?
- What do you think constitutes a satisfying sex life? How would you compare your own sex life to your idea of a satisfying or "perfect" sex life?
- How often do you masturbate? Do you ever have any trouble masturbating? What kind of trouble do you have?
- Are your sexual partners, both now or previously, primarily women or men? Have you ever had any same-sex sexual encounters? Have you ever had sex with

a prostitute? With an intravenous drug user? (Be *sure* to consider the patient's religious beliefs carefully when asking such questions regarding high-risk behaviors and masturbation.)

- How do you feel about your body? Are you happy with it or are there some things that worry you, or that you wish could be different? Have you ever tried to change anything about your body? How are your eating habits?
- How difficult is it for you to talk about sex with your partner? Have you ever talked about anything in particular?
- How difficult is it to get a full erection during sex? During masturbation?
- How often do you get erections? Are they as firm as you would like them to be, or as they used to be?
- Is it ever painful when having sex?
- Are you able to orgasm/cum/ejaculate?
- How interested or disinterested are you in sex? Is this level of interest different than before?
- Do you ever have trouble or pain when urinating? Do you need to go to the bathroom more often than you would like? Has your prostate ever been checked?
- What medical problems do you have? What have you been diagnosed with?
- What prescription medications are you currently taking? What over-the-counter medications do you take, even on an occasional basis?
- How often do you smoke, drink alcohol, or use drugs?
- Would you like to know more about the typical changes that take place with sexual functioning as a man becomes older?

Note that issues involving masturbation, quality of communication between partners, participation in high-risk behaviors, underlying medical conditions, concerns about one's overall appearance, and overall interest in sex are just as important as specific questions about the presence and firmness of an erection. Older male clients often benefit from the implicit message that their global sense of sexuality and self is at least as important, if not even more important, than the singular functioning of their penis.

Erectile Dysfunction

Definitions and Prevalence

Erectile dysfunction or ED represents one of the most feared symptoms of sexual dysfunction among men. Previously referred to as impotence, the currently used and presumably less pejorative term ED (promoted strenuously in advertising by its producer, Pfizer) is defined by the US National Institutes of Health as "an inability of the male to achieve an erect penis as part of the overall multifaceted process of male sexual function" (NIH, 1992). A similar definition is promoted by the United Kingdom's Sexual Dysfunction Association, in which ED represents "the persistent or recurrent inability to attain or maintain an erection sufficient to complete sexual intercourse or another chosen sexual activity" (Sexual Advice Association, 2011). The focus of these definitions appears to be upon a man's inability to maintain or sustain an erection sufficient enough to engage in some kind of penetrative sexual activity.

Estimates suggest that ED affects more than 150 million men worldwide, including those in European nations, the USA, Asia, and developing countries (Khoo et al. 2008; McKinlay, 2000; Moreira et al. 2006). In the USA, ED is identified as the most common source of sexual dysfunction among men. By midlife 40% of all men are expected to experience ED (Sand et al. 2008), whereas by age 70 up to 67% may experience ED (Laumann et al. 1999). Additional reviews suggest that the vast majority of men with ED never receive a diagnosis (Feldman et al. 1994). It also is important to note that by the time men reach age 40, more than 90% experience at least one erectile failure. Such transient inability to attain or sustain an erection represents a normal part of aging, and not a sign of ED (McCarthy, 2001). To complicate matters further, few men or their health care providers broach the topic on a regular basis (Hillman 2008a, b), and those men that do may not seek clinical treatment (Wentzell & Salmeron, 2009).

Clinical Stereotypes and Assumptions

Many men are reluctant to even discuss ED and are not likely to be cognizant of the underlying medical problems, surgical procedures, and side effects of prescription medications that can cause it. To compound the problem, many clinicians employ stereotypes that older or elderly men do not have a true or valid need for penetrative vaginal or anal intercourse, and that it simply becomes a matter of helping these men garner enjoyment from "other" sexual activities (Butler et al., 1994). Similarly, findings from various studies suggest that men from certain socioeconomic and ethnic groups are more likely to cease participation in all sexual activity if they become unable to achieve penetration with a partner (Cogen & Steinman, 1990; Wentzell & Salmeron, 2009). In other cases, a clinician's cultural or religious background may inhibit a discussion of the importance of the client's symptoms. It becomes the clinician's primary responsibility to help an older male patient discover the underlying cause of his ED and to look for both medical and evidence-based psychological treatments to help ameliorate his symptoms. Only as an absolute last resort should an older man be told, in effect, "Well, your inability to have [penetrative] sex is a significant loss that will have to be recognized, discussed, and mourned…There is really nothing you can do about it."

Mike

Mike was a 66-year-old single man who was involuntarily admitted to a psychiatric facility following a suicide attempt. He had been scheduled to go through a painful rectal procedure and took a bottle of sleeping pills the night before in order to avoid going through the surgery. At the time, he had been unemployed and homeless for the last 6 months and was staying with his financially successful older sister and her husband. Mike engaged reluctantly in psychotherapy, primarily because he acknowledged that if he were not willing to discuss his suicide attempt he was not likely to be released from the unit any time soon. He was diagnosed with major depression and dependent personality disorder. On the unit he was passive and had difficulty discussing his own needs and wants. Mike began to progress in therapy and recognized that his outbursts of yelling and cursing ostracized him from the family members he cared most about, and that he could find more assertive ways to deal with his problems. He began to explore his feelings of failure as a younger man and acknowledge that his suicide attempt was a way to gain attention from his family members, to make his family members feel guilty, and to avoid taking responsibility for his own health. He also agreed to take antidepressant medication at the request of the staff psychiatrist.

Three weeks into treatment, Mike discussed his difficulties in finding a "steady girlfriend." His suicide attempt was preceded by his previous girlfriend's abrupt termination of a year-long relationship after he lost his job as a masonry worker because of layoffs. He expressed a desire to be married, but feared that he was getting too old for anybody to want him, particularly with his sporadic work history and moderate level of income. He also told his therapist that he was having trouble with some things "down there, you know?" When his therapist said that she wanted to understand what he meant by that, Mike said that he "had trouble, you know, moving things along, getting things into place, when [he was] out on a date." He said that he didn't even want to go on dates when he knew that if "things started getting close…I couldn't deliver." His therapist assured him that these were very important concerns, and suggested that while they discuss the possible impact of his recent breakup with his last girlfriend on his ability to "be prepared," he simultaneously ask the staff psychiatrist about possible, underlying medical causes for his problem. (Only the staff psychiatrist could make other medical referrals.) The therapist also approached Mike's psychiatrist privately and informed her that Mike was gathering the courage to schedule an appointment to discuss a particularly sensitive issue.

A few sessions later, Mike seemed particularly angry and morose. When his therapist asked him if something triggered his abrupt change in mood, he answered bitterly, "That bitch said that it wasn't something I should be worried about—that I have more important things to work on like not being so depressed…What the hell does she know? Maybe I'm depressed because I can't get it up, you know?…She said that she might make a referral for me to see a urologist in 'a month or so if it's still bothering [me].' You know, what the fuck is that?" The therapist quickly mirrored Mike's frustration with his inability to get important information, and more importantly, his inability to be taken seriously. (The therapist also had to work carefully to avoid making a split in the treatment team by offering unprofessional comments regarding the staff psychiatrist.) Mike's initial response to his psychiatrist's snafu was to have his therapist approach her instead and "do it for me, would you?… You obviously know her better than I do, and she'll respect you more since you're a professional and she thinks I'm just some stupid bricklayer." In addition to not getting his concrete needs for information and validation met, he felt symbolically (and literally) castrated. With further discussion, Mike recognized that he was entitled to have his own needs and concerns addressed, and that his job description had nothing to do with his inherent rights as a person or a patient. In a very concrete way, he was prepared to approach his psychiatrist again, tell her that this issue was very important to him, and ask for an immediate referral to the hospital urologist.

After this session, the therapist brought up this issue during the next treatment team meeting. She presented what Mike told her and asked for clarification about "what really happened." Again, the therapist felt that she would have to proceed carefully in order to gather the appropriate information and to not ostracize the psychiatrist and promote even greater difficulties for her patient in the future. To the therapist's surprise, the psychiatrist responded unabashedly, "Oh, yes, I remember that...he's the little guy with dark hair...I don't see why he needs a referral now, anyway. He's not married and he doesn't even have a serious girlfriend...He doesn't need that problem treated right now; he's got more important things to work on." The occupational therapist on the team quipped, "So what if he doesn't have a serious girlfriend! It's his body and if he wants it to work right, why can't he get a referral?...Maybe he likes having one night stands or he likes to 'take care of things himself' but that's not any of our business to judge him."

To the staff psychiatrist's credit, a lively discussion ensued in which she acknowledged that her cultural and religious views prevented her from seeing the importance of being able to achieve erection and climax, even without the promise of a steady partner. Mike's psychologist also was able to articulate that his loss of potency resonated painfully with his overall feelings of passivity and hopelessness. The fact that his psychiatrist dismissed his concerns so easily also reinforced his feelings that he had no sense of agency or worth. The therapist then advised the psychiatrist to wait until Mike approached her to discuss the issue again (in order to provide him with positive reinforcement for a newly attempted, assertive behavior) before she made the appropriate referral. Working with a urologist, Mike and the psychiatrist then selected a different antidepressant medication that did not interfere with Mike's ability to have erections. Now that he was less preoccupied with his sexual functioning, Mike also was able to make more progress in therapy regarding his depression and more defensive interpersonal style.

Underlying Causes

Although it is vital for clinicians to address the potential psychological underpinnings of ED, as a man ages it becomes more likely that the cause of his ED is physiological (Beaudreau et al. 2011; Pommerville, 2006). Because the causes of ED become multifactorial with increasing age, educating and encouraging our clients about seeking appropriate referrals becomes vital. Rightly or wrongly, many men and their partners feel relieved to learn that medical problems may be responsible for their ED because most men attribute their inability to perform as a sign of personal failure or as a loss of their masculinity. Having a general knowledge base of specific medical disorders commonly associated with erectile difficulties can help speed this process. Medical problems associated with ED often include (e.g., Chun & Carson, 2001; Mayo Clinic 2011a, 2011b, 2011c, 2011d, 2011e). It also is important to note that men with various high-risk factors for ED (e.g., hypertension, high cholesterol, diabetes, and depression) may be unaware of their increased risk for ED (Shabsigh et al. 2010) and require appropriate education.

- Diabetes mellitus. The vascular changes associated with type II diabetes, including a reduction in blood flow and circulation, can lead to difficulties in achieving erection. Neurological damage also can be associated with poor sexual function. Estimates from various studies suggest that between 35 and even 90% of men with type II diabetes will experience some form of erectile dysfunction (Malavige & Levy, 2009).
- Vascular diseases including atherosclerosis, high cholesterol, hypertension (i.e., high blood pressure), sickle cell anemia, and Leriche syndrome. Plaque and occlusions in the blood vessels can extend from the coronary arteries to the penile arteries, making obtaining an erection difficult if not impossible. As noted in Chapter 5, other patients who have had heart attacks may fear the chest pain that often accompanies intercourse and limit their sexual activities.
- Endocrine and metabolic disorders such as hypothyroidism, hypogonadism, hyperprolactinemia, Addison's disease, and Cushing's disease.
- Systemic disorders such as renal failure, myotonia dystrophia, and chronic obstructive pulmonary disease.
- Neurological disorders including multiple sclerosis, Parkinson's disease, temporal lobe epilepsy, Alzheimer's disease, stroke, pelvic nerve lesions, and spinal cord injuries.
- Substance abuse and dependence. Tobacco, alcohol, amphetamine, heroine, and cocaine have been shown to cause difficulties in achieving erection and orgasm. Cirrhosis of the liver also has been associated with ED.
- Urinary tract infections (Tsai & Pan, 2010)
- Pelvic surgeries that lead to damage of the neuromuscular bundle of the penis (the nerves and blood vessels that promote erections). Surgery to remove prostate gland tissue, colorectal surgery, and bladder surgery are typically responsible for such damage, even though procedures are being improved to avoid disrupting these nerves and vessels. Estimates suggest that removal of the prostate may cause ED in up to 12% of all men who have this surgical procedure (Schiavi & Rehman, 1995).
- Obesity, with a waist circumference equal to or greater than 40 in. (Shabsigh et al., 2010).
- Psychological disorders such as depression and anxiety. Although it remains unclear to what extent ED may be caused by psychological or physiological factors in these cases, ED and overall loss of libido are common symptoms of depression in middle-aged and older men.

This list of health-related risk factors is certainly not exhaustive, and increased levels of overall stress, fatigue, and distraction can play a significant role in sexual dysfunction. It also is important to note the complicated etiology of ED in many cases. For example, a man might be clinically depressed and as a result develop ED, yet when his depression is resolved successfully with a prescribed selective serotonin reuptake inhibitor (SSRI) such as Prozac, he finds that an unwanted side effect of his medication itself is ED (Nurmberg et al. 2003).

Although many cases of ED can be addressed primarily through medical means, many older men also display problems with sexual performance as a result of underlying psychological conflicts. Widower's syndrome (e.g., Meston, 1997) is described as a commonly occurring phenomenon in which a man finds himself unable to consummate a new marriage or relationship, even many years after the death of his wife or partner. Unresolved issues of loss, grief, guilt about engaging in a new relationship, and fears of another potential loss are believed to manifest themselves psychosomatically and prevent the older man from fully engaging in a new romantic relationship. For these widowers, individual psychotherapy and participation in support groups can provide an outlet for these feelings and thus unblock channels to future intimacy. As with younger men, general performance anxiety also can lead to sexual dysfunction. The use of sensate exercises, in which both partners have the freedom to explore sexual pleasure without the need to engage in actual intercourse, can alleviate such anxiety and reestablish psychological, as well as physiological, intimacy.

Side Effects of Medication

As noted, in addition to medical problems inducing sexual dysfunction, a number of prescription medications can induce ED (e.g., Mayo Clinic 2011a, 2011b, 2011c, 2011d, 2011e; Nurmberg et al., 2003). It is important to acknowledge that some physicians themselves are unaware of these side effects, and they may not assess the pros and cons of prescribing one medication versus another when the effects of ED are particularly distressing for a male patient. Psychologists and other mental health care providers can employ general knowledge of these medications to help steer patients in the right direction to obtain appropriate referrals and consults. See Table 8.1 for a list of some of the more common medications that have been shown to produce sexual dysfunction, including ED.

Some classes of medications, such as antihypertensives for high blood pressure, produce sexual dysfunction in many male patients (Mayo Clinic 2011a, 2011b, 2011c, 2011d, 2011e). Other medications that treat depression are implicated in ED (Nurmberg et al., 2003), although some antidepressants such as bupropion have been shown to produce less sexual dysfunction than others. Clinicians also should inquire about the presence of such sexual side effects; low rates of medication compliance may result when ED emerges as an unexpected side effect. It also is important to note that treatment of an underlying chronic illness itself (e.g., engaging in lifestyle changes to reduce weight and high blood pressure) can also serve as a treatment for ED (Wentzell and Salmeron 2009).

Table 8.1 Drugs associated with sexual dysfunction

Medication class	Name	Generic or brand name
Prescription drugs		
Alpha blockers	Prazosin	Minipress
	Terazosin	Hytrin
Antidepressant	Clomipramine	Anafranil
	Fluoxetine	Prozac
	Paroxetine[a]	Paxil
	Sertraline[a]	Zoloft
	Venlafaxine	Effexor
Antimania	Lithium	Eskalith, Lithonate
	Topamarite	Topomax
Antianxiety	Alpraolam[a]	Xanax
	Lorazepam[a]	Ativan
	Temazepam[a]	(Generic)
Antiarrythmic	Disopyramide	Norpace
Anti-inflammatory	Naproxen	Anaprox, Naprelan, Naprosyn
	Indomethacin	Indocin
Antihypertensive & diuretics	Atenolol[a]	(Generic)
	Clonidine	Catapress
	Digoxin[a]	Lanoxin, Digibind
	Furosemide	Lasix
	Hydrochlorothiazide[a]	Lopressor
	Propranolol	Inderal
Antiparkinsonian	Biperiden	Akineton
	Benztropine	Cogentin
	Bromocriptine	Parlodel
	Levodopa	Sinemet
	Procyclidine	Kemadrin
	Trihexyphenidyl	Artane
Antipsychotic	Haloperidol	Haldol
	Mesoridazine	Serentil
	Trifluoperazine	Stelazine, Suprazine
Antiseizure	Phenytoin[a]	Dilantin
Chemotherapy	Tamoxifin	Nolvadex
Muscle relaxant	Cyclobenzaprine	Flexeril
	Orphenadrine	Norflex
Prostate cancer medication	Flutamide	Eulexin
	Leuprolide	Lupron
Sleep Aid	Zolpidem	Ambien
Miscellaneous	Celebrex	Celecoxib
	Donepezil[a]	Aricept
Over-the-counter and herbal preparations		
Antihistamine	Dimehydrinate	Dramamine
	Diphenhydramine	Benadryl
	Drixoril	Tavist-D

(continued)

Table 8.1 (continued)

Medication class	Name	Generic or brand name
Histamine receptor antagonist	Cimetidine	Tagamet
	Nizatidine	Axid
	Ranitidine	Zantac
Herbal	Alkaloids	Rauwolfia
	Glycyrrhiza glabra	Licorice
	Hypericum	St. John's Wort
Recreational and frequently abused drugs		
Analgesic	Opiate	Heroin, Methadone, Opium
Depressant	Alcohol	Beer, Liquor, Wine
	Barbiturates	Amytal, Downers, Yellowjackets
Hallucinogen	Cannibus	Pot, Marijuana, Hemp, THC
Stimulant	Amphetamine	Speed, Ice, Bennies, Crystal meth
	Cocaine	Coke, Crack, Blow
	Nicotine	Cigarettes, Cigars, Chew, Snuff

Notes: This list of drugs is meant to be illustrative rather than comprehensive
[a]Identified as one of the 50 most frequently prescribed oral drugs among older adult hospital patients (Steinmetz et al. 2005)

Ralph

Ralph was a 69-year-old man who was admitted to a psychiatric facility after being arrested for disorderly conduct. He had broken antennas off of parked cars after an alcohol-filled night out on the town. Ralph had been diagnosed with bipolar disorder and had apparently stopped taking his medication a month ago. He had become manic, unable to speak coherently because of pressured speech, and he manifested grandiose delusions and poor social judgment. Although he had a relatively high-paying job as a computer repairman, he spent most of his savings during the past few weeks on expensive clothing, dinners, and horse racing. His wife of 20 years threatened to leave him if he was not willing to stay in the hospital for extended treatment.

About 3 weeks after admission, Ralph had begun to make progress. He was prescribed lithium by the unit psychiatrist, and his mental status improved significantly. He still had some grandiose delusions, but his speech rate became normal and staff members could understand clearly what he was saying. Ralph was now able to plan for the future and recognize the consequences of his actions. He also had a healthy sense of humor and he used it to make friendly acquaintances with other patients. He participated actively in group and individual therapy, and the treatment team decided that he had made enough progress to have unrestricted visiting hours and run of the grounds during scheduled breaks.

The next week, however, a nursing assistant found Ralph sitting on a pool of blood on his bed around 4 a.m., clutching at his arm and hand. The window in his room had been broken, even though it was entwined with wire and sprayed with special coatings to make it impenetrable to all but the strongest blows.

After receiving more than 35 stitches, Ralph was asked what had happened. Through pressured speech, he admitted that he had not taken all of his medication. When asked what had caused him to stop taking his pills as prescribed, Ralph stopped talking and stared straight ahead. He then lifted his arm above his head and screamed plaintively, "I can't masturbate anymore. I mean, I can't get off!…You can't take that away from me—even in this hell hole!"

Apparently, a side effect of the lithium prevented Ralph from experiencing orgasm; he described masturbating in his room "for hours" to no avail. "You have no idea how frustrating that is, you just don't know, especially for someone as important as me." What Ralph meant symbolically was that no one cared to recognize how important the functioning of his body was to him, particularly on a sexual level. Neither his psychiatrist nor psychologist had discussed ED with him as a potential side effect of his medication, nor had they asked him directly about the presence of any such symptoms as he resumed taking his medication. Although Ralph's psychologist certainly was not responsible for his medication directly, the application of her knowledge in this area would have been quite relevant. If these professionals had intervened, or even made an appropriate inquiry, Ralph's serious injury could probably have been avoided.

Treatments for Erectile Dysfunction

A number of treatments are available for older men experiencing ED. They generally fall into two categories: medical and psychological approaches. (Alternative approaches such as high-potency vitamins, herbs, and acupuncture are not discussed here as available evidence for the effectiveness of these alternative treatments is limited; Mayo Clinic 2011a, 2011b, 2011c, 2011d, 2011e). As noted, ED among middle-aged and older men does not always have one focal, underlying cause. Rather, the cause of ED is often multidimensional, and may include multiple medical problems or some combination of medical and psychological difficulties. Clinicians must be careful to avoid a narrow focus upon one specific problem or factor. A current tendency in the field appears related to the medicalization of ED in which psychological factors tend to be downplayed or ignored. This trend appears heightened with the prevalence of advertising and mass marketing of oral medications.

PDE-5 Inhibitors as a First Line Medical Approach to Treatment

Medical treatments for ED can be grouped into four primary categories: PDE-5 inhibitors (e.g., Viagra; Cialis), vacuum devices, injectable vasoactive drugs, and penile implants. Each type of treatment has its own degree of invasiveness, effectiveness, financial cost, and observed rates of compliance. The most recent medical advances in the treatment of ED involve oral, prescription medications also known as PDE-5 inhibitors. These drugs, including sildenafil citrate or Viagra, have been touted as miracle pills because a man with ED can take a pill, and within 1–4 h before intercourse have an erection with purportedly few side effects. Because only a simple pill is required, no cumbersome equipment, surgery, or needles are involved, and a higher degree of privacy or secrecy can be maintained if the man so desires. Unlike other medical means of treating ED, Viagra appears effective in treating sexual dysfunction caused by a variety of underlying problems including vascular disease, hypertension, diabetes mellitus, pelvic surgery (e.g., removal of a diseased prostate), spinal cord injury, and most notably a number of drugs including antihypertensives, diuretics, antidepressants, and antipsychotics. The cost of these medications is somewhat prohibitive, but under a variety of circumstances, Medicaid will pay for its disbursement. In contrast, a number of private insurance companies refuse to pay for this relatively expensive treatment, stating that sexual intercourse at a certain age "is not medically necessary."

Viagra represents the first oral treatment for ED, introduced by Pfizer in 1998. Initially developed for the treatment of angina or chest pain, Viagra failed to reduce chest pain but was surreptitiously discovered to increase blood flow to the penis and significantly increase the likelihood of an erection. Viagra, as well as the later introduced Levitra and Cialis, all work by inhibiting the body's PDE-5 enzyme. A primary function of the PDE-5 enzyme is to metabolize or break down the neurotransmitter cyclic guanosine monophosphate (cyclic GMP), which helps relax smooth muscle tissue. When PDE-5 is inhibited, more cyclic GMP remains available in the body which causes the walls of smooth muscle tissue, including arterial blood vessels, to relax and expand. With PDE-5 located primarily in the nose, skin, and penis, blockage of PDE-5 leads to an increase in cyclic GMP and related blood flow to those areas, leading to an increased likelihood of erection. With Viagra, men are typically instructed to take the "little blue pill" approximately 1 h before intending to engage in sexual activity with the expectation that an erection can be obtained within the next four hours (Pfizer, 2011).

Levitra was introduced next by Bayer in 2003 as a more selective PDE-5 enzyme inhibitor requiring a smaller dose of medication when compared to Viagra. Erections can sometimes be obtained within 30 min (Levitra, 2011), and users typically do not have dietary restrictions as cautioned with Viagra. (Fatty foods and alcohol can slow the absorption of Viagra into the bloodstream; Pfizer, 2011.) Cialis was brought to the market shortly after in 2003, and this PDE-5 inhibitor from Lilly offers an extended half-life when compared to its competitors. Specifically, a man who ingests Cialis has a 36 h window in which the drug remains in his blood stream long enough to help obtain an erection. Thus, Cialis is sometimes referred to as the "weekend [sex] pill" (Berner et al. 2006; Carson, 2006).

So, which PDE-5 inhibitor works best? Although this question is complex and must truly be answered with individual examination, diagnosis, and consultation with an appropriate health care professional, some facts are available for general consideration. Viagra has been on the market for more than a decade, with long-term data available regarding its efficacy and overall safety, whereas Cialis is the newest to market and does not offer the same wealth of empirical data. Levitra requires a lower dose than Viagra and can be effective for men who have achieved only limited success with Viagra. Although Cialis offers the longest window of action for obtaining an erection, allowing for greater spontaneity in the initiation of sexual activity, the lengthiest terminal half-life of this PDE-5 inhibitor also means that any side effects also are likely to last longer (Carson, 2006). Similarly, side effects from Levitra may persist longer than those experienced with Viagra. A recent meta-analysis of clinical trials among the three available PDE-5 inhibitors taken at their maximum recommended dosages revealed that Viagra, Levitra, and Cialis each demonstrated significant clinical efficacy in improving erectile function when compared to placebo (Berner et al., 2006). However, not all men in all trials with these medications were able to achieve an erection; they were highly effective with most men *on average*.

Similarly, both large-scale clinical trials (Berner et al., 2006) and smaller clinical practice studies of Viagra (Marks et al. 2006) indicate that Viagra improved erections by 71–95%. Closer examination of the these results, however, reveal that men with ED with minimal dysfunction, who could occasionally have erections firm enough for penetration, had a significantly better response to the drug than men with more moderate or severe ED who had partial or no erections. In addition, men who had their prostate surgically removed did not respond as well to PDE-5 inhibitors, although some men who had the surgery in conjunction with a nerve-sparing procedure had greater success achieving an erection with treatment (40%; Marks et al. 2006). Overall, the greater the severity of ED, irrespective of its underlying etiology, the less likely the desired response to treatment with the drug became. Not surprisingly, the more severe the ED, the less likely the patient was to continue treatment with Viagra.

Interestingly, less than half of all men refill their prescription for PDE-5 inhibitors such as Viagra. Hypotheses for this failure of men to refill their prescriptions include general disappointment with the clinical effects, feelings that treatment is "unnatural," attempting to achieve an erection too soon after ingestion, having insufficient arousal or physical stimulation, and the experience or fear of side effects (Porst et al. 2003). Identified side effects among all of the PDE-5 inhibitors are similar, including headache, stuffy or runny nose, flushing, nausea, muscle aches, photosensitivity, vertigo, and visual disturbances (e.g., blue or green tinted vision; Berner et al., 2006). In some men, the incidence of such side effects may diminish with use over time (Pfizer, 2011). Only one dose of the medication can be taken safely per day. Among older men, healthy men over the age of 65 showed a 40% greater concentration of Viagra in their blood plasma after use. It remains unclear whether older men are more likely to experience side effects compared to their younger counterparts.

Men taking PDE-5 inhibitors are advised to seek medical help immediately if they experience a sudden loss of vision or hearing, a painful erection or one that lasts for more than four hours (i.e., priapism), heart palpitations, chest pain, or breathing difficulties. Contraindications for use of this class of drugs include the presence of heart disease, recent stroke or heart attack, and high or low blood pressure. Men taking nitroglycerine or other nitrate-based medications also are advised to avoid PDE-5 inhibitors (Porst et al., 2003). Various reports exist of men suffering heart attacks while using Viagra and other PDE-5 inhibitors (Goldstein et al. 1998), but researchers caution that statistically similar numbers of men suffer heart attacks during sexual activity without the use of the medications.

Anecdotal evidence indicates deaths from cardiac arrest or stroke have occurred in men who took Viagra without supervision from a physician, who lied about their current medical status in order to receive the medication, or who took Viagra in conjunction with recreational drugs such as amphetamines and cocaine. It is important for clinicians to remember that just because a man is "older," he may engage in the use of such recreational drugs and should be cautioned accordingly. In contrast, unrealistic and extreme fears about side effects from PDE-5 inhibitors may lead some men to avoid discussion of this potential treatment with their health care providers (Wentzell & Salmeron, 2009).

Myths and Misconceptions about PDE-5 Inhibitors

Misconceptions about use of Viagra and other PDE-5 inhibitors are common (Rubin, 2004) and can lead to lower clinical efficacy and cause frustration, discomfort, and even potentially dangerous outcomes. For example, many men and women are unaware that Viagra, Levitra, and Cialis have no (i.e., zero) impact whatsoever on sexual desire (Goldstein et al., 1998). In other words, PDE-5 inhibitors are not aphrodisiacs in and of themselves. A man must first become sexually aroused, which causes his brain to release nitric oxide from specialized cells. In turn, this nitric oxide causes the formation of cyclic GMP. Only after the man's sexual arousal is great enough to initiate the cascade of nitric oxide to the formation of sufficient amounts of GMP will the PDE-5 inhibitors help allow the blood vessels in the penis to relax and become filled with blood, causing an erection. If a man has low sexual desire or interest, Viagra and all the other PDE-5 inhibitors are essentially ineffective.

Even when a man who takes Viagra or other PDE-5s drugs becomes sexually aroused, additional physical stimulation, for a relatively long period of time when compared to previous sexual encounters before the occurrence of ED, may be required for the production of an erection. Because many health care providers fail to relay this vital information about the drug's method of action, requiring heightened emotional or psychological sexual arousal as well as direct stimulation of the penis in most cases, many consumers may become frustrated or disillusioned (McCarthy, 2001). Pfizer itself recommends that men try Viagra up to four different times before deciding to abandon use of the drug (Pfizer, 2011). It bears repeating that *when a man takes Viagra or any other PDE-5 inhibitor, an erection will not suddenly develop on its own.* Even with these powerful medications, sexual arousal and physical stimulation of the penis are typically required to produce an erection.

The quality of an erection associated with male-enhancement performing drugs also differs from that produced prior to the time a man experienced ED. Erections produced with PDE-5 inhibitors are generally not as firm as the erections men typically experience in young adulthood. In other words, the erections produced with the assistance of Viagra and other PDE-5 inhibitors typically become firm enough to engage in intercourse, but they are not the "rock hard" erections that many men (or their partners) might expect when taking the drugs.

Another misconception with dangerous and even fatal consequences is that Viagra and other performance-enhancing drugs provide protection against sexually transmitted diseases including HIV/AIDS. Understanding that PDE-5 inhibitors provide no protection whatsoever against STDs is particularly important for men and women over the age of 50, who typically do not see themselves as susceptible to infection. Because older adults are typically less familiar with HIV, less likely to have their health care providers talk to them about HIV, less likely to use condoms due to decreased fears of pregnancy, and more susceptible to HIV per exposure due to age-related declines in their immune systems, some researchers suspect that increased numbers of HIV infection among older adults can be linked, to some extent, with the use of PDE-5 inhibitors (Hillman 2008a, b).

Unfortunately, studies suggest that the recreational use of PDE-5 inhibitors among men who have sex with men (MSM) in urban cities is increasingly common and is associated with a greater likelihood of unprotected intercourse, illicit drug use, diagnosis of an STD in the past year, and sex with multiple partners (Sanchez & Gallagher, 2006). All of these aforementioned activities represent increased risk factors for contracting HIV/AIDS. Although such studies provide vital information about increased risk factors among certain segments of the population, it is important to note that virtually no data exists regarding prevalence or attitudes toward ED among aging gay, bisexual, and transgendered men, much less the use of PDE-5 inhibitors among MSM who are in long-term committed or monogamous relationships. Although some qualitative studies suggest that older gay men face considerable social pressure to maintain a more youthful physical appearance and consistent level of sexual functioning (Murray & Adam, 2001), it is essential not to generalize these emergent trends regarding recreational PDE-5 use and high-risk behaviors among promiscuous MSM in US urban centers members to the entire GBT community worldwide.

Another misconception is that men without ED who take Viagra and other PDE-5 inhibitors will heighten their sexual prowess, experience firmer erections, or avoid premature ejaculation. For men with normal erectile function, the recreational use of PDE-5 inhibitors does not produce any of these effects. Rather, the likely result of recreational use of these drugs is the development of a headache or becoming flushed. Taking above the maximum recommended dose of PDE-5 inhibitors in hopes of "extreme enhancement" or a more pronounced therapeutic effect also appears ineffective (Goldstein et al., 1998); the result appears to be even longer lasting side effects. Thus, the use of Viagra and related male-enhancement drugs among otherwise healthy men with normal erectile function for "extreme enhancement" or partying is certainly ill-advised.

Mass and Black Market Influences

Another confounding factor in the treatment of ED with Viagra and other PDE-5 inhibitors is the ability of many men (and their partners) to obtain the drug. Individual factors include ease of purchase, cost, and quality. Viagra and the other PDE-5 inhibitors can be obtained only by prescription in some countries (e.g., the UK and the USA), but can be readily purchased over the counter in others (e.g., Brazil). Even in countries that require a prescription for purchase at a pharmacy, some people turn readily to Internet pharmacies or other sources to obtain the medication. The approximate cost per pill of PDE-5 inhibitors at their maximum recommended individual dose is between $18 and $20 for Viagra, Levitra, and Cialis.

Additional concern exists regarding the presence of generic, knock off, or black-market versions of PDE-5 inhibitors, which cost significantly less than their prescription form. Counterfeit PDE-5 inhibitors now account for the majority of illegally produced pharmaceuticals; millions of these pills are seized annually, and most are sold over the Internet (Jackson et al. 2010). Although generic or nonbrand name drugs are expected to be of identical content and produced with similar levels of quality control as their brand name cousins, studies of confiscated drugs from domestic households, including those labeled as Viagra and Levitra, suggest that up to 25% of knock off or black market drugs deviate significantly from the expected content or concentration (Lown, 2000). Many such off market drugs for ED come in nearly identical packaging that is quite difficult to identify as counterfeit. These counterfeit PDE-5 inhibitors are typically contaminated with potentially hazardous substances including commercial paint and printer ink (Jackson et al., 2010). Even miscellaneous "herbal" or "natural" forms of Viagra fail to contain the necessary active chemical ingredients and may introduce their own series of adverse or side effects.

The mass marketing of Viagra and other PDE-5 inhibitors, with budgets approaching one million US dollars, and the formation of a "medical market" itself presents interesting subjects for analysis. One primary concern is that in a free market consumers are expected to be informed about the products available, be aware of differences in product quality, have bargaining power, and have the freedom to purchase what they desire. In a medical market, these assumptions are typically violated (Lown, 2000). Advertisers typically play upon men's fears about ED and associated perceived failures (e.g., inability to have an erection "on demand" either automatically or by strength of one's will) to please a stereotypically heterosexual partner. Academic symposia, including medical journal supplements, are often funded by the pharmaceutical companies who produce these drugs, offering what could easily be construed as a conflict of interest (Sigmund, 2002).

Direct-to-consumer advertising of these male performance-enhancing drugs only available by prescription in certain markets (e.g., the USA) also provides inherent challenges in the ability of potential consumers to arrive at the medically appropriate decision to seek treatment (Sigmund, 2002). Male patients may be more likely to breech the subject with their health care provider, but physicians may then spend most of their time with that patient explaining why that advertised medication is not an appropriate choice for them. In contrast, men (and women) who purchase PDE-5 inhibitors via the Internet or over the counter, without the benefit of a medical exam and consultation, may inadvertently miss a valuable opportunity to detect and treat a serious, underlying cause for ED such as prostate cancer or diabetes.

Advertisements for Viagra and other brand name PDE-5 inhibitors, which boast annual marketing budgets approaching hundreds of millions of US dollars annually, also tend to portray healthy sexual activity as a significant source of male prowess and identity. Linking Viagra with male spokespeople including presidential candidates and professional athletes in soccer, baseball, and NASCAR as well as "ordinary" men fosters stereotypes and inappropriate expectations that the ability to engage in heterosexual intercourse is inexorably linked to a man's value and worth. Such advertising also promotes misconceptions that even normal, occasional difficulties with an erection represent ED, a medical disorder that can afflict even the young, healthy, rich, and famous, and thus certainly the average man. The focus of these advertisements appears exclusively upon a singular, apparently malfunctioning body part rather than the status of an entire person, much less the emotional and dynamic interaction between two people (Hillman 2008a, b; Sigmund, 2002). In many cases, these drugs are portrayed as the only viable solution to ED, which ignores more potentially complex, underlying health and relationship difficulties.

Impact Upon Partners

A critically important yet typically overlooked aspect of PDE-5 use is that of male users' partners. Efficacy studies typically involve heterosexual couples and suggest that female partners generally concur with their male partners' and physicians' ratings of the drugs' effectiveness (Goldstein et al., 1998). However, little is known regarding female and male partners' perceptions of the drug and its effects within the context of their own experience or relationships. Qualitative analyses of older women's interview responses regarding ED and Viagra reveal recurrent themes including the desire to expand the focus to male and female sexual and pleasure, a sense of sexual obligation in long-term relationships with men, the unfortunate equation of sexuality with masculinity, and continued surprise about the prominence of sexuality in Western culture (Conaglen & Conaglen, 2009).

Many older women interviewed about Viagra use mentioned that if the drug was not obtained in consultation with them, as the primary partner, many felt angry or obligated to engage in sexual intercourse when it was no longer that important to them or their sense of the relationship (Conaglen & Conaglen, 2009). These comments are consistent with general advisories for older women to communicate openly with their male partners about their feelings when Viagra or other PDE-5 inhibitors are introduced into the relationship, and for men and women to employ additional foreplay, lubricants, and a slower pace if sexual intercourse has not been a typical part of their sexual relationship to avoid discomfort (Hillman 2008a, b).

Still other women expressed frustration when their male partner expected them to provide additional physical stimulation of the penis to achieve an erection, particularly when they had become accustomed to greater cuddling and nongenital foreplay during the untreated periods of ED (Conaglen & Conaglen, 2009). (Even though a male partner may "be ready" for sex within half an hour, older women, in particular, often require a longer period of foreplay and additional lubrication in order to enjoy sexual intercourse without pain or discomfort.) In contrast, other women were thrilled with the resultant changes in their partner's ability to engage in intercourse, as well as their sense that their male partners felt more positive about themselves. In sum, it appears that female partners have both positive and negative responses to Viagra as a treatment for ED, and that communicating clear expectations for both emotional and physical responses and expectations between male and female partners is essential for a more positive outcome.

Medical Vacuum, Injectible, and Implant Therapies

PDE-5 inhibitors such as Viagra are typically employed as the first line of treatment for ED due to their relatively noninvasive nature. For men who do not wish to take these drugs, or for whom these drugs have not delivered acceptable results, more invasive, alternative approaches are available, including vacuum pump devices, injectible vasoactive drugs, and penile implants. Of these three treatments, vacuum devices appear to be most effective in cases in which vascular (i.e., cardiovascular) problems underlie ED (Galindo & Kaiser, 1995). Vacuum devices also appear moderately effective for ED related to diabetes and neurological disorders (Hellstrom et al. 2010).

A physician can prescribe the use of an FDA-approved vacuum device at a cost between approximately $200 and $500. Essentially, a cylinder is placed over the penis and a vacuum draws blood into the organ to create an erection. A clamp or band (i.e., a cock ring) is then placed at the base of the penis to maintain the engorgement of blood and the rigidity of the penis. Patient education is essential. Side effects from the use of vacuum devices can include bruising, numbness, and pain upon ejaculation. From a psychological perspective, many men are hesitant to use these pumps because they appear invasive and artificial, and because it is difficult to hide the use of such devices from a partner. Some patients also find that a combination of PDE-5 inhibitors and a vacuum pump provides a more acceptable level of symptom relief than use of either treatment modality alone (Hellstrom et al., 2010).

Another, more invasive approach to treatment of ED is through the use of self-injected drugs. A variety of injectable intracavernosal drugs such as papaverine HCI, phentolamine mesylate, alprostadil, and prostaglandin E can produce an erection in as little as 5 min that can last as long as 30–60 min. A major problem with this approach is that few of these drugs are approved for penile use by the FDA. Other problems include cost (up to $30 per injection), burning and pain at the site of injection, and, rarely, bruising, prolonged erection, and liver problems. Because both a high level of manual dexterity and visual acuity are required for proper administration of these drugs, older men with arthritis or visual impairments may be unable to use this treatment. Some men are shocked to learn that the medication must be administered directly into the penis itself, rather than in the arm or buttocks; sometimes a partner can learn to give the injections.

Some men prefer these self-injectable vasoactive drugs over both vacuum devices and penile implants because their partners do not have to be involved in the treatment; they can maintain some level of privacy or secrecy in relation to their ED. Of course, it remains up to debate whether such willingness to hide ED from a partner is a simple matter of privacy or if it represents intra- or interpersonal difficulties. Studies regarding patient satisfaction indicate that among a sample of men with ED who did not achieve sufficient results with PDE-5 inhibitors, those who most satisfied with injectable vasoactive drugs included those who were older, had younger partners, and achieved a "fully rigid" penis after injection (Hsiao et al. 2011).

In cases of spinal cord injury, severe vascular or neurological disease, or when all other approaches have failed (Hellstrom et al., 2010), penile implant surgery may represent a last resort, as this form of treatment is most invasive. Penile implants are accomplished via expensive surgical procedures that may range in cost from $3000 to $5000. During the surgery, pockets are created inside the penis to allow silicone or saline-filled rods to be inserted along the length of the penis. Some of these rods require external inflation for erection to take place. A number of problems have been reported following these surgeries, including complications from anesthesia, scarring, and problems in operating the device. If an implant malfunctions and has to be removed, the odds of ED occurring as a result also are relatively high (Galindo & Kaiser, 1995). Middle-aged men have reported greater satisfaction with this procedure than older men, perhaps because older men are more likely to experience postoperative infection. Because this procedure is so invasive, urologists recommend that patients undergo psychological treatment both before and after the procedure (Mayo Clinic 2011a, 2011b, 2011c, 2011d, 2011e).

Integrated, Psychologically Based Approaches

Sexual intercourse or activity with or without any of the above medically based treatments to ED does not occur in a vacuum. The use of any medical treatment, including the typically noninvasive PDE-5 inhibitors, typically occurs within the context of a relationship. Improved communication between partners, as well as accepted changes to established rituals among long-standing partners regarding sexual initiation, often become necessary for both parties to resume and enjoy sexual relations.

Advertisements for male-enhancing drugs virtually always fail to mention that optimal treatment outcomes are more likely to occur when men with ED receive both pharmaceutical agents and couples-based sex therapy (Brooks & Levant, 2006; Rosen, 2000). Specifically, treatment of ED with both Viagra and sex therapy has been shown to provide increased erectile function and marital satisfaction when compared to treatment with Viagra alone (Aubin et al. 2009). Relapse prevention also appears to be enhanced when treatment with a PDE-5 inhibitor is coupled with sex therapy. With the help of a trained therapist, men and their partners can become educated about positive, realistic expectations regarding the sexual response cycle of each partner, both with and without the use of drug enhancement. Therapists can also employ cognitive-behavioral techniques to help both partners view sexual activity as occurring within the context of a relationship rather than an isolated physical event revolving entirely around the ability to produce and sustain an erect penis.

In sex therapy, men and their partners can be helped to learn that periodic erectile failure, experienced both with and without the use of Viagra and other PDE-5 medications, is natural and to be expected. Once both partners have realistic expectations and view occasional disruptions and changes in the sexual response cycle as merely variations rather than cause for anxiety or alarm, they can be helped to plan for flexible and variable options in an expanded repertoire of sexual activity including intimacy, eroticism, and mutual and self-stimulation. Men with ED can also use Viagra and other male-enhancement drugs as an aid to masturbation. Experimenting with self-stimulation both with and without these drugs can help men regain a sense of confidence that can then transfer to sexual expression with a partner. Viewing a partner as a source of enjoyment and various forms of pleasure rather than a simple demand for an erection also is advised. Such an expanded perception of realistic and enjoyable sexual activity lends itself to enhanced relapse prevention of ED (McCarthy, 2001).

Even if men with ED do not seek out formal sex therapy in conduction with PDE-5 treatment, general education about the sexual response cycle alone in relation to increasing age delivered through formal workshops and even simple written materials appears to significantly increase the likelihood of help seeking behavior, communication among partners, and increased sexual satisfaction (Berner et al., 2006, Phelps et al. 2004). Physicians, other health care providers, sex educators, and even advertisers can be advised to include access to at least written or online educational material in conjunction with PDE-5 use. For health care providers who feel they do not have sufficient time or the personal inclination to discuss such matters at length with their male patients, results from these studies suggest that offering written materials to their patients would provide at least some benefit. (Certainly, a primary goal is to encourage physicians and other health care providers to garner some level of comfort in discussing sexual activity regularly with their male and female patients; Chun & Carson 2001; Hillman 2008a, b).

Regardless of the medical interventions used to treat ED, clinicians also must remember that *none* of the aforementioned, medical approaches to treatment provide protection against sexually transmitted diseases. It often becomes the purview of mental health practitioners to educate patients about the consequences of engaging in high-risk behaviors, or in assisting them in ceasing their particular high-risk behaviors (e.g., having unprotected sex with intravenous drug users or prostitutes). With few exceptions, most medical treatments do not call for psychological intervention, nor do they necessarily involve partner participation.

More general, psychologically based approaches for treatment of ED include desensitization in the form of sensate focus exercises (Masters & Johnson 1966), anxiety reduction techniques, and general psychoeducation about sexuality and sexual performance. In many cases, fear of ED can become just as crippling as ED itself. In sensate exercises and anxiety-reduction techniques, couples are typically prohibited from engaging in genital stimulation and intercourse. Couples are given homework assignments in which they are to massage each other, take a lingering bath or shower, and explore each other's bodies without touching or stimulating their genitals. It is believed that by distracting the male from his internalized expectations about what is successful, he can become sexually aroused and paradoxically experience an erection (Cranston-Cuebas & Barlow, 1990). Such sensate focus therapies have been accepted and used widely as an effective means of reducing ED (Rosen & Leiblum, 1993).

Additional psychologically based treatments for ED that have garnered more empirical support, particularly among older adults, include cognitive therapies that incorporate basic sex education (Rosen & Leiblum, 1995). Sometimes simply identifying unrealistic expectations, highlighting normal age-related changes to the sexual response cycle (Wiley & Bortz, 1996), disavowing the central role of the penis in bringing pleasure (e.g., Zilbergeld, 1992), and normalizing the experience of help-seeking (Schover et al. 2004) allow couples to engage in more fulfilling sexual relations.

Fred

Fred was a 67-year-old man who believed that he should be able to achieve an erection and have sex with his wife at least two times a week. Mary, his wife of 43 years, agreed to come to couples therapy at the insistence of her husband. She did not quite understand why he asked her to come to "his sessions," since she believed that he had been coming to the geropsychology outpatient clinic for treatment of mild depression. She described their relationship as comfortable and easy going, and said that from the time they met at age 16, they "were destined to be sweethearts for life." Both husband and wife were community living and relatively free from illness. Fred did have mild hypertension, and his medication may have contributed to his inability to have erections consistently.

In one therapy session, Fred announced, "Last time I asked [my wife] if she was ready for her 'semi-annual,' she replied, 'Oh, you mean our annual semi?'" He took his wife's comment as an insult to his masculinity and a jab at his inability to consistently achieve erection. Fred thus felt compelled to try to have sex with her even more often to demonstrate his "abilities as a man," which further added to his performance anxiety. On discussion, he was quite relieved to learn that she did not base the quality of their sex life on his ability to penetrate her vaginally; she was quite happy to engage in vaginal intercourse on a more sporadic basis while maintaining their participation in foreplay and other activities. Mary also recognized that although her comment was "funny," it masked some of her previously unexpressed, angry feelings toward Fred. She had begun to feel "put upon" because Fred kept trying to initiate sex when she wasn't always ready or "in the mood…just petting and watching television together in bed would have been good enough for me that night."

In conjunction with a referral to a urologist, the couple decided that Fred would experiment with Viagra in order to alleviate some of his concerns regarding ED. One important issue that the couple was able to agree on in therapy was that Fred and Mary were both to agree on the use of the medication *before* Fred took it, and that Fred would be primarily responsible for initiating these discussions. Mary maintained that she was happy with Fred "the way he was," but that she understood if he wanted to tryout this "new fangled medication…heck, I guess it can't all be bad…a little more fun never hurt anybody." Both Fred and Mary reported that they were able to "talk about this kind of thing" like they never had before, even when they were much younger. They both noted independently that "coming to the doctor" had helped them significantly because this topic was certainly not a topic of conversation among their friends.

Prostate Issues

Just as any responsible discussion of female sexuality includes a review of breast and cervical cancer, any responsible discussion of male sexuality must include information about prostate enlargement and cancer. Only within the past decade has increased attention been paid to these disorders of endemic proportions. Enlargement of the prostate gland, or benign prostatic hyperplasia (BPH), has been described as a symptom of viropause or manopause. Prompted by age-related changes in the production of testosterone, the majority of men can expect to experience BPH at some point in their lifetime. Autopsy studies of men over the age of 80 suggest that nearly 90% of all men from this age group present with an enlarged prostate (Boyle 1994). Findings from epidemiological studies suggest that 1 in 6 American males will develop prostate cancer at some point in their lifetime, and their risk increases significantly with age (National Cancer Institute, 2009). Additional risk factors for prostate cancer include a family history of the disease, obesity, and Black ethnicity (Mayo Clinic 2010a, 2010b, 2010c), although Black Caribbean men have higher rates of incidence than US born Black men (Bunker et al., 2002).

In the early stages, prostate cancer may not generate any apparent symptoms. In the more advanced stages, symptoms may include difficulty urinating, blood in urine, blood in semen, swelling of the legs, and pain or discomfort in the pelvic area. Significant controversy exists regarding the most appropriate method of screening for prostate cancer (Barry, 2008). A commonly employed screening method is via a blood test to detect the level of prostate-specific antigen (PSA) in the blood. A level above 4.0 ng per liter of PSA typically leads to biopsy or ultrasound of the prostate in order to make a more definitive diagnosis. Although a level above 4.0 represents the typical cut off for ordering a biopsy for many physicians, the level of PSA itself is age dependent, with 6% of healthy men in their 60s, 21% of healthy men in their 70s, and 28% of cancer-free men in their 80s expected to have levels of PSA above 4.0.

In other words, the possibility of false positives in PSA testing significantly increases with age (Welch et al. 2005). The risk of complications in response to biopsies, including incontinence, ED, infection, and increased mortality, also increases significantly for men over the age of 75 (Begg et al. 2002).

An additional form of screening is via a digital rectal exam, in which a health care professional inserts a lubricated, gloved finger into the patient's rectum and palpitates the prostate gland to detect any potential abnormalities in texture, shape, or size (Mayo Clinic 2010a, 2010b, 2010c). The digital rectal exam does not produce any physiologically induced side effects after the procedure. Many physicians employ both PDA testing and a digital rectal exam to provide a more comprehensive screening.

One of the most insidious barriers to appropriate diagnosis and treatment of both BPH and prostate cancer is the hesitation of middle-aged and older men to seek treatment for symptoms. (Socioeconomic status, including access to health care, also represents a limiting factor; Kudadjie-Gyamfi et al. 2006). Many older men appear to recognize the symptoms associated with an enlarged prostate, including frequent urination (particularly at night), painful urination, difficulty in stopping urination, intense urges to urinate, urinary retention, incontinence, and difficulties in achieving an erection. However, many fail to seek medical attention until their sex lives are affected. One 70-year-old man cited, "I don't care if I have to get up at night five times to go to the john. I just don't want my 'other functions' to start going, if you know what I mean." Reflecting the narcissistic injury inherent in the disorder, one 81-year-old man retorted, "Who gives a shit anyway; I know I'm old. So I probably have [prostate cancer]…I don't need to have my nose rubbed in it, too."

Although the most effective test for assessment of BPH appears to be a rectal examination as compared to the antigen blood test (Lee and Oesterling 1995), some men decline to see their physician or other health care professional based upon significant anxiety and fear about the rectal examination itself. Some men fear that assent to the rectal exam suggests that they may be gay, or they may have fears that they will somehow "like it" or become aroused and embarrass themselves in front of their physician. A particular concern among Latino men, who have some of the lowest rates of screening among various ethnic groups (American Cancer Society 2011), is that they will lose their manliness (i.e., machismo) if they passively accept penetration of anything into their rectum, indicating that they are gay. For men without access to health care and accurate information about prostate screening, additional fears may stem from inaccurate beliefs that a rectal exam will cause both ED and incontinence. In recent qualitative studies, some Latino males have noted that they would consider the digital rectal exam only as an absolute last resort (Rivera-Romas & Buki 2011). The open discussion of men's fears and concerns, as well as the delivery of psychoeducation, appears essential in order to help minimize existing disparities in prostate cancer screening.

Another challenge to engage older men in screening for prostate cancer is the fallacy that treatment for prostate cancer leads to outcomes that are worse than living with the cancer itself. As noted by one 66-year-old who rationalized, "I heard that you can live with this for a long time…and that when they cut it out you lose it anyway [develop ED], so what's the point in knowing if I have it?" Unfortunately, the revised guidelines from the US Preventive Services Task Force, mandated by Congress, in which screening for prostate cancer via PSA testing is *not* recommended for men aged 75 and older (U.S. Preventive Task Force, 2008), will likely contribute to the challenge in getting older men to report symptoms and to have medical professionals respond. The use of such government-mandated guidelines in relation to PSA testing for prostate cancer represents one of the most controversial aspects of modern medicine (Barry, 2008).

Fortunately, a variety of treatments have been developed to combat both BPH and prostate cancer. These include coagulation therapies and laser therapy. For cancers in the early stages, radiation treatments and surgical removal of part or all of the prostate gland may be used. Additional approaches used by some medical professionals include "waiting it out" because the presence of BPH does not guarantee the emergence of prostate cancer (c.f., Barry, 2008). Treatment for advanced prostate cancer may include chemotherapy or androgen-deprivation therapy (ADT), in which specific hormones are prescribed to interrupt the supply of testosterone to growing cancer cells.

Unfortunately, common side effects of these aforementioned treatments and particularly chemotherapy and ADT, include osteoporosis, anemia, weight gain, loss of muscle mass, breast tenderness, hot flashes, fatigue, and depression (American Cancer Society, 2007), as well as a decrease in libido and ED (O'Connor & Fitzpatrick, 2006). To complicate matters, surgical removal of part or all of the prostate gland can induce significant side effects such as localized pain and swelling, a frequent need for urination, and even incontinence and ED (Mayo Clinic 2010a, 2010b, 2010c). Additional side effects from any of the aforementioned treatments also may include rectal difficulties including loose stools or pain during bowel movements (Mayo Clinic 2010a, 2010b, 2010c). Interestingly, a recent qualitative analysis of men living with prostate cancer indicates that the fatigue brought on by hormonally based treatment presented more of a challenge to their daily coping and life satisfaction than other treatment side effects such as ED (Jonsson et al. 2009).

Misinformation, fear, and denial play a significant role in poor compliance with both diagnostic and treatment protocols. The assistance of a mental health practitioner may be the key factor in allowing older men to cope with the anxiety surrounding the screening procedures and the potentially impending diagnosis of prostate cancer. Mental health practitioners can be invaluable in assisting older men and their partners in gathering accurate information and reviewing various options for treatment more effectively with their medical providers. For example, some types of prostatectomy (e.g., making an incision in the abdomen versus the perineum) are less likely to cause nerve damage and resulting incontinence and ED than others (Mayo Clinic 2010a, 2010b, 2010c). An interdisciplinary approach, including both medical and mental health professionals, would certainly help produce more positive patient outcomes. Even basic psychoeducation for men with prostate cancer appears to confer benefits, and men in distress, with lower levels of self-esteem and depression, appear to benefit just as much or even more from psychoeducation than their more well-adjusted peers (Helgeson et al. 2006).

It also becomes essential to treat prostate issues as a couple's issue (Bronner et al. 2010; Harrington et al. 2009) whenever a partner is involved. Although few empirical studies are available for review, martial satisfaction and greater daily positive mood was reported among both older husbands coping with prostate cancer and their wives when engaged in collaborative coping (i.e., joint problem solving) strategies (Berg et al. 2008). Therapeutic couple's work can be used to help prostate cancer survivors and their partners join resources to foster mutually productive strategies for coping with increased fatigue and other stressors in relation to the husband's cancer treatment and its side effects. If ED occurs as a result of the cancer or its treatment (as in the case of some invasive surgeries), various forms of treatment appear effective including traditional sex therapy, cognitive-behavioral approaches, PDE-5 inhibitors, vacuum devices, injectible vasoactive drugs, and penile implants (Mayo Clinic 2010a, 2010b, 2010c). Despite these advances, it also is important to note that virtually no studies are available that examine the impact of prostate screening and cancer upon gay, bisexual, or transgendered individuals or couples.

Body Image

Because women are traditionally viewed as sex objects who are supposed to remain thin, beautiful, and youthful in order to gain companionship and vital resources, men's concerns about body image, body functioning, and overall appearance are often overlooked as minimal or unimportant. The vast majority of empirical research on body image takes place with female participants (Pope et al. 2000), with only limited findings available from studies involving typically White middle-aged and older men (Peat et al. 2011). A commonly overlooked problem among older men is that of distortions in body image. Some of the more distressing age-related changes for men are likely to include an overall decline in physical strength and musculature, male pattern baldness, and gynecomastia (enlargement of the fatty tissue of the breast).

Empirical studies have suggested that although men view their bodies in a variety of ways, they often focus on three primary aspects of their body: upper body strength, physical stamina, and level of attractiveness (Brown et al. 1990; Franzoi and Shields 1984). The few empirical studies that exist regarding older men's body image suggest that throughout midlife and old age, men appear to place significantly less emphasis on their body image than women (Peat et al., 2011). Consistent with these findings, the vast majority of young and older people suffering from eating disorders are women. However, a small percentage of those afflicted are men, who are equally deserving of clinicians' and researchers' time and attention (e.g., Van Deusen 1997).

Like their female counterparts, men with anorexia and bulimia tend to choose an inappropriate ideal for their body image which is too thin for their normal height and body frame (Barry and Lippmann 1990). Many times older male patients who lose a significant amount of weight in a short period of time are assumed to be just depressed or "[physically] sick." Stereotypes persist that elderly adults simply lose interest in food and other activities as they age. It becomes vital for informed clinicians to assess for the presence of distortions in body image or a related eating disorder whenever conducting an initial interview.

Another common concern among older men is that of baldness (e.g., Morley 1996). Male pattern baldness has generated a multimillion-dollar industry that boasts realistic and unrealistic claims for treatment including toupees, hair transplants, and drug therapies. The fact that this is a multimillion-dollar industry highlights the psychological difficulties that many middle-aged and older men endure as they experience hair loss. One 76-year-old man commented, "Well, I can cover up the rest of me with nice suits and shirts and things, but I can't do too much with this chrome dome…I don't like wearing a hat inside, and besides, I'm afraid that wearing a hat will make me lose even more hair…My wife doesn't seem to mind, but it does get to me every now and then." It also remains unclear why some older men adjust well to hair loss whereas others do not. Some men indicate that baldness is a problem because it makes them unattractive to women, whereas others point out that every time they look in the mirror, they are given a painful reminder of their aging bodies, even if they had been feeling good about themselves the moment before they glanced in the mirror. On a practical level, other middle-aged and older men cite concerns about sunburn and the increased risk of skin cancer on their exposed scalps.

A significantly less discussed but equally problematic age-related change for some older men is that of gynecomastia (Morley 1996). The majority of men experience either a subtle or dramatic change in the fatty composition of their chest, specifically in their breasts (Carlson 1980). This increase in fatty tissue may first present itself in one breast only, and may require a visit to a physician to rule out breast cancer. Traditionally, this increase in fatty tissue takes place in both breasts and is, for better or worse, a sign of normal aging. The only available medical treatment is breast reduction surgery; hormonal treatments do not appear to be safe or effective (Carlson 1980). Some older men simply respond to this body change by noting, "Well, it's about time for me to get old and flabby, I guess." Others attempt to hide their chests by wearing large shirts or suits with heavily padded shoulders. One 83-year-old man articulated his distress by stating, "I mean, I look like a girl…it's embarrassing just to take my shirt off. I don't even know the last time I went swimming." Sometimes educating the patient about the normality of this event is enough to allow him to feel more comfortable with his body's changes. At other times, a discussion of these issues leads to important therapeutic work in differentiating male and female roles, identifying distortions in body imagery, and uncovering homophobic or homosexual tendencies.

Summary

An essential goal in work with men regarding sexual issues is for both patients and practitioners to recognize that male sexuality encompasses significantly more than the ability to have an erection or the number of times per month one has intercourse.

The multimillion-dollar industry of PDE-5 inhibitors such as Viagra and Cialis now fuels increased demand for prescription, black market, and over-the-counter treatment for ED. Unfortunately, lack of knowledge about these medications as well as lack of communication between medical providers and patients about exactly how they work can lead to inappropriate expectations for a miracle cure as well as disruptions in established relationships. Additional difficulties ensue when partners are not included in treatment.

Despite the increasing medicalization of male sexuality, mental health providers continue to have a pivotal role providing appropriate care and psychoeducation for our male patients, and the ability to help male clients to expand their own views of what represents healthy sexuality. Many men are unaware of the normative, age-related changes that occur with their sexual response cycle as well as the myriad of medical disorders (e.g., high blood pressure, obesity, diabetes), prescription, and over-the-counter medications that can lead to ED. Additional factors that deserve consideration in any discussion of male sexuality include body image (from concerns about musculature to male pattern baldness) and prostate health. In sum, an integrated approach that incorporates medical, psychological, and educational components, in both an individual and couples' context appears most effective in helping men cope with issues related to their sexuality and aging.

Chapter 9
Sexuality and Aging with LGBT Populations

Estimates suggest that more than 3.5 million Americans self-identify as lesbian, gay, bisexual, or transgender. LGBT elders face a variety of unique stressors associated with aging including discrimination in health care, caregiving burdens and fears, and legal challenges to domestic partnerships. Recommendations are offered to help clinicians make their practices more LGBT friendly.

Aging adults with an LGBT orientation are likely to face a variety of challenges when preparing to address issues such as retirement, long-term health care, and finances, which ultimately influence the expression of sexuality, potential discrimination, and the ability to live authentically. Successful treatment begins with a trusting relationship and a willingness to negate negative societal stereotypes on the part of both patient and therapist. Looking for the larger social issues behind potential sexual dissatisfaction remains an essential part of treatment planning and assessment. Qualitative analyses of older adults' perceptions of various LGBT orientations, including unexpectedly positive aspects also will be examined. With the aging of the Baby Boom generation, society will be increasingly accepting of LGBT relationships, late-life dating, and cohabitating when compared to traditional, long-term heterosexual marriage.

Challenges for Clinicians

More clinicians are finding themselves working with middle-aged and older adults who are involved LGBT as well as heterosexual relationships. Surveys of mental health professionals indicate that within the last decade, 99% of all psychologists have worked with at least one LGB client at some point in their career (Garnets et al. 1991), and more than of all therapists report working with a LGB client, of any age, within the past week (Murphy et al. 2002). Also within the last decade, even the terms typically used to describe various types of romantic and sexual relationships

J. Hillman, *Sexuality and Aging: Clinical Perspectives*,
DOI 10.1007/978-1-4614-3399-6_9, © Springer Science+Business Media New York 2012

have evolved from the more derogative "traditional and nontraditional" to the more inclusive lesbian, gay, bisexual, and transgender descriptor. (The more inclusive term LGBT will be used here, in this chapter, for example.) Unfortunately an older patient—whether LGBT or heterosexual—must typically cope with societal biases and stereotypes in response to any romantic or sexual relationship outside of "traditional," long-term heterosexual marriage. For example, an LGBT elder may face homophobia in their community and discrimination in health care, while a heterosexual older woman may face stigma for living with someone outside of marriage.

To complicate the situation, few clinicians have received much graduate training or continuing education regarding LGBT relationships, particularly within the context of sexuality and aging. Gaining an understanding of LGBT issues and stigma in general, as well as the vital resources now available for both patients and clinicians, can be quite beneficial. It also is important to note that although the terms LGBT and gay (indicating LGBT, collectively) will be used here for the sake of both ease and inclusiveness, it is critical to recognize that lesbian, gay, bisexual, and transgender elders should not be viewed as a homogenous group. (The terms men who love men, women who love women, two-spirit, on the down low, and others also are relevant as well as distinctive and important, but for purposes here they are subsumed within the LGBT umbrella label for the sake of inclusiveness and simplicity.) Acknowledging and appreciating differences among these distinctive sexual-minority aging populations remains essential. It also can be helpful to consider whether an LGBT individual came of age before or after the start of the gay civil rights movement in the 1970s.

Historical Context

LGBT individuals in the current Baby Boom generation have lived through significantly different historical events than those in older age cohorts (see Haber 2009, for an outstanding review). For example, the current cohort of LGBT older adults over the age of 65 (also referred to as pre-liberation LGBT individuals) grew up at a time in history when homosexuality was typically viewed as pathological, illegal, and sinful. Few LGBT individuals openly professed their status out of fear of discrimination and violence. Most people referred to those with a lesbian or gay orientation as homosexuals. Married couples were the only ones who could adopt children, and no laws were in existence to help protect LGBT individuals from victimization. Those in the medical community offered "treatment" for homosexuality via involuntary hospitalization and shock therapy. Some of the only refuges for LGBT individuals existed in certain urban or vacation areas (e.g., San Francisco) with large same-sex communities and businesses.

A turning point for the gay community occurred in 1969 with the Stonewall Inn Riots. During the 1960s in New York City, demonstrations of homosexuality were illegal in public, police raids were used to shut down gay bars, and the names of gay patrons were published in newspapers to evoke shame. When a popular gay bar in Greenwich Village, the Stonewall Inn, was raided by the police, a violent demonstration and riot ensued which lasted for days. This riot, in which more than 1,000 men and women filled the streets shouting, "Gay power!" is commonly identified as the start of the gay civil rights movement in the USA 1 year later, in 1970, the first gay newspapers were published, gay rights groups were founded in many US cities, and Gay Pride marches took place on the riot's anniversary in New York City, San Francisco, Los Angeles, and Chicago.

Within the lens of the scientific community, data was readily available by the 1950s that gay individuals were as similarly happy and well adjusted as their hetero-sexual counterparts (e.g., Hooker 1957). Despite these and other research findings, the American Psychiatric Association and the American Psychological Association took nearly two more decades, until 1973 and 1974, respectively, to remove homo-sexuality from their diagnostic and statistical manuals as a psychological disorder or disease (Spitzer 1981). Although social stigma and discrimination against gays were still pervasive in the 1970s, this change in medical status now meant that experts in the fields of psychology and medicine regarded gays and lesbians as psychologically healthy and normal. Baby boomers had the benefit of living through this monumen-tal social change, and are sometimes referred to as postliberation LGBT individuals. However, as evidenced by the "don't ask, don't tell" policy of the US military, the limited number of states that sanction same-sex marriage, and other factors, includ-ing the fact that the Family and Medical Leave act does not provide coverage for domestic partners, social equality is still not realized for LGBT individuals.

Emergent Research

A general lack of empirical research regarding older LGBT individuals persists, even though even more research findings are available now than even a decade ago. An accurate population count of LGBT individuals in the USA is not even available due to difficulties in sampling and self-disclosure. Most studies of LGBT elders are based on smaller convenience samples and are typically oversampled with White, well-educated, middle-class participants (Haber 2009). General estimates suggest that there are between 1 and 3.5 million older gays (men) and lesbians in the USA By the year 2030, this number is expected to double to nearly 7 million people (Jackson et al. 2008). As noted, due to challenges with sampling and self-disclosure, it remains unclear what proportion of the population identify themselves as older bisexual and transgender individuals, as well as what proportion of the population are older gay minority group members. Research on gay minority group elders is essential as existing research suggests that older lesbians and gay men of color have reported experiencing significantly more victimization than their White counter-parts (Balsam and D'Augelli 2006).

Although some excellent studies exist regarding older gays and lesbians (e.g., Brotman et al. 2003; Goldberg et al. 2005; Martinez 2005; Slusher et al. 1996) and older heterosexual adults who date or live together instead of becoming married (von Sydow 1995), they are few in number and have significant methodological limitations. For example, most research regarding older LGBT individuals has been generated from financially secure, well-educated, urban populations. Although these studies' findings are legitimate and add a wealth of information to our knowledge base about older LGBT individuals, the selective and small sample sizes make it more difficult to generalize from these findings to other samples of gay elders who may be from different ethnic groups, from a lower socioeconomic status, or from a rural background. (Virtually no empirical research is available regarding older bisexual and transgender individuals; Addis et al. 2009.)

For these reasons, among others, it also remains virtually unknown what proportion of LGBT elders can be found living among the institutionalized, oldest-old, or rural older populations. Virtually nothing is known about older lesbian, bisexual, or "two-spirit" women in Native American cultures, for example (c.f., Lehabot et al. 2010). It also is difficult to fault researchers for their skewed samples, however. Because of fears about disclosure of one's sexual orientation, recruiting LGBT participants can be both difficult and expensive. Urban areas are among some of the more accommodating places to recruit LGBT participants because of their more liberal and accepting climate, and simply because greater numbers of people are available in one place at one time.

In response, a recent MetLife study of LGBT baby boomers offers the first representative profile of this population in a national context (MetLife Mature Market Institute 2010). This study's findings of more than 1,000 LGBT baby boomers' online responses indicate that the majority of those sexual-minority boomers are middle-income group, well educated, employed, and in a long-term committed relationship. LGBT boomers also are more likely to serve as a caregiver for a parent or other family member, partner, or friend when compared to their heterosexual counterparts. Twenty percent also report having at least one child. This MetLife study is unique in that it provides some of the first empirical information about LGBT baby boomers and because its online format allowed for more representative population sampling.

Just as it appears dangerous to generalize from empirical studies about older sexuality within the context of LGBT relationships, generalizing from clinical anecdotes may be just as potentially damaging to patients and clients. When faced with a lack of concrete information, it is easy to make assumptions that one clinical case or presentation will be similar to another. Such inclinations are natural, as clinicians (and patients) want to avoid confusion and feelings of inadequacy in response to incomplete information. A common defense against such feelings of loss of control is to rationalize and assume that one already has the knowledge needed to work with such patients. Another equally damaging stance is to assume that if little empirical knowledge is available about the topic, it must not be of much professional importance. Fortunately, such therapeutic work can be among the most fulfilling and challenging, if clinicians can accept this lack of general empirical knowledge, and simply adopt the necessary focus on their patient's own, unique experience.

Age Cohorts Within LGBT Populations

The clinical issues among older and younger gay adults appear to have some similarities. Common themes have been described as: concerns about coming out to others; whether or not to affiliate with the gay community; the maintenance of meaningful, interpersonal relationships despite societal pressures to the contrary; familial conflicts including family secrets and parental guilt; difficulties with prohibitive religious beliefs; anxiety about HIV and AIDS, including fears of contraction and the loss of loved ones to the disease; fear of discrimination that may include fears of physical as well as emotional harm; concerns about available social supports such as informal friendship networks and formal organizations; and legal problems related to the exclusion of same-sex marriage and partnerships. Related legal issues that come to the fore in the LGBT community include a lack of shared insurance coverage, retirement benefits, and visitation rights.

Despite these similarities between the two age cohorts, important differences also appear to exist. Gerontologists suggest that, compared to their younger baby boom counterparts, LGBT elders:

- Are significantly less likely to receive vital information about HIV education, treatment, and prevention (see Chap. 6).
- Have fewer family members available to tend to arising instrumental needs, particularly regarding financial and long-term health care support. Older LGBT adults do appear to gather significantly more social support from established friendship networks (e.g., a "family of choice") than from biological family members (MetLife Mature Market Institute 2010).
- Are more likely to have children, grandchildren, spouses, and ex-spouses.

Gay elders were raised in a generation in which one was supposed to ignore or put aside homosexual tendencies in order to ascribe to traditional family values and relationships. Thus, older gay adults are more likely to have married early in life or to have had children to "try to change" or to "do the right thing." The family dynamics involved have the potential to become challenging and, at worst, problematic.

- May have more difficulties forming informal social support groups or romantic relationships because they are less likely to easily identify one another out of concerns about revealing LGBT identities to others. To compound this problem, few formal support groups or social organizations for gay elders are available outside of large urban areas (Slusher et al. 1996).
- Are more likely to have suffered persecution by the community at large. For those in this older age cohort, gays who came out and revealed their sexual identity often suffered significant, negative consequences that included physical and emotional abuse at the hands of predominantly prejudicial, heterosexual peers (e.g., Lehabot et al. 2010). The related clinical presentations of these unfortunate experiences in later adulthood could present themselves in later life as diffuse anxiety and somatic reactions to outright posttraumatic stress disorder.

Such issues can be expected to impact, both directly and indirectly, upon a gay elder's sexuality. Clinicians must be aware that LGBT individuals who present with concerns about sexual dysfunction in late life also are likely to be coping with more broadly based interpersonal and societal issues. Specifically, negative societal attitudes and pressures can be found, in some instances, to catalyze or even generate some of the problems associated with older sexuality. For example, one older gay man suddenly developed problems with ED after his daughter (from a prior marriage in his early 20s) failed to ask both him and his partner of 13 years to an important family function. Recognizing that sexual dysfunction does not occur within a vacuum is essential in work with older gays and lesbians.

Myths About LGBT Elders

Historically, gay elders have been believed to be depressed, poor, and lonely (e.g., Slusher et al. 1996). Logically, it follows that older LGBT individuals would be either unable or unwilling to engage in meaningful or satisfying sexual relationships. In his classic article, Kelly (1977) outlined some of the specific stereotypes regarding gay elders, particularly older gay men. The myth of the older gay man is of a man who is obsessed with sex, particularly with younger men, but who is unable to develop an emotionally or physically satisfying long-term relationship with anyone. He purportedly rambles from gay bar to gay bar looking for one-night stands while becoming increasingly depressed, unattractive, and fearful of being labeled as gay by other members of the community. The myth further states that the older gay male evolves into a paranoid, effeminate queen who seeks out friendships almost exclusively with heterosexuals.

The research findings regarding older gay men and women, however, dispel virtually all of these stereotypes and myths. In various samples, older gays have reported that while they may have some concerns about disclosure of their sexual orientation to others, particularly in occupational settings, the majority describe themselves as moderately involved in gay culture. The majority also report a moderate or high interest in sexual relations and a high level of satisfaction in current sexual relations, primarily with members of their own age group. There is no evidence to suggest that gay elders suddenly begin to seek out younger partners. Those older gay men and women who do not have sexual partners have cited the loss of a long-term partner, physical illness, or a conscious choice to avoid participation in a long-term relationship as the reason for their abstinence (e.g., Berger 1980). Other studies reveal that older gays and lesbians are not any more or less depressed (Dorfman et al. 1995) or more or less obsessed with their appearance than their heterosexual counterparts (Kelly 1977).

Contrary to popular belief, some theorists suggest that the developmental and demographic changes associated with aging may actually work to the advantage of LGBT individuals (McDougall 1993). As stigma persists in many segments of the US society, it remains more socially acceptable for two older men or older women to live together as roommates or purchase a house than it is for a younger, same-sex couple. It seems that society's recognition that people want and need companionship, coupled with its agist assumptions that older adults do not engage in sex and that older gays and lesbians do not even exist as a group (because older single people must naturally be widowed or divorced), allows older same-sex couples to live together without causing any undue distress or homophobic anxiety among heterosexual members of the community. Because older adults in general are falsely assumed to be in poor health, heterosexual members of the community also do not appear to consider homosexuality as an issue when they see two older women or two older men walking arm in arm as they are ostensibly "just helping each other out."

Additionally, many segments of society assume that aging gay and lesbian adults are physically disabled (i.e., harmless) and that they do not pose the same psychological threat as younger gay or lesbian adults. This bias actually may benefit older LGBT individuals in some respects. For example, one 68-year-old gay patient in a rural outpatient treatment program cited, "I love being old now! No one bothers me or my lover anymore, even if we sit close when we go out walking or out to eat. It's like they think we have run out of steam or something, and that no one will bother us because we certainly can't bother them...I wish it was this simple when we were younger." Another older lesbian patient mused, "When [Lauren] and I bought our house together 20 years ago at 45 and 42, a few people in the neighborhood gave us a hard time. Nothing that bad happened, but there was no welcome wagon or cookies for us either. But now that we're older, it's like the neighbors know that we are there for them if their grandchildren fall off their bikes and need someone to look after them. I think they like that we are good neighbors who don't throw wild parties and play loud music...I wouldn't say we are everyone's favorite, but compared to some of the rough ones down the street, they could care less anymore if we are two older lesbians or not." Another older lesbian patient remarked, "I guess I don't even look like a [stereotypical] lesbian anymore. No one looks at me twice with my short hair and lack of makeup. No one expects some old woman to dress to the nines anymore. *So,* I can really blend in if I want to."

Unexpected Benefits and Resilience

Social support is an important aspect of life that allows someone to adjust more easily to life changes and crises. This function is no different among gay elders, who have been found to have similar levels of social support when compared to their heterosexual counterparts. One essential difference that has been observed between LGBT and heterosexual elders, however, is that LGBT elders typically garner more social support from friends than family members (e.g., MetLife 2010). Although long-standing social norms suggest that family members are expected to provide more instrumental services and support (e.g., money, long-term care) than friends, LGBT individuals have a history of developing meaningful, supportive friendship networks (i.e., family by choice), in part because their own family may have been resistant to or unsupportive of their sexual orientation. In this way, LGBT elders may be better equipped to gather different types of social support from more than one source when compared to typical heterosexual, older adults.

As an additional boon to gay elders who establish peer-related, social support networks, theorists suggest that friends may be a significantly better source of social support than family members. Compared to family members, friends are more likely to engage gay and lesbian elders in important life review discussions including difficult or sensitive topics such as personal disappointments, fears of aging, and current and prior sexual relationships (Lewittes 1988). Friends also are more likely to offer support and friendship because "they want to," whereas family members may offer assistance because they feel obligated or expected to. It often is easier to accept help from someone who gives it freely than from someone who extends help only because of perceptions that he or she has to.

Support groups often help LGBT individuals focus upon coping skills and resilience. One potentially positive aspect of living in a society that tends to be homophobic is that LGBT elders have had to overcome many obstacles in their younger days, compared to their heterosexual counterparts living in mainstream society, leading them to possess a variety of coping skills that allow them to adjust more easily to the challenges often associated with aging (MetLife Mature Market Institute 2010). Many older heterosexuals cite discrimination and stigma (i.e., agism) as one of the most distressing aspects of aging, and because LGBT individuals typically face prejudice and discrimination (e.g., homophobia) at much earlier ages, they may be better equipped to adapt to the stigma inflicted on them as they become older themselves. Thus, they may better avoid internalizing negative self-views or succumbing to negative, self-fulfilling prophesies.

Consistent with this theory, a qualitative analysis of narrative interviews with more than 100 participants suggested that older gays and lesbians often possess the following intra- and interpersonal strengths: a strong sense of independence and self-sufficiency; the establishment of interests outside of one's family and career; the ability to foster and preserve both romantic and nonromantic relationships; an increased sense of personal autonomy and inherent worth; and the capacity to adjust to living with a stigmatized identity (Wolf 1982). Consistent with this finding, nearly half of the LGBT participants in the MetLife (2010) study reported that their experience of being LGBT in a predominantly heterosexual culture helped them prepare for aging in some way. Additional potential benefits of a lesbian or gay identity may include the development of empathy and compassion, involvement in social justice and activism, freedom from gender-specific roles, and the potential to serve as positive role models (Riggle et al. 2008). Resilience among LGBT men and women must be acknowledged and recognized and can provide considerable benefits within the context of any psychotherapy.

The Need to Differentiate Among LGBT Populations

As noted previously, it is inappropriate to make blanket generalizations about lesbian, gay, bisexual, and transgender elders; these four populations must be viewed as uniquely distinct (Brotman et al. 2003). Within the context of the literature, some empirical findings are available regarding older gays and lesbians, with significantly less information available about older bisexual and transgender individuals. A brief review of some of these differential research findings will follow.

Lesbian Elders

For older women who are lesbians, empirical studies suggest that among this cohort the participants indicate that although the majority recognized their attraction to other women by the age of 18, the majority also did not have an intimate sexual or emotional relationship with another woman until they were in their mid-20s (e.g., Goldberg et al. 2005). The majority of lesbian respondents report favoring monogamous same-sex relationships, typically with women within 10 years of their own age. Approximately 40% of older lesbians report having being married at some point in their lives, with estimates suggesting that up to 33% have had children (Jones and Mystrom 2003; MetLife 2010). For the majority, friendship with other women and participation in a variety of lesbian-only or lesbian and gay groups appears central to their well-being and identity (Goldberg et al. 2005). Compared to older gay men, older lesbians are reported to have lower incomes, are more likely to have committed partners, and have larger social networks (Fredriksen-Goldsen, and Muraco 2010).

Although the majority of older lesbians report feeling positive about their sexual orientation, a significant number also report feeling lonely and socially isolated. Some researchers suggest that older lesbians face "triple jeopardy" in relation to their increased age, gender, and sexual orientation (Quam and Whitford 1992). These difficulties are obviously likely to be compounded for older lesbians who also are ethnic minority group members. Some primary concerns of aging for older lesbians include the potential lack of income and adequate savings in later life, concerns about giving and receiving long-term health care, and for many, especially those in rural areas, fears about social isolation and loneliness (Goldberg et al. 2005; MetLife 2010). Studies suggest that up to 80% of older lesbians have experienced discrimination of some sort due to their sexual orientation (Deevey 1990).

It also is important to note that despite myths maintained by both older lesbians and service providers themselves that older lesbians are immune to the transmission of sexually transmitted diseases (STDs), older lesbians should receive both routine screenings such as pap smears and appropriate preventative care messages (see Addis et al. 2009). For example, nearly half of older lesbians have had heterosexual intercourse at some point in their lives and 20% of women who have never had heterosexual intercourse have the HPV virus (the primary cause of cervical cancer). Although rare, female-to-female transmission of the HIV virus also has been documented (Hughes and Evans 2003). In sum, older lesbians, their health care providers, social service agencies, and public health care policy makers all need to be attentive to aging lesbians' sexual as well as general physical and mental, health issues.

Gay Male Elders

Estimates suggest that more than 2 million male Americans may identify themselves as gay (MetLife 2010). Obtaining an accurate estimate of older gay men in the population, however, is challenging. For example, some men from various ethnic groups define the term gay quite differently than those from the majority White culture. Specifically, some African-American men do not identify with the term gay, but rather refer to themselves as "men who have sex with men" or "men who love men." Latino and African-American men also may refer to sexual behavior between two men not as gay per se, but as "being on the down low." Among Latino men, in particular, a man who engages in penetrative sexual activity (a.k.a. a top) is not viewed as gay, regardless of whether his partner is male or female. Only a passive or receptive (i.e., a bottom) male partner is perceived as gay. In addition, among some Native American tribes, men (and women) who seek out same-sex relationships are not viewed (negatively) as gay, but are elevated to a special status (Hinrichsen 2006). Thus, such cross-cultural and ethnic issues must be taken into account in any potential therapeutic setting as well as from an epidemiological perspective.

Although it is important not to generalize about any LGBT group, results from various studies suggest that, on average, older gay men maintain a variety of social networks including partners, friends, family members by choice, and children and partners from previous marriages or other relationships (Shippy et al. 2004). Despite these strong social networks, gay men are two times more likely to live alone and four times less likely to have adult children available for help with caregiving in later life (Anderson 2008). Older gay men also may prefer the term homosexual when compared to younger men who prefer the term gay (Adelman et al. 2006). Although myths that older gay men are obsessed with their looks and severely depressed in advanced age have been dispelled, some studies suggest that older gay men do face significant pressure regarding cultural expectations for both male musculature and thinness. A recent study indicates that gay men were more likely to be dissatisfied with their bodies than heterosexual men, and that this level of dissatisfaction, with their weight in particular, increased with advancing age (Tiggemann et al. 2007).

In relation to historical events, the HIV/AIDS epidemic likely had greater impact upon older gay men than older lesbians (Haber 2009). Some theorists posit that living through the initial AIDS epidemic may have led to increased participation of gay men in advocacy and community outreach (Oswald and Masciadrelli 2008), as well as increased interest in monogamous, long-term relationships. More than 10% of all new HIV/AIDS cases are among Americans over the age of 50, and older gay men are at increased risk (CDC, 2008b; Linley et al. 2007). Unfortunately, no national HIV/AIDS education programs exist for older gay men, or for older LGBT individuals in general. Older gay men face significant stigma based upon their advanced age, their sexual orientation, and their HIV positive status. (Please see Chap. 6 for a detailed analysis of HIV/AIDS and other STDs as they impact upon LGBT and heterosexual elders.)

Perhaps in part because they are more likely to live alone than older lesbians (Heaphy 2007), older gay men routinely express concerns that they will experience discrimination by health care providers and in long-term care based upon their sexual orientation (Hughes 2009). To compound these challenges, older gay men are more likely than their heterosexual peers to encounter issues related to the diagnosis and treatment of prostate cancer due to coping with their own prostate issues, their partner's, and even both. More than 20,000 gay men may be diagnosed with prostate cancer each year, and limited information is available about what gay men know about this cancer and its sexual challenges (Asencio et al. 2009).

What information we do have suggests that because gay men often place significant emphasis upon ejaculation during sexual activity, various treatments for prostate cancer may cause significant distress as the ability to ejaculate may be diminished or stopped all together. Although the ability to experience an orgasm may remain intact with certain nerve-sparing procedures, the inability to cum or ejaculate may cause considerable challenges for any man's sexual identity, and particularly for gay men (e.g., Martinez 2005). Treatments for prostate cancer, with potential side effects such as ED, incontinence, and rectal pain, may be seen as more problematic than the disease itself.

Available empirical studies of older gay men's knowledge and attitudes about prostate health suggest that older gay men, like their heterosexual peers, have limited and even incorrect knowledge about prostate disease. In Asencio et al.'s (20009) study of middle-aged and older gay men, some men believed (incorrectly) that participation in rough anal sex either increased or decreased their risk of prostate cancer, and that using sex toys for penetration rather than one's penis would further decrease their risk. Older gay Latino men in the study indicated that receiving a digital rectal exam, commonly used for screening, would be seen as problematic, particularly if a Latino engaged only in penetrative, rather than receptive, sex. Overall, gay men from a lower socioeconomic status appeared to have lesser knowledge of prostate screening and treatments than those from a higher socioeconomic status, and gay African-American men maintained the least accurate knowledge. All gay men in the study expressed concerns about the effect of potential prostate cancer treatments upon their current partnerships and their ability to develop or sustain future relationships. Mental health care providers can encourage older gay men to discuss their concerns and fears about sexuality and prostate health, as well as help them communicate more effectively with health care providers, who are generally unlikely to discuss concerns related to both anal and oral sex.

Bisexual Elders

Very limited empirical data is available regarding older bisexual individuals, and most of these studies draw their findings from small, opportunistic samples (Haber 2009). In the MetLife (2010) study of LGBT baby boomers, only 15% of the 1,000 participants self-identified as bisexual. Some researchers theorize that the proportion of bisexual elders in the population is consistently underestimated (i.e., they become invisible) because many only report being heterosexual, gay, or lesbian based upon the gender of their current or most recent partner (Kingston 2002). The few studies available suggest that older bisexual individuals, on average, report being married or divorced, and are likely to have continued relationships with their children and ex-spouses (Dworkin 2006).

Another challenge for bisexual elders is that some segments of society, which may include heterosexual, gay, and lesbian individuals, express general distrust toward bisexual men and women, maintaining that they simply "choose" to be open to both male and female partners out of convenience, and that bisexual individuals may simply choose an opposite-sex partner to avoid being labeled gay or lesbian. Heterosexuals also may express fears that being involved with an older bisexual could significantly increase their risk of contracting HIV (Haber 2009). Such stigma and distrust could certainly contribute to an older bisexual person's concerns about aging, identity, and social isolation.

Transgender Elders

Probably even less is known about older transgender individuals. In the MetLife study, only 1% of the 1,000 participants identified themselves as transgender (2010), and it remains unknown what proportion of older transgender individuals have transitioned in their youth, are currently considering a transition, and who have transitioned in middle-age or older adulthood (Cook-Daniels and Munson 2010). Even the use of the term transgender is a relatively new event. Significant historical and cohort differences also exist for transgender individuals, compared to their lesbian and gay counterparts. When homosexuality was removed from diagnostic and treatment manuals as a psychological disorder in 1973 in the USA, men and women struggling with gender identity issues were labeled as having gender dysphoria syndrome. The diagnostic label "gender identity disorder" was not adopted until decades later. Individuals in the USA were first exposed to sex reassignment surgery in the mass media when Christine Jorgensen, an American citizen, underwent the surgery in Denmark in 1953.

Because most citizens were unaware of gender identity issues, in which individuals feel that they are "trapped" in a body of the wrong biological sex (leading typically to depression and other emotional distress), most Americans incorrectly assumed that gay men used sex reassignment surgery to literally turn themselves into women in order to present a mainstream heterosexual identity (Cook-Daniels 2006). Throughout the 1950s, significant controversy ensued about whether US physicians should perform such sex reassignment surgeries. One faction of the American Medical Association wanted to pass a law that would prohibit sex reassignment surgery (Meyerowitz 2002). In 1966, surgeons at Johns Hopkins University began to offer an experimental series of hormonal treatments and sex reassignment surgeries and a New York Times article featured this treatment advance in a front page article.

Currently, for an individual to undergo this surgery (at least in the USA), a psychiatrist or psychologist must verify a diagnosis of gender identity disorder. Most insurance companies do not provide coverage for the expensive series of surgeries and the hormonal treatments required both before and after surgery. (Some European countries fund sex reassignment surgeries for women and men diagnosed with gender identity disorder.) Additional challenges ensue when a transgender individual is in between biologically male or female states (sometimes referred to as intersex) and is in need of health care without fears of discrimination. Even finding an appropriate public bathroom or locker room can be problematic.

The majority of transgender individuals who undergo sex reassignment procedures do not do so until they are middle-aged or older. Some hypotheses for this postponement of treatment include the increased cost of medical procedures (one may need to work and save for decades to accumulate the required tens of thousands of dollars) and decisions to wait until retirement to avoid coming out at work and to wait until one's parents die to avoid familial disclosure (Haber 2009). Some transgender individuals align themselves with the LGBT community, whereas others do not, citing discrimination from among the gay community itself. Older transgender individuals may face challenges related to increased risk of diabetes or heart disease due to years of hormonal treatments, and a lack of regular screening for prostate issues (Asencio et al. 2009). Within the last decade, options for support groups and health care information are increasingly available via the Internet. It also remains unclear, however, what proportion of older transgender individuals use or even have access to this informational system.

Underutilization of Services

As emphasized previously, LGBT populations must be viewed as distinct from one another. One factor that can be viewed as a commonality among the individual cohorts of older lesbian, gay, bisexual, and transgender adults without argument, however, is that of discrimination. A well-documented finding is that LGBT elders are 80% less likely to access *needed health and social services* when compared to their heterosexual counterparts (US Department of Health and Human Services 2001), and negative reactions from social service workers and health care providers appear to be at least a partial culprit.

Some of the specific social service programs underutilized by LGBT elders include food stamps, senior centers, domestic violence shelters, counseling programs, and medical care itself (Haber 2009). For example, alarming findings from a national survey of the Area Agencies on Aging (the federally funded agency that sponsors and oversees US senior centers) indicates that more than 50% of all senior centers would not welcome someone who identified themselves as LGBT. It is no wonder, then, that less than 20% of LGBT seniors actively participate in programs at their local senior center, and more than 70% feel hesitant about even visiting one (Knauer 2009).

In relation to needed medical care, estimates suggest that less than 25% of LGBT elders reveal their sexual orientation to their health care providers (Gelo 2008). Apparently, coming out to some health care professionals can be rife with harmful consequences. Studies show consistently that upon disclosure of their sexual orientation, LGBT individuals have encountered a variety of negative reactions from health care providers including physicians, nurses, home health aides, nursing home staff, psychologists, and other mental health providers. These reported negative responses range from outright hostility, avoidance of physical contact, condescending remarks, refusal to provide treatment, and breaches of confidentiality to emotional detachment, embarrassment, anxiety, pity, and excessive curiosity (e.g., Barranti and Cohen 2000; Brotman et al. 2003).

Approximately 20% of the LGBT baby boomers in the MetLife (2010) study indicated that they had "little or no confidence that medical personnel will treat them with dignity and respect" (p. 52). To compound this lack of confidence, the typical 4-year medical school curriculum includes less than 4 h of instruction on LGBT issues (Bonvicini and Perlin 2003). Even though some states require diversity training as part of their continuing education requirements for psychologists and other mental health care professionals, none require specific coverage of LGBT and aging issues. Fortunately, some organizations now offer medical and mental health care providers a variety of educational opportunities regarding LGBT issues. A number of these organizations are noted in the resources section at the end of this chapter.

Transgender individuals face unique issues in relation to accessing health care and other services. Studies suggest that transgender men and women are more likely to be unemployed and live in poverty than their heterosexual and LGB counterparts, particularly if they are members of a minority group (Xavier et al. 2004). Findings from a rare sample of transgender elders revealed that nearly one-third reported experiencing at least one episode of healthcare discrimination. A specific incident reported was one in which a military doctor refused to treat a transgender patient for a post-sex reassignment surgical infection (Cook-Daniels and Munson 2010). Because the transgender individuals in the study were predominantly White and middle class, it remains unclear what proportion of older minority group or impoverished transgender elders have faced such discrimination or abuse.

Issues in Caregiving

Related to problems with accessing needed health care services are the often unique caregiving roles assumed by older LGBT individuals. As noted in the MetLife study (2010), older LGBT individuals are significantly more likely to serve as caregivers than older heterosexuals. In addition, gay men and lesbian women were similarly likely to assume the role of caregiver. (Among heterosexuals, women are significantly more likely to shoulder the role.) In essence, a gay man is significantly more likely to serve as a caregiver than a heterosexual man. Unlike their heterosexual counterparts, nearly half of all LGBT caregivers provide assistance to someone outside of their family of origin including partners, friends, and neighbors. This expanded circle of caring suggests that LGBT women and men have "families of choice" that expand well beyond legal and biological boundaries.

Some definite challenges and concerns exist for LGBT elders. Although 75% of the LGBT in the MetLife study expected that they would provide care for a partner, spouse, friend, or family member at some point in their lives, nearly 20% had no idea who would provide care for them, if needed. Older LGBT caregivers also appear to provide significantly more hours per week providing care than their heterosexual counterparts. Nearly 20% of LGBT caregivers provide care as a "full time job," spending more than 40 h per week providing assistance. In addition, nearly 20% of those caregivers have full-time employment outside of the home. The significant economic contribution of, as well as the emotional, physical, and financial demands placed upon, this LGBT demographic cannot be overlooked (MetLife 2010).

Challenges also exist for LGBT elders within the context of long-term care and other institutional settings. For example, very few nursing homes provide accommodations or support for the healthy expression of heterosexual behavior among consenting adults, much less that of healthy LGBT sexual behavior. (Also see Chap. 4). Empirical studies suggest that nursing home staff maintain significantly more negative attitudes toward same-sex than heterosexual activity among residents (Hinrichs and Vacha-Haase 2010). The available nursing homes that are oriented toward LGBT populations are few and certainly represent the rare exception rather than the rule. Although more assisted-living facilities and retirement communities cater to LGBT residents than nursing homes, these facilities are few in number, tend to be quite expensive, and are generally limited to larger urban centers. Transgender elders have cited fears of entering care settings in which assistance with personal care (e.g., bathing; dressing) would force them to come out if they chose to keep their sexual identity hidden (Williams and Freeman 2007) and have reported various instances in which health care providers met with their transsexuality with discomfort, hostility, and substandard levels of care (Drabble et al. 2003).

Some LGBT elders go to great lengths to live without fear of discrimination in institutional settings. For example, some older lesbians legally change their last name to match that of a partner so that they can "pass" and live together in the same nursing home room as sisters. Others report that even though they were "out" before entry into a nursing home, they were so afraid of potentially negative treatment that they suddenly dressed and acted as though they were straight, even to the point of hiding personal photos and mementos that would "give their [LGBT] status away." The majority of older LGBT individuals report a desire to live independently within their own homes as long as possible, rather than face an asexual or discriminatory institutional living situation (MetLife, 2010). For some LGBT individuals, this could mean receiving inferior or even abusive care in their own homes from relatives or others who may be uncomfortable with their sexual orientation (Cook-Daniels and Munson 2010; Williams and Freeman 2007). In more than 40 states, no laws exist that prohibit discrimination against sexual orientation in either public or private housing.

Lifestyle Changes Across the Life Span

A number of theorists have speculated that single heterosexual women may begin to adopt gay or bisexual orientations in late life in order to assuage their needs for an otherwise unavailable, male sexual partner. In terms of general economic principles of supply and demand, this adoption of a more flexible sexual orientation would appear to mediate the pervasive lack of available male heterosexual partners in later life (von Sydow 1995). Anecdotal stories of older women who develop such a lesbian or bisexual orientation in late life seem to abound, and often receive a great deal of attention in informal discussions of this topic. For example, one older woman remarked, "I really missed having a companion after my husband died. Judy and I do everything together now…Sometimes we even sleep in the same bed and hold hands…I really like having someone there." Unfortunately, empirical research regarding this phenomenon is practically nonexistent. It also is unclear if women who engage in same-sex romantic or domestic relationships had lesbian interests earlier in life that were never reported or acted upon due to fears of discrimination.

One pioneering survey of women between the ages of 50 and 91 years suggested that up to one-third of their community-living sample "expressed interest" in women, but that only 4% actually engaged in recent lesbian activity. Further, the study found that those older women who did engage in a same-sex intimate relationship in later life had previous experience with same-sex sexuality in adolescence or midlife. (It also is important to note that the majority of the older women who reported some lesbian activity throughout life reported having positive, satisfying sexual relations with women.) In sum, these limited empirical indicators suggest that this notion of older heterosexual women blossoming into bisexuality in late life is not as common as perhaps believed. What is more likely is that younger lesbians who grew up in oppressive social conditions were unable to explore or express their orientation openly, and now have the proclivity or perceived freedom to do so in their advanced years. Ultimately, it appears vital that these theorists' hypotheses, although supported by general principles of "supply and demand" and some clinical anecdotes, may not represent the experience of the majority of older adult heterosexual women. Once again, it becomes essential to view each patient's case in the most unbiased and open-minded manner possible.

Georgette

Georgette was an 87-year-old, twice-divorced woman who was seen in therapy for treatment of narcissistic personality disorder and major depression. She was a resident of a nursing home because arthritis now prevented her from tending to her activities of daily living, and because she had no immediate family members to tend to her daily care. Consistent with the external focus of someone suffering from narcissistic personality disorder, Georgette was obsessed with money, material possessions, and her appearance. Life review therapy allowed her therapist to initiate a trusting relationship with Georgette, and it assisted in important information gathering for an otherwise withholding reporter. More importantly, this work in therapy appeared to reduce Georgette's depression and anxiety.

In her course of life review, Georgette discussed at length her divorce from her second husband at 55. Although he was a wealthy businessman, Georgette managed to make him so angry that he refused to share his money with her (the court system at that time was not of much assistance). With what finances she did have, Georgette spent it on extravagant spa vacations and clothing instead of making any attempt to save for the future. At age 62, Georgette's best friend, Bette, took her in after she could hardly make the rent payments on her small garden apartment. Bette, a long-time friend from high school who lived in a nearby town, also liked to attend parties and other important social functions. In contrast to Georgette, Bette was a widow whose husband left her with a large estate and a massive inheritance. Georgette said that the two of them did everything together. "Bette was so terrific. She wanted me to have the best of everything, just like she had…When we would go to the jewelry store, she always insisted on buying me something, too…I wish I hadn't given that emerald necklace to my niece…I always said 'no' to Bette about her buying me things like that, but she insisted that I have fun, too."

For more than 10 years Georgette and Bette lived in Bette's estate, each in separate rooms in her large residence. After a bout with heart disease, Bette passed away and left a sizable inheritance to her children and to Georgette. Although Georgette said that she was very sad and depressed when discussing her best friend's death in therapy, she did not shed any tears or appear overly distraught. On discussing the issue further, Georgette's therapist asked her if there was anything that she didn't like about Bette, noting that it was OK to have both good and bad feelings about the same person. Georgette became very quiet and said, "Well, there was that 1 day after the big party. I didn't like it…I didn't like at all."

According to Georgette, one evening after a large social function, Bette had a few more cocktails than usual. After the guests left, Bette followed Georgette into her bedroom and sat on the couch in the nearby sitting area. Georgette was sitting on the bed when Bette began to cry and talk about how much she missed having someone special in her life, and how much she liked having Georgette to talk to and to spend time with. Georgette said that she went over to sit next to Bette on the sofa and put Bette's head on her shoulder saying, "It's OK, friend. We have each other, right? Forget about those men!" A few seconds later, Georgette said that "Bette reached up-and grabbed my breast! I just stared at her and then I started screaming, 'What the hell are you doing? Get out of my room, right now! If you bring this up I'm leaving.' And, we never discussed it again." Georgette told her therapist that from that moment on "we just pretended that it never happened…I decided it was still worth it for me to live there…she was my best friend even if she was funny like that…But it always did sort of bother me." Apparently, Bette felt trapped in a love-less marriage with her husband. She was a lesbian who had gotten married because it seemed to be her only option at the time.

Despite her narcissistic concerns and her obvious desire to remain in a situation in which her material needs were amply met, Georgette was able to process her confusion, fear, and anxiety about the entire incident in therapy. Her therapist worked carefully to assess whether Georgette had her own issues related to potential, latent homosexuality, and Georgette and her therapist felt that she did not have underlying homosexual tendencies. Through this discussion, Georgette also was able to move beyond her homophobic anxiety ("You mean it's OK to talk about all of this?") and recognize that Bette had faced significant adversity in her life; "It must have been awful to be married to someone, and have to sleep with them and do all of that, even if you didn't really want to…It was bad enough for me when my husbands wanted to get in bed with me and I didn't want them to because I was mad at them or something." Georgette was able to acknowledge both her own discomfort at Bette's attempt to become intimate with her, and her own sorrow for Bette's unsuccessful search for romantic companionship. "That must have been lonely. I know I got lonely always trying to look for dates…I guess it was a lot harder for her to look, too…And then I never even wanted to talk about it."

Although Georgette never fully acknowledged her sometimes selfish reasons for living with Bette (she did admit that she might have wondered if Bette "liked her" when she bought her all of that expensive jewelry), she began to appreciate their friendship even more. She even recognized that, on some level, she was flattered by Bette's advance, but that out of unrealistic fears of becoming "homosexual" herself, she could hardly allow herself to consider it as a compliment at the time. Although Bette had passed away and could not benefit from Georgette's increased understanding and acceptance, Georgette's work in therapy allowed her to widen her acceptance of other residents from various backgrounds at the nursing home. This reduction in LGB stigma certainly benefited all residents, as well as Georgette herself.

Recommendations for Clinicians

To help facilitate social support among LGBT older adults, various groups and programs have been implemented successfully. Some of these nationally known activist and support groups include SAGE in New York City, GLOE in San Francisco, and GLOW in Ann Arbor, Michigan. Although the vast majority of these organizations are located in urban areas, some LGBT elders and their therapists have formed support groups in suburban and rural areas through the use of local papers, local gay bars, and the Internet. (Please see the resources section at the end of this chapter for additional organizations.) One way that clinicians can assist their LGBT clients is to provide links to such groups, as well as provide information about these organizations and LGBT-related topics in their waiting rooms and in other public areas. Providing literature and visible information about LGBT issues also can help provide important gay-friendly messages about service provision (Pugh 2005).

The few available studies in the literature suggest that additional treatment factors are relevant in creating more positive psychotherapy outcomes with older LGBT patients and clients. For example, empathy, positive regard, warmth, and trust in the therapeutic alliance appear essential (David and Cernin 2008). Although based upon responses from young adult participants, LGB psychotherapy clients have identified a variety of therapist characteristics as beneficial. These characteristics include having a therapist of the same sex, who is LGB themselves or who has LGBT friends or family members, who is "generally familiar" with LGBT support and social groups, who is aware of the unique stressors for LGBT community members, and "who seems to like [them]" (Burckell and Goldfried 2006, p. 38).

Additional desirable therapist qualities include therapists who are cognizant of the stigma facing LGBT individuals, particularly in relation to coming out to friends and relatives, who identify the value of family members of choice, and who believe that one's specific sexual label (e.g., lesbian, bisexual) is less important than one's feelings of personal happiness and adjustment. In contrast, therapist characteristics that could be identified as turnoffs included therapists who assumed initially that their client was heterosexual, that people can change their sexual orientation, and that asking questions about their client's sexual orientation would be offensive or infer prejudice (Burckell and Goldfried 2006).

These aforementioned therapist characteristics parallel those recommended in the American Psychological Association's Practice Guidelines for LGB Clients (2011a). Additional considerations from the APA Practice Guidelines include mandates that psychologists recognize the unique experience of bisexual clients, and consider the impact of socioeconomic, spiritual, cultural, and historical (i.e., age-related) influences upon sexual-minority clients. (The fact that the APA Guidelines do not include evidence-based recommendations for work with transgender clients reflects the lack of available empirical research findings regarding effective psychotherapy with transgender clients, and not a therapeutic or organizational bias per se.) Consistent with APA guidelines that therapists seek out professional education and training in LGB issues, therapists themselves can contact both local and national LGBT-oriented groups for access to both local and national resources, and for professional continuing education and sensitivity training. Please also refer to the Resources for Professionals provided at the end of this chapter.

A specific, easily accomplished recommendation for clinicians that has emerged consistently from focus groups of LGBT elders is that of providing "LGBT friendly" or "LGBT inclusive" intake forms. Simply asking new patients and clients to indicate their sexual orientation from a list of modifiers including heterosexual, lesbian, gay, bisexual, and transgender can offer a sense of potential cultural competency and safety (Landers et al. 2010). Additional suggestions offered for establishing and maintaining an LGBT-friendly therapy or support group, particularly in a geographic area that is rural, suburban, or generally discriminatory toward LGBT individuals, can include the selection of a sexually neutral (i.e., not an exclusively gay) locale, an emphasis upon social support as well as education, and the provision of sensitivity training for involved therapists and other staff members (Slusher et al. 1996), including staff members with therapeutically peripheral receptionist and housekeeping duties.

Whether or not a clinician offers exclusively LGBT or gay-friendly services, it is vital that clinicians obtain the requisite education and supervised experience when working with LGBT patients and clients. For example, it is essential to give partners, friends, and other "family members of choice" of (both LGBT and heterosexual) clients equal parity with biological family members (e.g., Landers et al. 2010). As noted, intake forms can also allow for potential patients and clients to indicate "family members of choice" as well as strictly biological family members.

It also is essential that clinicians not treat LGBT clients "as if" they were hetero-sexual. Knowledge of the unique issues facing LGBT patients and clients, including historical and legal issues, stigma, and potential discrimination, is vital.

Legal Issues

Clinicians can familiarize themselves with many of the legal challenges that LGBT elders may face. Although a mental health provider certainly cannot and should not serve as an expert in civil rights and elder law, acquiring general knowledge in order to make client referrals to relevant organizations can be readily accomplished. (As noted, LGB psychotherapy clients have indicated that they prefer therapists with general knowledge of LGB issues; Burckell and Goldfried 2006). Although some Fortune 500 companies make exceptions, the federal Family and Medical Leave Act does not cover time off from work for care of domestic partners. Social Security Supplemental Security Income (SSI) and Disability benefits are not extended to LGBT domestic partners, employers are not obligated by law to provide bereave-ment leave for domestic partners, and if a partner dies leaving property behind, states typically do not allow the surviving partner to inherit the property without paying property and other taxes, even if the property is jointly owned. The rules for Medicaid spend-downs, designed to keep a healthy spouse from losing their home and becoming bankrupt in order to obtain nursing home care for an ill spouse, do not apply for domestic partners (Hash and Netting 2007). Immigration laws also provide no protection for same-sex couples. A US citizen with a same-sex partner cannot sponsor their partner for citizenship, no matter how long they have been committed to each other (Hinrichsen 2006).

Although a 2010 federal mandate now calls for all Medicare and Medicaid funded hospitals (and other care facilities) to guarantee visitation rights for both LGBT and heterosexual domestic partners, LGBT patients can be informed that most hospitals offer Ombudsman services in case problems or challenges to visita-tion and other rights do arise. General legal recommendations also can be offered to LGBT individuals in terms of establishing living wills, medical and durable (legal) powers of attorney, and wills and trusts. For transgender individuals in particular, the Americans with Disabilities Act "[denies] protection for conditions related to gender dysphoria," which prompts insurance companies to routinely deny hormone therapy and sex reassignment surgery as treatment (Hong 2003, p. 1).

Summary

Clinicians working with older adults can expect to encounter a wider variety of rela-tionship, health care, and legal issues with LGBT elders than once thought. Successful treatment begins with a trusting relationship and a willingness to overlook societal stereotypes on the part of both the patient and therapist. Clinicians can make their practices more LGBT inclusive by modifying intake forms to include LGBT orientations, requesting and valuing information about family members by choice, and by providing general information and referrals, when needed, for legal issues related to caregiving and other end of life issues. Helping LGBT elders harness their likely well-honed coping skills also can provide insight into personal strengths as well as a focus upon resilience.

Resources for Professionals

American Psychological Association
Division 44, Society for the Psychological Study of LGBT issues
http://www.apadivision44.org

LGBT issues and aging bibliography
http://www.apadivision44.org/resources/lgbt_aging_bibliography.pdf

Gerontological Society of America (GSA)
"Rainbow Research" Special interest group for LGBT issues among older adults
http://www.geron.org/Membership/interest-groups#rainbow

GSA: 2010 Pre-Conference Workshop:
"Developing and Implementing Research on LGBT Older Adult Populations"
http://www.geron.org/component/content/article/70-2010-annual-scientific-meeting/544-pre-conference-workshops

Academic Journals

International Journal of Transgenderism
Journal of Bisexuality
Journal of Gay & Lesbian Mental Health (formerly Journal of Gay & Lesbian Psychotherapy)
Journal of Gay & Lesbian Social Services
Journal of GLBT Family Studies
Journal of Homosexuality
Journal of Lesbian Studies
Journal of LGBT Health Research
Journal of LGBT Issues in Counseling
Law & Sexuality: A Review of Lesbian, Gay, Bisexual & Transgender Legal Issues

Continuing Education

Gil Gerald & Associates, Inc.
Cultural Responsiveness in Serving LGBT Individuals and Families (3 h, BBS/
 CAADAC)
Working with Sexual Minority Youth (6 h, BBS/CAADAC)
http://gilgerald.com/lgbt-access/2010/9/11/online-continuing-education-courses.html

Greater Washington Society for Clinical Social Work
Offers a variety of continuing-education programs, some focused on LGBT and aging.
http://www.gwscsw.org/education.php

The Kinsey Institute
Focused on research in Sex, Gender, and Reproduction, provides lists of continuing
 education opportunities for continuing education and postgraduate training.
http://www.kinseyinstitute.org/resources/cont-education.html

LGBT: Clinical Work with Lesbian, Gay, Bisexual and Transgender Clients;
 Providing culturally competent therapy with sexual minorities
Continuing education course through Zur Institute
(5 credit online course)
http://www.zurinstitute.com/lgbtcourse.html

Education for the Workplace

Pride Center of Western New York
Sensitivity trainings for businesses and organizations towards LGBT clients, cus-
 tomers, and coworkers.
http://www.pridecenterwny.org/site/st.asp

An outline of the infamous USDA workplace sensitivity training (Developed
 January 2010)
http://www.la.nrcs.usda.gov/about/LGBT/GLBT_Training_January_2010.pdf

Pacific Pride Foundation
LGBT Sensitivity Training for Service Providers. Based in Santa Barbara, CA
http://www.pacificpridefoundation.org/LGBT_Services/lgbt12_Sensitivity_
Training.htm

Brian McNaught, "the godfather of gay sensitivity training," who presents at col-
 leges and universities and also offers corporate trainings, books, and DVDs
http://www.brian-mcnaught.com

National Resource Center on LGBT Aging
Offers free training for agencies working with LGBT and/or older populations.
http://www.lgbtagingcenter.org/about/training.cfm

Resources for Consumers

The following LGBT-focused organizations provide information and assistance in relation to aging, legal issues, health care (including long-term care), and social support.

AIDS Health Project
San Francisco
415.476.3902
www.ucsf-aph.org

American Society on Aging, also known as Lesbian and Gay Aging Issues Network (LGAIN)
http://www.asaging.org/lain

Association of Lesbian and Gay Affirmative Psychotherapists (ALGAP)
New York, NY
646.486.3430
www.algap.org

Bay Area Lawyers for Individual Freedom
San Francisco
415.865.5620
www.BALIF.org

Gay and Lesbian Association of Retired Persons (GLARP)
Los Angeles
(310)-722-1807
www.gaylesbianretiring.org

Gay and Lesbian Elder Housing
1602 North Ivar Avenue
Hollywood, CA 90028
(323)-957-7200
www.gleh.org

Gay and Lesbian Medical Association
San Francisco
415.255.4547
www.glma.org/

Gay & Lesbian Outreach and Engagement (GLOE)
Part of the Washington DC Jewish Community Center
Washington DC
(202)-777-3253
www.washingtondcjcc.org/social-networks/gloe/

GLBT National Hotline
888.843.4564
www.glbtnationalhelpcenter.org

Human Rights Campaign (HRC)
Washington, DC
800.777.4723
www.hrc.org

Lambda Legal
New York, with regional offices in the Midwest, South, and West
212.809.8585
www.lambdalegal.org/nhg

Lavender Seniors of the East Bay
San Leandro, CA
510.667.9655
www.lavenderseniors.org

National Academy of Elder Law Attorneys
Tuscon, AZ 85716
520.881.4005
www.naela.org

National Center for Lesbian Rights
San Francisco
415.392.6257
www.nclrights.org

National Resource Center on LGBT Aging also known as Service and Advocacy for
   GLBT Elders (SAGE)
National Headquarters in New York, NY
212.741.2247
www.lgbtagingcenter.org
e-mail: info@sageusa.org
www.sageusa.org

Transgender Aging Network
A partner of National Resource Center on LGBT Aging and FORGE
http://forge-forward.org/aging/
Transgender Law Center
San Francisco
415.865.0176
www.transgenderlawcenter.org

The following organizations do not espouse a specific LGBT mission, but may provide services in relation to caregiving, research, advocacy, or social support

Family Caregiver Alliance
San Francisco
800.445.8106
www.caregiver.org

The FCA offers an online LGBT caregiver forum, the LGBT Caring Community
    Online Support Group
http://caregiver.org/caregiver/jsp/content_node.jsp?nodeid=347

Medicare Rights Center
Hotline: 1.800.333.4114
National Council on Area Agencies on Aging
Elder Care Locator
800.677.1116
www.eldercare.gov

Chapter 10
Emergent Issues in Sexuality and Aging

The field of sexuality and aging is still in its infancy. Emergent issues include the role of the Internet, dating, and divorce among older adults, and sexuality at the end of life.

Once considered a virtual societal myth, it has been established consistently via research and clinical findings that older adults can and do engage in a variety of sexual behaviors, and express their sexuality in a variety of ways. However, new forces in society including drugs such as Viagra for treatment of ED, marketing campaigns that promote a highly sexualized, physically healthy, and ageless ideal, and opportunities on the Internet can bring significant change to the expression of older adults' sexuality. Dating has changed for older adults, particularly in relation to the limited number of available, older single heterosexual men compared to women, increased rates of divorce, and various online opportunities.

This chapter is designed to review emergent areas of clinical interest and research in adulthood and aging. With increased rates of divorce, cohabitation, and second (and other) marriages among aging adults, issues in couple's therapy may emerge related to the sharing of resources as well as allegiances to adult children from previous relationships. Although limited empirical research is available regarding its use, the relative anonymity of the Internet offers older adults unique access to sexual health information, erotica and pornography, and social networking including dating. Another emergent area of importance is that of sexuality at the end of life. Although research indicates that sexuality remains important to older men and women in palliative care, few professionals acknowledge or even broach the subject.

The Internet

Older adults represent one of the most rapidly growing segments of Internet users. Adults over the age of 65 account for approximately 13% of the US population, and in similar proportions, represent nearly 10% of all active US Internet users.

The number of older adults who identify themselves as active Internet users has recently doubled, with more than 17 million online users over the age of 65 spending an average of more than 48 h a month online (Nielson 2009). Although more men than women over the age of 65 report using the Internet, the number of older women users is increasing steadily each year. The AARP (2010) reports that nearly half of all adults over the age of 50 feel "extremely or very comfortable" going online. In contrast, only 1 in 4 adults over 50 reports that they never used the Internet. Additional findings suggest that baby boomers place greater value upon the ability to access the Internet via a smart phone or other handheld device than younger adults (Yang and Jolly 2008), and that older adults profess only minimal anxiety about the use of technology (Niemela-Nyrhinen 2007). These findings consistently dispel myths that older adults are uncomfortable with, and do not like, using computers or the Internet.

In terms of specific online activities, the Nielsen company (2009) reports that nearly 90% of older adults use the Internet primarily for e-mail. Nearly half of those surveyed reported that they used the Internet to search for health-related information, which could include information related to sexuality (Nielson 2009; Nimrod 2009; Wyatt et al. 2005), and more than one-quarter reported using social networking sites to keep in touch with friends and family (AARP 2010). The third most commonly visited site for users over the age of 65 was Facebook. (Notably, in the year prior to the 2009 Nielson report, Facebook was only the 45th most commonly visited site among this older age group). More than 70% of older users report that their online social networks are very important to them (AARP 2010). Nearly 10% of all visitors to social networking and blog sites were over the age of 65, with slightly smaller numbers of teenagers following suit (Nielson 2009). In other words, older adults are more likely to use social networking than teens. Although more than half of all older adults access the Internet via a desktop computer rather than with a laptop (26%) or smartphone (8%; AARP 2010), social networking and online interaction with others is clearly important to older adults and is becoming increasingly popular.

Evidence also exists of a significant digital divide within the older adult population. Additional findings from the AARP's (2010) study of Internet use revealed that Latinos over the age of 50 were significantly less likely to use the Internet, and felt significantly less comfortable using the Internet, than their White counterparts. For example, only 20% of Latinos, compared to 47% of Whites, reported being "extremely or very comfortable" using the Internet. Although more than three-quarters of Whites report using the Internet regularly, more than half of all Latinos report that they have never gone online. It is unclear whether the discrepancies observed between Internet use among older Whites and Latinos can be explained by differences in socioeconomic status, culture, or even the availability of certain Web sites in Spanish. It also remains unclear if there are other significant differences in Internet use among middle-aged and older adults from different ethnic and cultural groups. It is clear, however, that both clinicians and researchers need to be aware of these differences in Internet access and comfort.

The Internet offers a variety of benefits for older adults. In terms of the expression of sexuality, it is a powerful medium for a number of reasons (Adams et al. 2003). The relative anonymity of users is paramount. An older adult can, from the relative privacy of their own home, visit a variety of Web sites and participate in various groups and chat rooms without having to reveal personal information about themselves. Unless an Internet user chooses to submit online photos or use a Web cam, older adults can interact with others without being discriminated against for having gray hair, wrinkles, being over or underweight, having a different skin color, sitting in a wheelchair, or for having some other physical disability (Adams et al. 2003). Curiosity about potentially embarrassing or stigmatizing topics can be explored without being conspicuous. Pornography, sex toys, lubricants, and condoms can be viewed or purchased without having to go out in public to an adult bookstore. Most sites on the Internet are free or at low cost, so for those who have Internet access, most online material is quite affordable.

Certainly, each of these factors poses various pros and cons. Like individuals of any age, older adults can become so involved with online pornography that it begins to interfere with their own personal relationships. Other older adult Internet users, particularly those with some degree of cognitive impairment or a lower level of education, may be more likely to misinterpret health information found on the Internet. Multiple accounts exist of older adults being duped or scammed by online parties who are less than honest or forthcoming. To help address some of these practical issues, the AARP offers valuable information for people who want to explore the Internet with a greater degree of online safety and awareness (e.g., www. aarp.org/technology/how-to-guides).

Online Dating

Although large increases in the absolute number of older adults has occurred as a result of the aging of the baby boom generation, additional changes appear when one considered the sheer number of older women compared to older men. Specifically, by age 75–84, women outnumber men by at a ratio of nearly 2–1. For Americans aged 65–74 there are more than 100 women for every 82 men (Gist and Hetzel 2004). A number of factors may help account for this discrepancy. Women have a longer life span than men, and women tend to marry older men. Thus, married women can typically expect to become widows and live as a widow for many years to come. As a result, the odds are stacked poorly against an older heterosexual woman who wishes to find an available male partner. This discrepancy in population size between older men and women also means that older, heterosexual men will find themselves with multiple opportunities for available partners. (When the author presented some of these findings at a local community presentation, a middle-aged male audience member shouted out from the back of the room, "Woo-hoo! Now you've given me a reason to live!") This virtual glut of single women in the older adult population certainly influences the experience of dating for both older heterosexual men and women.

More than one million older adults have turned to Internet dating sites and services, perhaps in part to the demographic challenges of finding an older, single, male partner. Baby boomers typically hold more liberal attitudes toward sexuality and dating then their previous older adult cohorts, and the Internet can also provide outlets for companionships and dating for older adults who may have physical challenges to their mobility or for those who are socially or geographically isolated. In recent years on Match.com, one of the largest online dating sites, the number of new members over the age of 65 has more than doubled (Los Angeles Times 2004). Although generalist dating sites such as Match.com, eHarmony, and Yahoo!Chat accept adults over age 65 as members, additional online dating sites designed specifically for middle-aged and older adults including SeniorPeopleMeet.com, SeniorMatch.com, SeniorFriendFinder.com, ThirdAge.com, BabyBoomerPeopleMeet.com, SilverSeniors.com, SeniorNet.org, and LavaLife.com, among others, now exist with hundreds of thousands of members.

Most dating services, including SeniorNet.org and Sassyseniors.com offer links or membership to LGB elders. Some sites also are available for middle-aged and LGBT elders including cafmoscommunity.com for middle-aged and older gay men and onlinegaydating.co/uk. However, stigma against LGBT adults exists, even online. eHarmony, one of the largest online dating sites, which initially accepted only heterosexual members, was court ordered in 2008 to offer a separate online dating service for LGBT individuals, including LGBT seniors (Axon 2010). Due to the stigma typically faced by LGBT older adults, especially those who may be physically or socially isolated, online opportunities for friendship, dating, and social support can offer significant benefits.

One of the few studies available that provides information about older adults' online dating profiles and preferences reveals similarities to those observed among younger adult users (Alterovitz and Mendelsohn 2011). An analysis of 600 heterosexual personal ads on Yahoo! revealed that men aged 60–74 and those aged 75 and older were seeking female partners younger than they were. Older female users aged 60–74 were seeking male partners who were similar in age or older, whereas female users aged 75 and older sought male partners younger than they were. All of the males in the study, from age 20 to 80 were seeking physically attractive female partners, and provided information about their status (e.g., income and educational level). All of the female users, including those aged 60 and older, were more selective about selecting a male profile, and placed emphasis upon a potential partner's level of income and education. So even in advanced age, men appear to value physical attractiveness and youth in a romantic partner, whereas women appear to value personal resources and status over youth. However, beginning at age 75, women appear to deviate from this pattern, and place similar value upon status and relative youth.

Individual reports suggest that many older adults find companionship, dating partners, love, and even committed partners and spouses through various online services and activities. Actual data on member satisfaction is lacking, so most reports remain anecdotal. Individual reports of negative outcomes from online dating also appear. For example, a 64-year-old widow began chatting online with a "Mr. John Smith" from Canada on a Christian Internet dating site. After a few months of chatting, Mr. Smith proposed and asked the widow to send various household goods and money to his address to help him prepare for an online business before their wedding. The woman complied for more than 2 weeks before one of her adult children became concerned and alerted her to the scam. Local authorities then opened an investigation. Another adult daughter wisely prevented her 63-year-old, recently separated mother from sending her new out-of-state, online boyfriend the money to buy an airline ticket to visit for the upcoming holidays. Once the 63-year-old informed her boyfriend that she was unable to send the money for his ticket, all communication from him suddenly ceased. The AARP offers additional sound advice for seniors who wish to engage in safer online dating (e.g. http://www.aarp.org/money/scams-fraud).

Pornography and Online Purchases

It is notable that none of the aforementioned studies, including the large-scale Nielsen (2009) and nationally representative AARP (2010) studies of Internet use, assessed the extent to which older adults used the Internet to view pornography. Virtually no information is available regarding the extent to which older adults communicate in online chat rooms or forums dedicated specifically for sexual activity (and whether those communications are obtained for free or pay), or share sexualized pictures or videos via "sexting." It is unclear whether the researchers who designed these surveys were told not to include such measures, or whether the researchers failed to consider that older adults might use the Internet for such activities.

Some of the only empirical data available regarding older adults' use of pornography exists regarding both print and online forms (Kontula and Haavio-Mannila 2009). This nationally representative survey of Finnish adults found that more than half of middle-aged and older Finnish men found pornography arousing. Specifically, 64% of men aged 45–54, 57% of men aged 55–64, and 55% of those aged 65–75 found pornography arousing. In contrast, Finnish women were significantly less likely to report that they were aroused by pornographic material. Approximately one-third of middle-aged Finnish women aged 45–54 found pornography arousing, whereas less than one-quarter of Finnish women aged 55–64 and 64–75 found it arousing. It remains unclear how well these results could be generalized to other countries' adult populations, and to what extent these middle-aged and older adults sought pornography from online versus print sources.

Online pornography accounts for a sizeable proportion of Internet revenue (Koerner 2000), and the most popular queries on search engines are for adult material (Miller 2000; Ropelato 2011). More than four million adult Web sites are available online, and America leads worldwide for video production. A new adult video is created every 39 min in the USA. For every four Americans who view online pornography, three are male and one is female. More women visit chat rooms for sexual content than men. Estimates of online viewers suggest that the majority (26%) viewing porn are men aged 35–44, with men aged 45–54 and those aged 55 and older both accounting for 20% of all viewers (Ropelato 2011). Virtually no empirical data is available regarding the numbers of older adult men and women who view online pornography or chat rooms.

It also is unclear what kind of pornography older adult men and women opt to view. For example, it is unclear if older adults tend to view or prefer younger or older adults in their sexually explicit material. Significant cultural differences may also arise. Despite common Western stereotypes that elderly sexuality is negative, distasteful, or otherwise uncouth, Japanese culture is demonstrating a wave of interest in "older adult" film stars. In fact, the fastest growing segment of their adult film industry features elderly adults. For example, a 74-year-old identified only by his screen name Shigeo Tokuda is regarded as a highly successful and sought after porn actor. As touted in Time magazine, Shiego's production company, Glory Quest, launched a series of "old man" or "grandfather" films starring him in 2004. In his more than 350 films, including "Forbidden Elderly Care," the 74-year-old is featured having sex with both younger and older women. Shigeo notes that as long as his country's interest in such "old man" pornography holds, he plans to continue working for at least 6 more years (Toyama 2008). Some women over the age of 70 are now being cast to star "mature woman" Japanese sex films (Sparrow 2008). Conversely, no popular adult film stars in the USA are over the age of 65.

There are potential benefits and harms related to the use of online pornography for older adults. For an older adult who may have difficulty with mobility (e.g., who may be wheelchair or bed bound), who live in predominantly rural areas, or who are embarrassed about going to a public, brick and mortar store to purchase pornography, the privacy afforded by the Internet can be very important. For older adults without a partner, with a partner who is unable to engage in sexual activity, or for older adults who wish to incorporate the viewing pornography into their sexual activity, and for older adults in general, online erotica and pornography can provide helpful fantasies and release. Online pornography comes in a myriad of varieties, including those of interest to straight and LGBT individuals. Of course, problems arise when any person's use of online or other forms of pornography begin to interfere with their off-line sex life, or someone develops beliefs that healthy sexual activity includes coercion or the persistent objectification of others. The 24-h, 7-day-a-week access of online pornography may make some individuals, including older adults who typically do not work full time outside of the home, more susceptible to some of these problematic situations.

Older adults also can purchase sex toys, vibrators, books, videos, condoms, lingerie, and lubricants from a variety of reputable online merchants, including Amazon.com (the author has no affiliation with this entity), with relative privacy and ease. And, Internet merchants are always open. If an older woman learns from her health care provider and therapist that masturbating with a vibrator could help increase her natural levels of lubrication and ability to orgasm, she may be hesitant to take the bus into town to purchase these items at the local pharmacy or adult bookstore. She may be significantly more likely to purchase one online and have it delivered surreptitiously to her front door. In other words, an older adult's "nosy neighbor" in their apartment complex or assisted-living facility does not need to know anything about such private but often essential purchases.

The relative anonymity of the Internet also unfortunately allows individuals of all ages, including older adults, to engage more freely in certain illegal or paraphilic sexual activities. Although virtually no empirical prevalence data is available, older adults do engage in illegal sexual activities including exhibitionism and the production and viewing of child pornography. Although federal sentencing guidelines allow older adults to receive shorter or reduced sentences when convicted, especially if alternative forms of punishment could be considered less costly (e.g., giving a shorter prison term if the elderly defendant is seriously medically ill), a recent appeals court decision ruled that in cases of child pornography, age alone does not provide grounds for a reduced sentence. So, elderly adults convicted of possession and use of child pornography can no longer expect leniency due to their advanced age (Arias 2010).

Sexual and Health Education

Adults over 50 report that one of the primary reasons they use the Internet is to obtain medical information (AARP 2010; Nielson 2009), which obviously includes that of sexual health and functioning. In the USA, more adults seek e-health information per day than those who visit their doctor in person (Fox and Rainie 2002). Used in conjunction with advice and discussion with a health care provider or other clinician, the Internet can provide individuals with free, easily accessible sex education as well as vital information about sexual dysfunction, menopause, and prostate and breast cancer, among other health-related topics. Some therapists find it helpful to assign certain Web pages, or online exploration, as homework for their clients if they are struggling with a certain issue. Other therapists ask their clients to print out relevant pages they have been searching, and review them together.

Unfortunately, if an Internet user is surfing the Web alone and is not well educated or familiar with evaluating Web sites for their potential accuracy (e.g., government and university sites are generally reliable sources of information, and end in .gov and .edu, respectively), problems with misinterpreting and receiving incorrect information can easily occur. It could be expected that many older adults are unaware that Wikipedia entries are not necessarily provided or approved by professionals for their accuracy, and that blogs and other Web postings can be generated by anyone, including those who hide or distort their identities and credentials. Simply finding relevant health-related information online can be challenging. Findings suggests that for Internet users looking for health information about cancer, for example, find what they want only 60% of the time, and typically after a lengthy search (Berland et al. 2001).

Even when individuals find appropriate Web sites with accurate medical information, it is important to keep in mind that older adults in the USA possess only basic or below basic levels of health literacy, on average (Kutner et al. 2006). It also is important to note that many people who are literate in standard or the everyday use of English, their health literacy may be significantly lower. Findings show that even when taking community-living older adults' years of schooling, cognitive ability, number of chronic health conditions, and visual acuity into account, older adults maintain significantly less familiarity with health-related terms and information than younger and middle-aged adults (Baker et al. 2000). Clinicians must keep in mind that their older adult clients may have limited health literacy, and that other clients may not even have access to the Internet. Many older adults turn to adult children or friends, sometimes referred to as warm experts, for assistance in locating and reviewing medical information on the Internet (Wyatt et al. 2005). These warm experts may or may not be able to properly evaluate the validity of the source material (e.g., does it come from a government sponsored Web site or from a personal blog) or provide adequate interpretations or explanations of the online health content itself.

As noted, the Internet can also allow homebound or potentially embarrassed elders the opportunity to purchase various items in relative privacy, including vibrators, condoms, and lubricants. It also remains unclear how many older adults make online purchases for untested and potentially unsafe "sexual enhancement products," designed to mimic Viagra and other prescription drugs. Other products with questionable efficacy that may be purchased online by older adults include various remedies for hot flashes, night sweats, vaginal dryness, and other symptoms associated with menopause. Some older adults may also have difficulty differentiating between advertising, personal opinions, and professional information on the Internet. Clinicians can provide valuable assistance to older adults who may seek out the assistance of the Internet in finding health and sexually related information.

Issues Related to Divorce and Remarriage

Single heterosexual middle-aged and older adults are more likely to date than marry, consistent with contemporary trends (e.g., Cooney and Dunne 2001). More than one-third of Americans over 50 are divorced, widowed, separated, or have never married, with the majority being divorced (AARP 2003). For older adults over age 65 who lose a spouse, up to one-third become interested in dating within 18 months after becoming single (Carr 2004). This marked growth in the number of older adults who are dating or living together is recognized among savvy businesspeople as a burgeoning market, primarily because the dating behavior of older adults appears more varied than that of younger adults (Schewe and Balazs 1992). Activities for dating among adults over 50 range widely from midnight walks, cooking, bird watching, camping, and attending the opera to square dancing, art classes, movies, mountain biking, and exotic vacations to attending church, visiting children and grandchildren, and making dinner at each other's homes. For international travel, Elder Hostels offer unique opportunities. The pace at which older adults date and develop emotionally and sexually intimate relationships also appears to be more rapid than among their younger counterparts. Although companionship is cited as the primary reason why older men and women seek romantic partners, studies suggest that up to 90% of these men and women seek love and sexual satisfaction as well (Bulcroft and Bulcroft 1991).

A nationally representative survey of more than 1,000 middle-aged and older adults (AARP 2004) revealed that among individuals aged 50 and older, wives were more likely to initiate divorce proceedings than husbands. Both men and women reported that going through a divorce at midlife was more emotionally devastating than they would expect if they lost their job, and only somewhat less traumatic than what they would expect if their spouse died. Although 45% of the participants indicated that their greatest fear after their divorce was that of being alone, the majority (80%) reported having a positive outlook on their lives. Seventy-five percent of female and 81% of the male respondents reported that they started dating and became involved in a sexually monogamous relationship within 2 years after their divorce.

Among middle-aged Americans, there are three single women available for every two single men (AARP 2003). The greater shortage of single older men among older age cohorts has forced many women over the age of 65 to forgo dating or to engage in a very competitive market. (As noted, some turn to online dating to help broaden their choices.) Women over 65 who do date often receive social prestige in their communities (McElhaney 1992) and tend to be healthier with greater physical mobility. Men over the age of 65 who date tend to be involved in social organizations and to own their own home. It remains unclear whether the women and men over age 65 who date are more psychologically healthy than those who do not, or whether dating itself fosters the development of coping skills and life satisfaction (Bulcroft and Bulcroft 1991).

Despite the apparent benefits of dating in later life, the high ratio of single older women to single older men has important interpersonal implications. One only has to visit a nursing home, retirement complex, or other institution to observe some of the associated dynamics. At one retirement community, an older man complained that his "dance card was always full," and that he never had a chance to relax with all "of those womenfolk around." Another older man's girlfriend lamented, "Every time he goes out to walk the dog, the divorced woman in next unit asks if he would like to come over and have some coffee…I can see that she's not wearing anything much under her housecoat, for goodness sake…Sometimes it makes me feel very unsure about [our] relationship. I mean, wherever we go, he's got women fawning all over him." Another single older woman noted, "If I don't ask the men to go for coffee, I don't think they will be asking me. I might as well take a number before they get around to it." Some of these very literal issues of supply and demand in terms of dating can prove quite challenging, particularly for older single women.

Justine

Dating can be difficult to accomplish, or even define, for some older adults. Justine was a 75-year-old, community-living widow of 8 years. She sought individual therapy for depression and for help in coping with problems related to her adult children. Her depression emerged soon after the death of her husband, with whom she appeared to have a loving, caring relationship. Justine spoke candidly and warmly about their fun times in which they took warm baths together, walked in the rain (with umbrellas so they wouldn't catch colds), and dancing in the living room to their favorite old records. Justine and her husband also monopolized most of each other's time; they worked together in a family business and had few outside friends or interests. After 2 months in treatment, Justine had begun to mourn her husband's loss and to seek out other social contacts in her immediate neighborhood. For the first time, she began to get to know her neighbors. She also took a taxi to the local senior center twice a week and started to make some same-sex friends. Justine also admitted that although she had considered dating other men, her adult son seemed rather opposed to the idea: "You and dad were the perfect couple…How could you ever top that?"

Approximately 1 month later, Justine's therapist began to worry about her patient's apparent change in mental status. Justine began to show significant signs of paranoia, and she described hiding things in her own apartment. She also talked about sneaking in and out of the apartment after "making sure that the coast was clear." Justine began to forget her appointments at the hairdresser, geriatrician, and ultimately with her therapist. Her therapist began to consider dementia or psychotic depression as potential diagnoses for Justine, especially when she claimed that a man was following her to most of her appointments. Justine claimed that she did not forget about them; she just wanted to "go to them alone in peace!"

Because Justine either could not or would not describe anything about the man following her, it was assumed that her mental status was so impaired via delirium that she was recalling prior events improperly, or that she was so depressed that she began to create psychotic delusions in which she was the subject of romantic interest from another man. Her therapist scheduled a neuropsychology examination to rule out dementia and asked Justine's geriatrician to run additional tests to see if any underlying medical problems could account for her apparent decline in mental status. The results of the neuropsychology testing were ambiguous. Justine was described as having only a mild memory impairment. Justine's geriatrician felt that she was quite healthy except for her arthritis, and that she was "just imagining things" to feel needed and important.

During her next appointment, Justine lamented that her mystery man had followed her to the hospital. Unwilling to ignore the possibility that Justine was relaying some aspect of truth in her convoluted stories, her therapist called security and asked Justine if she would accompany her to the waiting room to see this man for herself. Justine hesitatingly complied, and when they turned the corner into the waiting area, an older man stood up with his cane and said buoyantly, "Justine, I thought you wouldn't be done yet! Can I walk you home?" After Justine's therapist was assured that her patient was in no immediate danger or physical threat from this man, they returned to discuss the issue in their session.

Justine admitted that she had been "keeping secrets…that is my neighbor, Rob… He can't get enough of me. He follows me everywhere…Sometimes I like it and sometimes I don't…It's nice because he helps me put my garbage out when it's heavy. He shovels my walk when it snows. He brings in my mail for me when it rains and helps me carry my groceries up the stairs. And, sometimes I just like having coffee and talking with him." Justine began to wring her hands and continued,

"Of course, my son and…his wife don't like it…Rob's a married man. We, we haven't really DONE anything. Well, one time he tried to hold my hand and I pulled it away. But his wife sure is steamed. I'm afraid to come out of my house because one day she was waiting on her steps to yell at me and call me names…I haven't really done anything, don't you think? I can't help it if he doesn't love her anymore, can I? He told me that she doesn't even bother to take a bath anymore or try to look nice. He gets tired of her nagging and moaning."

A number of sessions followed in which Justine was able to speak openly about her relationship with Rob, her fears about becoming emotionally involved with another man, her concerns about betraying her departed husband and her son, and her feelings of ambiguity about getting involved with a married man. At one point, Justine considered a restraining order against Rob, but soon realized that she did enjoy his attention, even if it was a bit excessive. For the first time in 8 years, she felt "desirable and pretty…like [she] was courting again." At the same time, she feared that her relationship with Rob would generate a storm of gossip in the neighborhood in which she "would never be able to make friends, with what all of the other women would think." Justine's therapist had to monitor her own countertransference carefully so that she did not interject her own beliefs about marriage and dating into the treatment dynamic.

Ultimately, Justine tried to speak to Rob and his wife about the nature of their relationship. She spelled out that she and Rob "were just friends because she didn't want to be responsible for breaking up anybody's marriage." Rob's wife remained angry about the relationship, and she forbade Rob to even speak to Justine. For whatever reasons, Rob's wife tolerated his lack of attention and did not appear to pursue a divorce, separation, or counseling. So, Justine continued their relationship under more clandestine circumstances. She felt that even though Rob had amorous intentions toward her, she did not act on them, making them "just friends, after all." Justine said she sometimes felt guilty about all of the time they spent together, but she rationalized it as "Rob's choice to make his wife angry; I didn't ask him to do any of this."

In the next few months of therapy, Justine also came to recognize that by being friends with Rob, she had selected a companion who could not, by definition, compete with her dearly beloved husband or upset her son because she "couldn't marry [him] anyway." Justine became better able to acknowledge her own needs for love and companionship, and she began to spend more afternoons at the local senior center in hopes of finding a single male companion. She appeared to recognize, on some level, that she wanted a relationship in which she took responsibility for her own actions and in which she did not have to hurt anybody else (i.e., Rob's wife). She also felt that she had been able to "say good-bye" to her husband, which allowed her to seek a genuine, romantic relationship, rather than a pseudofriendship. She also recognized that her son would have to come to accept her choice in companions; she was the one living alone without a spouse, not him. Like most of the older women in her situation, however, the only single men who went to the senior center already had a steady girlfriend.

Sexuality in Long-Term Relationships

Couples therapy for older adults is essentially similar between older and younger adults. Vital aspects for exploration often include how the couple relates as romantic partners and friends; how they divide responsibility for various chores, tasks, and duties at home; how they experience subjective feelings of love and emotional intimacy; how satisfied they are with sexual expression; to what degree they share activities and leisure time; how they argue, and so on. However, because of age differences, younger therapists may feel somewhat uncomfortable discussing issues related to sexual satisfaction and fidelity. Although the therapist's tendency may be to discuss those issues more gingerly, they comprise just as important a part of the couple's clinical assessment as in any work with younger couples. Even more importantly, this may be the first time the members of an older couple have been given the opportunity to openly discuss their sex life, and their satisfaction or dissatisfaction with it.

The developmental challenges associated with aging also can pose some unique problems for older couples. These difficulties may be age or cohort related and can include:

- Disagreements with adult children over money, wills, finances, power of attorney, and lifestyle choices.
- Visitation rights with grandchildren, especially if a divorce has introduced stepparents and step-grandparents.
- The assumption of financial and emotional responsibility for grandchildren; anxiety about financial security, especially if on a fixed income.
- A move to a new community or care facility.
- Adjustment to a significant increase in leisure time or the loss of identity as a valued employee through retirement.
- The loss of friends and family members through illness or geographic relocation.
- Adjustment to new roles as caregiver for a spouse who may be acutely or chronically ill.

All of these changes and transitions can translate easily into sexual dysfunction or dissatisfaction.

In couple's therapy with older adults, it also is vital to assume that an older couple's initial presentation may not be accurate. This is not meant to say that older adults are devious, mischievous, or lacking in insight; many older adults may present initially as a gentle, happy, loving couple because they were socialized to "be on their best behavior" when dealing with a professional. The notion of discussing family problems with outsiders may appear foreign to many older adults, and may result in an initially positive, skewed presentation. In fact, many times couples do not initially seek out couple's therapy. One member of the pair may present to a clinician with problems of depression, anxiety, or substance abuse, which leads to a trial of couples therapy.

Clinicians should avoid subscribing to stereotypes that frail older adults are "perfect or sweet" because they have stayed together for so long. Even though older couples do tend to remain married, compared to younger couples who are more likely to divorce, older couples are just as likely to have serious problems in their relationships as younger couples. For better or worse, older couples may be more likely to overlook, deny, or tolerate spousal substance abuse or infidelity in their relationship in order to remain married.

The Hortons

Don and Winny, 72 and 67 years old, respectively, arrived together at a treatment team meeting. Winny was excited that she was given permission to take her husband home from the Alzheimer's unit. Don had been misdiagnosed by a previous practitioner as having Alzheimer's disease, when in actuality he had a mild case of vascular dementia from sleep apnea. He could be maintained with an alarm monitor and oxygen supply at night, and his dementia was deemed mild enough for him to resume life with his wife at home. Winny beamed, "Oh, we love each other, and get along so well...I can't wait to take my husband back home. Thank god!" The couple had been married for nearly 47 years and never reported any significant problems in their marriage.

Three weeks later, Winny suddenly came back to the hospital and wanted counseling because she "hated" her husband. She complained that Don was so upset that they had not had sex since he returned home, and she was furious about him "pawing all over me." When asked if anything else was bothering her, Winny exploded, "He drives me nuts! He wants me to do this, to do that. He tried to change the oil in the lawn tractor and dragged it all over the house and now I have to clean it up. And I have to make his favorite meals, and do the laundry. I am so sick of this!" With Winny's permission, the intake therapist consulted with members of Don's previous assessment team, and Don's original therapist was asked to engage the couple in marital therapy. The therapist's initial goal was to provide both Winny and Don with more education about Don's dementia and the changes that might follow in their relationship and household. Both parties also were educated about changes in sexual interest and response related to dementia.

During their second session, Don's sentiments mirrored those of Winny's. "The woman makes me crazy. I try to do something, and I guess I don't do it quite right. She ends up screaming at me for something or other, no matter what I do. I can't take it anymore. I just want to be with her, you know. I'm her husband and I'm back home now." The therapist pursued the relationship between Don and Winny, and about the realistic and unrealistic expectations that each held regarding Don's cognitive abilities, and their mutual concerns about resuming sexual relations after his diagnosis. For some reason, the therapist felt that this discussion was falling on deaf ears, and neither party seemed open to discussing their feelings or fears about Don's mild dementia. The use of more concrete attempts to engage the couple in developing a homework assignment also appeared to fail.

Reasoning that Don and Winny's apparent lack of motivation masked some deeper problem, the therapist decided to take a different tact. In order to assess their prior conflicts and coping skills, the therapist asked, "What is the worst thing that has happened between the two of you, either lately or in the life of your marriage?" Winny looked at the floor and began to tremble. Don said, "Well, I mean, I don't know…I guess when I almost ran the mobile home off the road, right? I did have a few to drink, but…" Winny interrupted, "Oh, yeah, you can't do shit right, can you Don? That's just it though, and you want me to just laugh and say, 'OK, that's all right,' isn't it?" Don's face contorted in anger and he leaned forward in his chair to invade her body space, "Yeah, just like you are always nagging, nagging, nagging! Goddamn woman. Don, do this. Don, do that. Don, don't touch me, blah, blah, blah, blah, blah."

The therapist turned to Winny, "What usually happens after Don talks to you like that?" Instead of Winny answering, Don blurted out, "Oh, I tell her, doc. I get right up in her face with my fist and tell her, 'I'm going to knock you a good one, woman, if you don't shut the hell up right now! I'm sick of your shit.' That usually shuts her up pretty good." Winny sunk in her chair, and Don started with a hmph and sat back confidently in his chair, staring at her. Immediately, the therapist established firm limits and boundaries, stating that violence, or even threats of physical violence, would never be tolerated and were unacceptable no matter what the circumstances. She informed Winny and Don that being upset and angry was a normal part of any couple's relationship, but that acting on those feelings with physical violence was completely unacceptable.

On further exploration, Winny indicated that Don rarely hit her, but that things "started getting worse" after his dementia was diagnosed and his impulse control had lessened. Dan hated feeling like Winny's "pathetic, can't do nothing right, child" instead of her husband, and his anger was expressed more and more often in inappropriate ways. It also appeared that Winny was accepting of Don's abuse over the years out of a sense of obligation and rationalization; "He never hit me in the face, and he always came home on time, didn't gamble, paid the bills, and never cheated on me…and I do love him." The couples therapy took on a radically different direction when it was obvious that their lack of sexual satisfaction was related to more dire, underlying problems.

Even though older adults are from an age cohort that espouses the virtues of marriage and committed relationships, older adults now often elect to live together rather than legalize their relationship through marriage. Older couples may avoid marriage because of pressure from adult children who have concerns about loyalty to deceased parents or about potentially dwindling inheritances, out of respect for their first spouse who they promised to cherish forever, or out of practical concerns regarding income tax problems and estate planning. For those who do have second (or multiple) marriages, problems may arise regarding all of the aforementioned issues. In clinical work with such couples, it becomes vital to identify the intersection between the emotional and practical issues underlying their presenting problem.

The Albertsons

Beverly Albertson, aged 67, was accompanied to the geriatric outpatient clinic by her husband, Dan, aged 71. Their marriage of 13 years was a second marriage (through divorce) for both of them. Beverly came to the clinic seeking therapy for major depression. She had lost interest in leisure activities, in seeing friends, in cooking, in eating, and in sex with her husband. Dan lamented that they once had been "very close," but that seemed to change about a year ago. Because Beverly and Dan did not appear to be well educated about depression, its origins, or treatment, couples therapy was proposed as an adjunct to Beverly's individual therapy.

During one couple's session, it became apparent that Beverly's depression emerged when she felt that things between them had so badly deteriorated that "it feels like there is no going back." When asked what she thought had happened, Beverly started to cry. Although Dan did not make any overt attempt to comfort her, he appeared anxious and distressed. Beverly continued, "I just don't feel special anymore…He doesn't even touch me anymore. I don't even want him to touch me anymore. What for, anyway? I don't count. I don't matter to him, anyway. I'm just not worth anything to him anymore." When asked to illustrate a specific time in which she did not feel special, Beverly spent more than 15 min relating a story in which she and Dan went out to eat at a family style pizza pub. Beverly would not even look at her husband when speaking.

Beverly: When he takes me out to eat there, we get the lunch buffet because it's cheaper. That's OK, and I can understand that, even though it's supposed to be like a date, or our special evening out. We don't have that much money between the two of us. But, he won't even let me order a soda. I can only get water unless I pay for the soda myself!

Dan: Beverly, we've been through this a hundred times! That's how they make all of their profit! They probably spend ten cents for that soda, and they get over a dollar from me! It's the principle of the thing. I want her to feel special, but it makes me sick to jack up their profit margin that much. I just can't do it. We can always get soda at home, so why do you have to have it when we go out?

Beverly: But I want it. Why can't you just get it for me. It's one stupid soda!

Dan: Well, it's one stupid dollar!

Beverly: Well, maybe I'm worth one stupid dollar.

Dan: Yes, but not when it goes in some scumbag, rich guy's pocket!

The conversation continued and the situation escalated. The therapist allowed for the argument to continue because she wanted to observe their typical pattern of interaction during conflict. Beverly and Dan continued to bicker and yell. They turned physically away from each other, but crossed their arms on their chests in a similar way.

Beverly: Well, the last time [your daughter] came to visit, you sure as hell bought soda when we went out to eat!

Dan: Well, she doesn't come to visit often from California, and I, I thought it was the right thing to do.

Beverly: The right thing for WHO?

After both parties were instructed to calm down in order to process the interaction, Beverly was able to admit that she always felt as though she played "second fiddle" to Dan's daughter from his first marriage. She wanted to feel like she was his primary love, and that she did not have to compete to get his attention. More importantly, she was able to admit that she felt that Dan did not love her as much as his daughter. Her lips trembled when she said this, and on seeing her tears, Dan put his arm around his wife, reportedly for the first time in over 3 months.

In a subsequent session, Dan admitted that he felt in constant competition with Beverly's first husband. He was a wealthy businessman who "allowed" Beverly to maintain a large investment account. Dan admitted he was upset because his pension check could never compare with the kind of money Beverly had in her account. When asked why he called it "her account," Dan said that he felt slighted that she did not offer to have him as a cosigner in her financial dealings.

On the one hand, Dan said he felt he could understand Beverly's desire to maintain separate investment, savings, and checking accounts "for her children, for later, if something bad happened," but that he still felt disdained and overlooked. After all, Beverly did a lot of shopping and bought herself expensive clothes, and Dan lived off of his own, somewhat meager pension. Dan said, "I mean, one dollar is a lot for me, and it is nothing for her…every time she asks me to spend that dollar, it's like… it's like I'm just giving it back to her first husband or something."

In addition, physical intimacy for the Albertsons had become tied to one of the most troublesome aspects of marriage for couples of all ages—money. Specifically, the distribution of valued resources in the marriage appeared tied to perceived competition with spouses from previous marriages, actual responsibilities to children from prior marriages, and the absence of trust between partners. Only after concrete planning and decision making was made about the distribution of money in the marriage, including money that was to be shared and separate, were Beverly and Dan able to resolve issues of trust. Beverly was able to understand that Dan sometimes felt that because he rarely saw his daughter, buying her things was one of the ways he felt he could be connected to her.

Dan became able to understand that since Beverly's first husband had an affair prior to ending their relationship, she felt" safer" hoarding her money in case her fears were realized and Dan decided he was going to leave her as well. Once Dan was able to set aside his competitive urges, to empathize with Beverly's fears, and to let her know that he was a different man than her first husband, Beverly decided to share some of her money in a joint account, and the couple began to feel secure enough in their relationship to engage in meaningful, satisfying sexual relations. In a more symbolic display of their ability to trust and give pleasure to one another, they also began to order two sodas and pay for dinner out of their joint account on their excursions to the local pizza parlor. More importantly, if the couple's therapist had pursued issues related only to their sexual dissatisfaction, the Albertsons' true, underlying problems related to their prior marriages would never have emerged.

Sexuality at the End of Life

For most individuals, thoughts about the end of life, hospice, and palliative care typically include pain, suffering, and sadness. Other associations to hospice may include hospital beds, oxygen masks, the administration of pain medication, and soft music and prayers. Although some individuals in hospice may not have the need or ability to engage in intercourse per se, the need and desire for other expressions of sexuality including holding, touching, fondling, kissing, massaging, and caressing may remain strong. For many patients, physical contact with a loved one can significantly improve their quality of life (Cort et al. 2004). Unfortunately, most people regard any expression of sexuality at the end of life as taboo, and many social workers, psychologists, nurses, physicians, and other health care providers fail to address hospice patients' sexuality (Katz 2011; Kutner et al. 2001).

Benefits

Sexuality obviously includes a variety of activities, ranging from physical activities that include others (e.g., intercourse, fondling, hugging, kissing, holding hands, cuddling) to those that are often solitary in nature (e.g., masturbation; bathing). Sexuality can also include nonphysical manifestations in the form of sex talk (or texting) with others, private thoughts, dreams, and fantasies, and the viewing of pornography either alone or with others. In terms of palliative and end-of-life care, adopting the broadest view of sexuality and sexual expression becomes essential. Clinicians can help dispel general societal myths that "sexuality equals sex," which only limit the ability of patients and their partners and health care providers to receive vital assistance (Redelman 2008).

A primary goal in palliative care is to improve patients' quality of life (Richards and Ramirez 1997). In one of the few empirical studies to examine hospice patients' sexuality, more than half of the 348 patients assessed, with an average age of 78 years, experienced significant problems with pain, lack of appetite, fatigue, drowsiness, concentration, and feelings of sadness. More than 60% of the patients assessed experienced problems with their sexual interest and activity, and the social workers in the study treating them reported that they felt unprepared to deal with their sexual problems (Kutner et al. 2001). In other words, nearly half of those older adult patients observed did sustain interest in sexual activity through the end of life. Other empirical studies suggest that terminally ill women and men desire sexual or physical contact with their partners, but feel too ashamed or "abnormal" to make their wishes known (Cort et al. 2004). It is as though these patients internalized general societal beliefs that sexuality is a privilege that belongs only to those who are healthy and young.

The benefits of sexual expression and intimacy in hospice and palliative care cannot be understated (Redelman 2008). Many dying individuals want to strengthen their relationships with partners and loved one, and sexual expression and physical intimacy can certainly be a significant part of that process (Singer et al. 1999). The simple act of physical touch or contact can be associated with significant changes in autonomic nervous system activity. The act of pleasant physical touch can release various neurotransmitters, leading to feelings of warmth and muscle relaxation (Weeks 2002). Participation in sexual activity including masturbation also has been associated with pain relief (Komisaruk and Whipple 1995) and improved quality of sleep (Weeks, and James 1998). When hospice patients' needs for sexual expression are not met, whether that means not being able to hold hands or to have intercourse, the results can be quite negative, including depression and significant declines in self-esteem (Leviton 1978).

Barriers to Expression

Significant barriers exist in relation to the expression of sexuality in hospice and palliative care. Scars from surgical procedures and even removal of various body parts including breasts and testicles often lead to feelings of embarrassment, loss, anger, and resentment (Rabow et al. 2004). The resulting declines in self-esteem and negative feelings about one's body image certainly do not bode well for the expression of sexuality even under optimal circumstances (Cagle and Bolte 2009). Concerns and issues related to pain, fatigue, and lack of privacy also prohibit participation in many sexual activities. Many hospice settings do not offer double beds, and many staff members come and go without knocking. Even when hospice services are offered in a patient's own home, many patients are moved to a single hospital bed, often on the first floor in a living room to provide more room for staff. This arrangement may have certain benefits, but the patient's privacy obviously becomes severely limited.

Even within the professional literature there appears to exist a dual stereotype or taboo in which sexuality among older adults as well as sexuality among the terminally ill becomes suspect. In a 2008 article appearing in the *American Journal of Hospice and Palliative Medicine* the author notes, "Many dying individuals will be in the older age group where the sexual needs between the couple may have found a nongenital/physical sexual expression, where sensuality expressed by hugging and kissing has become the norm…This may allow for an easier transition to the socially/institutionally acceptable interaction in [hospice care]…this may contrast sharply in a younger couple where an active sexual relationship is lost suddenly [leading to] anger…resentment…and frustration" (Redelman, p. 366). From this passage, it is as though only young adults have an active sex life and could be expected or entitled to mourn the loss of their sexuality. When professionals themselves discount the clear empirical evidence that older adults in intimate relationships can and do engage in sexual activity including penetrative intercourse, oral sex, genital fondling, and other forms of physical intimacy throughout their lives (e.g., Fisher, 2010), it becomes clear that there are more than institutional barriers to sexual expression within the context of hospice.

The literature is rife with portrayals of clinicians and health care providers who fail to recognize, acknowledge, and discuss issues related to sexuality with their end-of-life patients (e.g., Lemieux et al. 2004; Redelman 2008). One of the only available empirical studies of health care professionals' attitudes toward the discussion of sexuality with their patients in palliative care suggests that patient sexuality is medicalized, which limits discussions of sexuality to fertility, menopause, and erectile dysfunction (Hordern and Street 2007). This overreliance upon medical terms and diagnostic criteria also appears to allow health care providers to emotionally distance themselves from thinking about their patients as sexual beings.

The social workers, nurses, and physicians in Hordern and Street's study (2007) also avoided the topic by viewing their patients as asexual (especially if they were older), by assuming that someone else on the treatment team had already discussed sexual issues, by fearing that such a discussion would offend patients from various ethnic and religious backgrounds (e.g., Muslim) and by believing that they simply could talk about sexuality without becoming too personally vulnerable. Those clinicians who did attempt to broach the topic often tried to use humor to dispel their discomfort with the topic. Other researchers suggest that time constraints and fears that other professionals will regard discussions of sexuality with dying patients as inappropriate or over-attentive provide additional barriers to assessment (Cort et al. 2004). With pervasive social taboos about the open discussion of sexuality, aging, and death and dying, coupled with the generally limited amount of training and education available to clinicians about these topics, these findings sadly do not appear that surprising.

Talking with Patients and Clients

Although clinicians tend to avoid discussing sexuality at the end of life, individuals undergoing hospice care typically report that they want information about sexual functioning (Ananth et al. 2003). Patients also expect that their health care providers will broach the subject first (Gamel et al. 1993). The National Hospice and Palliative Care Organization regards an assessment of patient sexuality as a core competency for those working with individuals at the end of life, but their national curriculum provides no such detailed instructions (Hay and Johnson 2001). Fortunately, a number of researchers and clinicians offer outstanding recommendations about how to communicate more effectively with hospice and palliative care patients to address sexual issues. Hordern and Street (2007) offer some "opening lines" to help discuss the topic. Normalizing the topic is essential, as is placing control of the discussion with the patient or client.

- Many of the patients and clients I see express concerns about how treatment may affect their sex lives. How has this been for you?
- How has this experience affected intimate or sexual aspects of your life?
- Has your role as parent, partner, spouse, or intimate friend changed since you were diagnosed or treated?
- Is this the right time and place to discuss these issues further?
- Am I the right person for you to discuss these issues?
- In my experience, many people find that this disease or treatment has made a major impact upon their sexual activity or intimacy.
- How can I best provide you with information, support, or practical strategies to help you?

Taking a sexual history is essential (Stausmire 2004) regardless of a patient's marital status. Assumptions that the patient is heterosexual (or even male or female versus transgender) also should be avoided. For example, it is important to inquire about intimate partners rather than husbands or wives. Consideration of a couple's previous level of sexual and emotional functioning is essential (Cort et al. 2004), and will likely be obtained from the sexual history. Additional considerations for clinicians when speaking with patients about sexuality at the end of life include having a warm, empathic, and open attitude, providing information that dispels myths and misconceptions (e.g., many people who are terminally ill have sexual feelings and desires; radiation cannot be passed to a partner through sexual contact), speaking without technical jargon and terminology, using open-ended questions, and respecting the values of the patient (Cagle and Bolte 2009). Reinforcing the confidential nature of these discussions also remains essential (Cort et al. 2004).

It also is critically important to ask about permission to speak with significant others because many partners of those in hospice or palliative care are afraid to physically interact with their loved one out of fear of causing them discomfort, pain, or injury (Stausmire 2004). Giving patients and their partners time to come up with questions, or to be able to write them down and give them to the clinician can help provide an increased sense of confidence and control. It also is important to note that some patients may seek closeness with their partner, in a variety of forms, whereas the partner may wish to emotionally (and physically) distance themselves to avoid facing feelings of loss and abandonment (Redelman 2008). Working with partners individually and as a couple can provide a wealth of information that can ultimately benefit patients and their relationships.

Practical Suggestions

A number of practical recommendations also can be offered in work with patients or clients in palliative care who wish to engage in sexual activity, either alone or with a partner. In addition to the emotional challenges that present themselves in the face of illness at the end of life, clear medical, physical, and institutional challenges also present themselves. For example, indwelling catheters, IVs, and oxygen masks can pose clear obstacles for patients who wish to engage in sexual activity whether it involves intercourse to holding hands and cuddling. Psychologists, social workers, professional counselors, chaplains, and other mental health providers can work as part of an interdisciplinary team to help provide this assistance. For example (Cagle and Bolte 2009; Cort et al. 2004; Katz 2011; Lemieux et al. 2004; Stausmire 2004):

- Give patients and couples private time. Provide the use of "do not disturb" signs to patients, and enforce rules that staff and caregivers knock and receive permission before entering a patient's room.
- When possible, remove extra medical devices and equipment when couples wish to engage in sexual activity to make the setting feel less clinical. Encourage

patients and their partners to bring along things from home such as pillowcases, photos, and music.

- If a patient wants help with grooming, provide assistance or schedule visits with hairdressers and manicurists.
- Consider the importance of oral care, particularly if a patient expresses a desire to engage in kissing or other oral activities with their partner.
- If a patient experiences fatigue primarily in the afternoon or evening, consider suggesting that sexual activity take place in the morning to take advantage of increased energy.
- For patients who have surgical scars or who experienced a disfiguring surgical procedure, normalizing negative feelings is essential. For example, "After a mastectomy it is not unusual for women to report higher levels of dissatisfaction with their body image" (Cagle and Bolte 2009, p. 229). Some women may elect to wear a camisole or bra with an insert if that allows her to feel more attractive or to focus more upon her own sensual experience. Similarly, if a man is concerned about scars or any surgical change to his genitals, normalizing any negative feelings and offering suggestions for wearing underwear or shorts during sexual activity may provide similar relief.
- Discuss alternative positions for sexual activity, including those that place the patient on their side or back, supported by pillows.
- If a partner is afraid of causing pain by touching, consider bathing and other sensual activities involving massage, stroking, hugging, and applying body lotion.
- If possible, provide patients with a double bed so that their partner can lie next to them. Sometimes a lounge chair can be placed next to a hospital bed in the upright position so that partners can sit next to each other and touch and hold hands.
- Consider that some patients do not want to relate to their partners as caregivers per se, or that they may feel embarrassed or humiliated by engaging in certain activities in front of their partner. For example, some patients do not want their partners to be involved in certain activities of daily living and personal care including dressing or toileting.
- Consider timing the administration of pain medication to coincide with sexual activities.
- For patients who are short of breath, use of an inhaler or bronchodilator before sexual activity, if prescribed, can be helpful. Sitting upright or propped up with pillows can help make breathing easier.

Working with patients at the end of life is obviously challenging. However, clinicians can give patients and their partners the gift and freedom of communicating openly about sexual issues and concerns, rather than have them suffer in silence (Katz 2005). Because hospice and palliative care typically involves an interdisciplinary team, it is important that mental health providers such as psychologists, social workers, professional counselors, and chaplains take the lead in discussing sexual issues and concerns, as research suggests that most clinicians avoid the discussion or assume that other team members have already broached the subject. It also is important to remember that for those partners left behind upon the death of a hospice patient, the mourning process continues just as their sexuality and sexual needs continue. Sexuality remains a fundamental human need, even at the end of life, and psychologists and other mental health providers can play a lead role in ensuring its continuing value.

The Medicalization of Sexuality and Aging

Another challenge that can be expected within the context of sexuality and aging is the medicalization of sexuality. In other words, the medical model is likely to take center stage in the assessment and treatment of sexual dissatisfaction and dysfunction among aging adults. In our current culture of fast food, fast transit, and instant messaging, individuals often want a "quick fix," typically in pill form. They may lean exclusively on medical treatments to remedy sexual dissatisfaction that actually stems from underlying psychological problems such as marital discord, distortions in body image, or emotional difficulties in adjusting to disability. Simultaneously, however, medical knowledge and expertise play an essential role in helping middle-aged and older adults remedy many sexual problems. Mental health professionals need to become generally aware of these treatment options and make appropriate referrals. For example, all male clients concerned about ED should be referred to a urologist or geriatrician to rule out underlying medical problems. Accordingly, mental health providers will necessarily, if not preferably, find themselves enmeshed in this medical culture and will probably find themselves working within the context of interdisciplinary treatment teams.

As noted, the medicalization of sexuality offers significant challenges. Many clients and some medical professionals adopt a unidimensional approach to sexuality and aging, in which psychological issues and behavioral treatments (e.g., sensate focus; biofeedback) are either devalued or dismissed for the sake of a quick fix (e.g., a pill). Physicians typically command higher fees than psychologists and other mental health providers, and clients and physicians may see this difference as representative of the increased value or efficacy of medical treatment. Managed care companies, which tend to devalue psychological treatments, are more likely to pay for a visit to a geriatrician or urologist than to a psychologist. Clients who rely on insurance may be afraid to seek out treatment when they learn that their own insurance company does not value sexuality within the context of aging (e.g., certain managed care companies report that they will not pay for Viagra because "sex is not medically necessary" for older men).

Summary

With the aging of the baby boomers, American society will place increasing demands upon clinicians to be trained in both issues related to aging and sexuality when, unfortunately, the field of mental health is already behind in providing appropriate numbers of professionals trained in gerontology. Fortunately, with the American Psychological Association's newly approved recognition of geropsychology as a professional specialization, a variety of professional organizations are now available to clinicians seeking related training and education. The medicalization of sexuality and aging is likely to continue, and clinicians will likely find themselves working within interdisciplinary settings. Relatively unexplored areas such as sexuality among minority group and LGBT elders, HIV and aging, sexual consent capacity, and sexuality in institutional settings and at the end of life require increased clinical and research attention. Mental health professionals are in a unique position to shape the future of sexuality and aging, and the education of both clients and clinicians represents the first step in this meaningful and distinguished process.

References

AARP. (1999). *Sexuality survey*. AARP and Modern Maturity. Retrieved from http://assets.aarp. org/rgcenter/health/mmsexsurvey.pdf

AARP. (2003). *Lifestyles, dating, and romance: A study of midlife singles*. Washington, DC: Author. Retrieved from http://assets.aarp.org/rgcenter/general/singles.pdf

AARP. (2004). *The divorce experience: A study of divorce at midlife and beyond*. Washington, DC: Author. Retrieved from http://www.aarp.org/relationships/love-sex/info-2004/divorce.html

AARP. (2010). *Social media and technology use among adults 50+*. Washington, DC: Author. Retrieved from http://assets.aarp.org/rgcenter/general/socmedia.pdf

Abner, C. (2006). *Graying prisons: States face challenges of an aging inmate population* (pp. 8–11). Nov/Dec: State News.

Adams, M. S., Oye, J., & Parker, T. S. (2003). Sexuality of older adults and the internet: From sex education to cybersex. *Sexual and Relationship Therapy, 18*, 405–415.

Adams, M. A., Rojas-Cameroa, C., & Clayto, K. (1990). A small group sex education intervention model for the well elderly: A challenge for educators. *Educational Gerontology, 16*, 601–608.

Addis, S., Mavies, M., Greene, G., MacBride-Stewart, S., & Shepherd, M. (2009). The health, social care and housing needs of lesbian, gay, bisexual and transgender older people: A review of the literature. *Health and Social Care in the Community, 17*, 647–658.

Adelman, M., Gurvitch, J., de Vries, B., & Blando, J. (2006). Openhouse. In D. Kimmel, R. Rose, & S. David (Eds.), *Lesbian, gay, bisexual, and transgender aging* (pp. 247–264). New York: Columbia University Press.

Administration on Aging. (2001). *Older adults and mental health: Issues and opportunities*. Washington, DC: Department of Health and Human Services.

Administration on Aging. (2011). Aging statistics. Retrieved from http://www.aoa.gov/aoaroot/ aging_statistics/index.aspx

Aizenberg, D., Weizman, A., & Barak, Y. (2002). Attitudes toward sexuality among nursing home residents. *Sexuality and Disability, 20*, 185–189.

Akkus, Y., Nakas, D., & Kalyoncu, U. (2010). Factors affecting the sexual satisfaction of patients with rheumatoid arthritis and ankylosing spondylitis. *Sexuality and Disability, 28*, 223–232.

Allen, R. S., Petro, K. N., & Phillips, L. L. (2009). Factors influencing young adults' attitudes and knowledge of late-life sexuality among older women. *Aging and Mental Health, 13*, 238–245.

ALS Association. (2011). *Caregiving*. Retrieved from http://www.alsa.org/assets/pdfs/brochures/ caregiving.pdf

Alterovitz, S. S., & Mendelsohn, G. A. (2011). Partner preferences across the life span: online dating by older adults. *Psychology of Popular Media Culture, 1*(S), 89–95.

Altschuler, J., & Katz, A. D. (1999). Methodology for discovering and teaching countertransference toward elderly clients. *Journal of Gerontological Social Work, 32*, 81–93.

Altschuler, J., & Katz, A. D. (2002). Clinical supervisors' countertransference reactions toward older clients: Addressing the unconscious guide. *Journal of Gerontological Social Work, 39*, 75–87.

Altschuler, J., & Katz, A. D. (2010). Keeping your eye on the process: Body image, older women, and countertransference. *Journal of Gerontological Social Work, 53*, 200–214.

American Academy of Neurology AIDS Task Force. (1991). Nomenclature and research case definitions for neurologic manifestations of human immunodeficiency virus-type 1 (HIV-1) infection. *Neurology, 41*, 778–785.

American Bar Association, & American Psychological Association. (2008). *Assessment of older adults with diminished capacity: A handbook for psychologists.* Washington, DC: American Bar Association Commission on Aging & The American Psychological Association.

American Cancer Society (2011). Cancer facts and figures for Hispanics/Latinos 2009–2011. Retrieved from http://www.cancer.org/acs/groups/content/@nho.

American Cancer Society. (2007). *Prostate cancer overview.* Retrieved from http://www.cancer.org/Cancer/ProstateCancer/OverviewGuide/prostate-cancer-overview-diagnosedhttp

American College of Obstetricians and Gynecologists. (2009). ACOG practice bulletin: Cervical cytology screening. *Obstetrics and Gynecology, 114*, 1409–1420.

American College of Rheumatology. (2010). *Yes, you can have an active sex life with arthritis.* Retrieved from http://www.health.com/health/condition-article/print.

American Psychiatric Association. (2000). *Diagnostic and statistical manual of mental disorders* (4th ed., text rev.). Washington, DC: Author.

American Psychological Association. (2011a). *Practice guidelines for LGB clients.* Retrieved from http://www.apa.org/pi/lgbt/resources/guidelines.aspx

American Psychological Association (2011b, October 19). *Broad professional issues of concern to Psychologists working with older adults.* http://www.apa.org/pi/aging/resources/guides/practitioners-should-know.aspx. Accessed 19 Oct 2011.

Ames, T., & Samowitz, P. (1995). Inclusionary standard for determining sexual consent for individuals with developmental disabilities. *Mental Retardation, 4*, 254–268.

Ananth, H., Jones, L., King, M., & Tookman, A. (2003). The impact of cancer on sexual function: A controlled study. *Palliative Medicine, 17*, 202–205.

Andelloux, M. (2010). *Safer sex.* Retrieved from www.womensweb.ca/health/repro/safesex/lube.php.

Anderson, L. (2007, November 18). Alzheimer's: intimacy found after all is lost. *Chicago Tribune.* Retrieved from http://articles.chicagotribune.com/2007-11-18/news

Anderson, L. (2008, October 21). Aging even tougher for gays and lesbians. *Chicago Tribune.* Accessed 21 Oct 2008. http://www.chicagotribune.com/news/nationworld/chi-gay-elderly,0,2552870.story.

Annon, J. S. (1976). The PLISSIT model: A proposed conceptual scheme for the behavioral treatment of sexual problems. *Journal of Sex Education and Therapy, 2*(2), 1–15.

Apfel, R. J., Fox, M., Isberg, R. S., & Levine, A. R. (1984). Countertransference and transference in couple therapy: Treating sexual dysfunction in older couples. *Journal of Geriatric Psychiatry, 17*, 203–214.

Arias, M. L. (2010, March 29). No downward sentence departure for elderly who inappropriately use the internet. *Internet Business Law Services.* Retrieved from https://ibls.com/internet_law_news_portal_view.aspx?s=latestnews&id=2286

Armstrong, L. L. (2006). Barriers to intimate sexuality: Concerns and meaning-based therapy approaches. *Humanistic Psychologist, 23*, 281–298.

Asencio, M., Blank, T., Descartes, L., & Crawford, A. (2009). The prospect of prostate cancer: A challenge for gay men's sexualities as they age. *Sexuality Research and Social Policy, 6*(4), 38–51.

Aubin, S., Heiman, J. R., Berger, R. E., Murallo, A. V., & Yung-Wen, L. (2009). Comparing sildenafil alone vs. sildenafil plus brief couple sex therapy on erectile dysfunction and couples' sexual and marital quality of life: A pilot study. *Journal of Sex and Martial Therapy, 35*, 122–143.

Axon, S. (2010, January 28). eHarmony settles lawsuit, will merge gay and straight websites. *Mashable Business*. Retrieved from http://mashable.com/2010/01/28/eharmony-lawsuit/

Bachmann, G. A. (1995). Influence of menopause on sexuality. *International Journal of Fertility, 40*, 16–22.

Badeau, D. (1995). Illness, disability, and sex in aging. *Sexuality and Disability, 13*, 219–237.

Baker, L., & Gringart, E. (2009). Body image and self-esteem in older adulthood. *Ageing and Society, 29*, 977–995.

Baker, D. W., Guzmararian, J. A., Sudano, J., & Patterson, M. (2000). The association between age and health literacy among elderly persons. *The Journals of Gerontology: Series B, Psychological Sciences and Social Sciences, 55*, S368–S374.

Ballard, E. L. (1995). Attitudes, myths, and realities: Helping family and professional caregivers cope with sexuality in the Alzheimer's patient. *Sexuality and Disability, 13*, 255–270.

Balon, R. (2006). SSRI-associated sexual dysfunction. *The American Journal of Psychiatry, 163*, 1504–1509.

Balsam, K., & D'Augelli, A. (2006). The victimization of older LGBT adults. In D. Kimmel, T. Rose, & S. David (Eds.), *Lesbian, gay, bisexual, and transgender aging* (pp. 110–130). New York: Columbia University Press.

Bancroft, J. H. J. (2007). Sex and aging. *The New England Journal of Medicine, 357*, 820–822.

Barbach, L. (1996). Sexuality through menopause and beyond. *Menopause Management, 5*, 18–21.

Barclay, T. R., Hinkin, C. H., Castellon, S. A., Mason, K. I., Reinhard, M. J., Marion, S. D., et al. (2007). Age-associated predictors of medication adherence in HIV-positive adults: Health beliefs, self-efficacy, and neurocognitive status. *Health Psychology, 26*, 40–49.

Bargh, J. A., Chen, M., & Burrows, L. (1996). Automaticity of social behavior: Direct effects of trait construct and stereotype activation on action. *Journal of Personality and Social Psychology, 71*, 230–244.

Barksdale, J. (1999). American urological association guideline on the management of erectile dysfunction: diagnosis and treatment recommendations. *Journal of Pharmacotherapy, 19*, 573–581.

Barranti, C., & Cohen, H. (2000). Lesbian and gay elders: An invisible minority. In R. Schneider, N. Kropt, & A. Kisor (Eds.), *Gerontological social work: Knowledge, service settings and special populations* (2nd ed., pp. 343–367). Belmont, CA: Wadsworth/Thompson Learning.

Barrett, L. L. (2005). *Prescription drug use among midlife and older Americans*. Washington, DC: AARP.

Barry, M. J. (2008). Screening for prostate cancer among men 75 years of age or older. *The New England Journal of Medicine, 359*, 2515–2516.

Barry, A., & Lippmann, S. B. (1990). Anorexia nervosa in males. *Postgraduate Medicine, 87*, 161–165.

Bauer, M. (1999). The use of humor in addressing the sexuality of elderly nursing home residents. *Sexuality and Disability, 17*, 147–155.

Beaudreau, S. A., Rideaux, T., & Zeiss, R. A. (2011). Clinical characteristics of older male military veterans seeking treatment for erectile dysfunction. *International Psychogeriatrics, 23*, 155–160.

Beaulaurier, R. L., Craig, S. L., & Mario, D. L. R. (2009). Older Latina women and HIV/AIDS: An examination of sexuality and culture as they relate to risk and protective factors. *Journal of Gerontological Social Work, 52*, 48–63.

Beck, A. T. (1976). *Cognitive therapy and the emotional disorders*. Oxford, England: International Universities Press.

Begg, C. B., Riedel, E. R., Bach, P. B., Kattan, M. W., Schrag, D., Warren, J. L., et al. (2002). Variations in morbidity after radial prostatectomy. *The New England Journal of Medicine, 346*, 1138–1144.

Bell, J. (1992). In search of a discourse on aging: The elderly on television. *The Gerontologist, 32*, 305–311.

Berg, C. A., Wiebe, D. J., Butner, J., Bloor, L., Bradstreet, C., Upchurch, R., et al. (2008). Collaborative coping and daily mood in couples dealing with prostate cancer. *Psychology and Aging, 23*, 505–516.

Berger, R. M. (1980). Psychological adaptation of the older homosexual male. *Journal of Homosexuality, 5*, 161–175.

Bergeron, S., Morin, M., & Lord, M. (2010). Integrating pelvic floor rehabilitation and cognitive-behavioural therapy for sexual pain: What have we learned and where do we go from here? *Sexual and Relationship Therapy, 25*, 289–298.

Berland, G. K., Elliott, M. N., Morales, L. S., Algazy, J. I., Kravitz, R. L., & Broder, M. S. (2001). Health information on the Internet: Accessibility, quality, and readability in English and Spanish. *Journal of the American Medical Association, 285*, 2612–2621.

Berner, M. M., Kriston, L., & Harms, A. (2006). Efficacy of PDE-5 inhibitors for erectile dysfunction. A comparative meta-analysis of fixed-dose regimen randomized controlled trials administering the International Index of Erectile function in broad-spectrum populations. *International Journal of Impotence Research, 18*, 229–236.

Bhasin, S., Enzlin, P., Coviello, A., & Basson, R. (2007). Sexual dysfunction in men and women with endocrine disorders. *The Lancet, 369*, 597–611.

Bildtgard, T. (2000). The sexuality of elderly people on film—visual limitations. *Journal of Aging and Identity, 5*, 169–183.

Biskupic, J. (2007, November 13). A new page in O'Connors' love story. *USA Today*. Retrieved from http://www.usatoday.com/news/nation/2007-11-12-court_N.htm

Black, B., Muralee, S., & Tampi, R. R. (2005). Inappropriate sexual behaviours in dementia. *Journal of Geriatric Psychiatry and Neurology, 18*(3), 155–162.

Blackmore, D. E., Hart, S. L., Albiani, J. J., & Mohr, D. C. (2011). Improvements in partner support predict sexual satisfaction among individuals with multiple sclerosis. *Rehabilitation Psychology, 56*, 117–122.

Blinik, Y. M. (2010). The DSM diagnostic criteria for vaginismus. *Archives of Sexual Behavior, 39*, 278–291.

Blum, H. P. (1973). The concept of erotized transference. *Journal of the American Psychoanalytic Association, 21*, 61–76.

Bodenheimer, C., Kerrigan, A. J., Garber, S. L., & Monga, T. N. (2000). Sexuality in persons with lower extremity amputations. *Disability and Rehabilitation, 22*, 409–415.

Bogner, H. R., Gallo, J. J., Sammel, M. D., Ford, D. E., Armenian, H. K., & Eaton, W. W. (2002). Urinary incontinence and psychological distress in community-dwelling older adults. *Journal of the American Geriatrics Society, 50*, 489–495.

Bonvicini, K., & Perlin, M. (2003). The same but different: Clinician-patient communication with gay and lesbian patients. *Patient Education and Counseling, 51*, 115–122.

Bostwick, J. M., Hecksel, K. A., Stevens, S. R., Bower, J. H., & Ahlskog, J. E. (2009). Frequency of new-onset pathologic compulsive gambling or hypersexuality after drug treatment of idiopathic Parkinson Disease. *Mayo Clinic Proceedings, 84*(4), 310–316.

Botwinick, J. (1984). *Aging and behavior* (3rd ed.). Berlin: Springer.

Bouman, W. P., & Arcelus, J. (2001). Are psychiatrists guilty of 'ageism' when it comes to taking a sexual history? *International Journal of Geriatric Psychiatry, 16*, 27–31.

Bouman, W. P., Arcelus, J., & Benbow, S. M. (2006). Nottingham study of sexuality and ageing (NoSSA II). Attitudes of care staff regarding sexuality and residents: A study in residential and nursing homes. *Sexual and Relationship Therapy, 22*, 45–61.

Boyle, P. (1994). New insights into the epidemiology and natural history of benign prostatic hyperplasia. *Progress in Clinical and Biological Research, 386*, 3–18.

Bradburn, N., & Sudman, S. (1979). *Improving interview method and questionnaire design*. San Francisco: Jossey-Bass.

Branaghan, R. J., & Gray, R. (2010). Nonconscious activation of an elderly stereotype and speed of driving. *Perceptual and Motor Skills, 110*, 580–592.

Brockett, D. R., & Gleckman, A. D. (1991). Countertransference with the older adult: The importance of mental health counselor awareness and strategies for effectiveness management. *Journal of Mental Health Counseling, 13*, 343–355.

Bronner, G., Royter, V., & Korczyn, A. D. (2004). Sexual dysfunction in Parkinson's disease. *Journal of Sexual and Marital Therapy, 30*, 95–105.

Bronner, G., Shefi, S., & Raviv, G. (2010). Sexual dysfunction after radical prostatectomy: Treatment failure or treatment delay? *Journal of Sex and Martial Therapy, 36*, 421–429.

Brooks, G. R., & Levant, R. F. (2006). Is Viagra enough? Broadening the conceptual lens in sexul therapy with (heterosexual) men: A case report. *International Journal of Men's Health, 5*, 207–216.

Brotman, S., Ryan, R., & Cormier, R. (2003). The health and social service needs of gay and lesbian elders and their families in Canada. *The Gerontologist, 43*, 192–202.

Brotto, L. A., Heiman, J. R., Goff, B., Greer, B., Lentz, G. M., Swisher, E., et al. (2008). A psychoeducational intervention for sexual dysfunction in women with gynecologic cancer. *Archives of Sexual Behavior, 37*, 317–329.

Brown, T. A., Cash, T. F., & Mikulka, P. J. (1990). Attitudinal body-image assessment: Factor analysis of the body-self relations questionnaire. *Journal of Personality Assessment, 55*, 134–144.

Buckingham, S. L., & Van Gorp, W. G. (1988). Essential knowledge about AIDS dementia. *Social Work, 33*, 112–115.

Bulcroft, R. A., & Bulcroft, K. A. (1991). The nature and functions of dating in later life. *Research on Aging, 13*, 244–260.

Bullough, V. (1976). *Sexual variance in society and history*. New York: Wiley.

Bunker, C. H., Patrick, A. L., Konety, B. R., Dhir, R., Brufsky, A. M., Vivas, C. A., et al. (2002). High prevalence of screening-detected prostate cancer among Afro-Caribbeans: The Tobago Prostate Cancer Survey. *Cancer Epidemiology, Biomarkers and Prevention, 11*, 726–729.

Burckell, L. A., & Goldfried, M. R. (2006). Therapist qualities preferred by sexual-minority individuals. *Psychotherapy: Theory, Research, Practice, and Training, 43*, 32–49.

Bureau of Labor Statistics. (2011). *Consumer expenditure survey anthology, 2011*. Retrieved from http://www.bls.gov/cex/csxanthol11.htm

Burgess, A. W., Prentky, R. A., & Dowdell, E. B. (2000). Sexual predators in nursing homes. *Journal of Psychosocial Nursing and Mental Health Services, 38*, 26–35.

Burrow, J. A. (1986). *The ages of man: A study in medieval writing and thought*. Oxford: Clarendon Press.

Butler, R. N. (1969). Age-ism: Another form of bigotry. *The Gerontologist, 9*, 243–246.

Butler, R. N., & Lewis, M. I. (1976). *Sex after sixty*. New York: Harper & Row.

Butler, R. N., Lewis, M. I., Hoffman, E., & Whitehead, E. D. (1994). Love and sex after 60: How physical changes affect intimate expression. *Geriatrics, 49*, 20–27.

Cagle, J. G., & Bolte, S. (2009). Sexuality and life-threatening illness: Implications for social work and palliative care. *Health and Social Work, 34*, 223–233.

Cain, V. S., Johannes, C. B., Avis, N. E., Mohr, B., Shockien, M., Skurnick, J., et al. (2003). Sexual functioning and practices in a multi-ethnic study of midlife women: Baseline results from SWAN. *Journal of Sex Research, 40*, 266–277.

Call, V., Sprecher, S., & Schwartz, P. (1995). The incidence and frequency of marital sex in a national sample. *Journal of Marriage and the Family, 57*, 639–653.

Calvert, H. M. (2003). Sexually transmitted diseases other than human immunodeficiency virus infection in older adults. *Aging and Infectious Diseases, 36*, 609–614.

Campbell, J. M., & Huff, M. S. (1995). Sexuality in the older woman. *Gerontology and Geriatrics Education, 16*, 71–81.

Carlson, H. E. (1980). Gynecomastia. *The New England Journal of Medicine, 303*, 795–799.

Caro, J. L. L., Lozano Vidal, J. V., Vicente, J. A., Roca, M. A., Bravo, C., Sanchez Zamorano, M. A., et al. (2001). Sexual dysfunction in hypertensive patients treated with Losartan. *The American Journal of the Medical Sciences, 321*, 336–341.

Carpenter, L. M., Nathanson, C. A., & Young, J. K. (2006). Sex after 40: Gender, ageism, and sexual partnering in midlife. *Journal of Aging Studies, 20*, 93–106.

Carr, D. (2004). The desire to date and remarry among older widows and widowers. *Journal of Marriage and the Family, 66*, 1051–1068.

Carson, C. C. (2006). PDE5 inhibitors: Are there differences? *The Canadian Journal of Urology, 13*, S34–S39.

Catania, J. A., Turner, H., Kegeles, S. M., Stall, R., Pollack, L., & Coates, T. J. (1989). Older Americans and AIDS: Transmission risks and primary prevention research needs. *The Gerontologist, 29*, 373–381.

Centers for Disease Control and prevention. (2000). *Behavioral risk factor surveillance system survey data*. Atlanta, GA: Author.

Centers for Disease Control and Prevention. (2001). *Sexually transmitted disease surveillance 2000*. Atlanta, GA: Author.

Centers for Disease Control and Prevention. (2006). Revised recommendations for HIV testing of adults, adolescents, and pregnant women in health care settings. *Morbidity and Mortality Weekly Report, 55*(RR-14), 1–17.

Centers for Disease Control and Prevention. (2007). HIV/AIDS surveillance report, 2005. *Department of Health and Human Services, 17*, 1–54.

Centers for Disease Control and Prevention. (2008a). *Diagnoses of HIV infection and AIDS in the United States and dependent areas*. Atlanta, GA: Author.

Centers for Disease Control and Prevention. (2008b). *Persons age 50 and over: Centers for disease control and prevention*. Atlanta, GA: Author.

Centers for Disease Control and Prevention. (2008c). QuickStats percentage of adults aged >18 years who had ever been tested for human immunodeficiency virus (HIV) by age group and sex: National Health Interview Survey, United States, 2007. *Morbidity and Mortality Weekly Report, 58*(3), 62.

Centers for Disease Control and Prevention. (2010). *Breast cancer statistics*. Retrieved from http://www.cdc.gov/cancer/breast/statistics.

Centers for Disease Control and Prevention. (2011). *Healthy aging: at a glance 2011*. Retrieved http://www.cdc.gov/chronicdisease/resources/publications/aag/aging.htm.

Chandra, V. (1996). Cross-cultural perspectives: India. *International Psychogeriatrics, 8*(S3), 479–481.

Charney, D. S., Reynolds, C. F., III, Lewis, L., Lebowitz, B. D., Sunderland, T., Alexopoulos, G. E., et al. (2003). Depression and bipolar support alliance consensus statement on the unmet needs in diagnosis and treatment of modd disorders in late life. *Archives of General Psychiatry, 60*, 664–672.

Chen, C., Booth-LaForce, C., Park, H., & Wang, S. (2010). A comparative study of menopausal hot flashes and their psychosocial correlates in Taiwan and the United States. *Maturitas, 67*, 171–177.

Chiao, E., Ries, K., & Sande, M. (1999). AIDS and the elderly. *Clinical Infectious Diseases, 28*, 740–745.

Chrisler, J. C., & Ghiz, L. (1993). Body image issues of older women. *Women and Therapy, 14*, 67–75.

Chun, J., & Carson, C. C. (2001). Physician-patient dialogue and clinical evaluation of erectile dysfunction. *The Urological Clinics of North America, 28*, 249–258.

Clarke, J. (2009). Women's work, worry, and fear: The portrayal of sexuality and sexual health in US magazines for teenagers and middle-aged women, 2000–2007. *Culture, Health, & Sexuality, 11*, 415–429.

Clayton, A. H., & Montejo, A. L. (2006). Major depressive disorder, antidepressants, and sexual dysfunction. *The Journal of Clinical Psychiatry, 67*(S6), 33–37.

Cogen, R., & Steinman, W. (1990). Sexual function and practice in elderly men of lower socioeconomic status. *The Journal of Family Practice, 31*, 162–166.

Cohen, B. L., Barboglio, P., & Gousse, A. (2008). The impact of lower urinary tract symptoms and urinary incontinence on female sexual dysfunction using a validated instrument. *The Journal of Sexual Medicine, 5*, 1418–1423.

Coker, L. H., Espeland, M. A., Rapp, S. R., Legault, C., Resnick, S. M., Hogan, P., et al. (2010). Postmenopausal hormone therapy and cognitive outcomes: The Women's Health Initiative Memory Study (WHIMS). *Journal of Steroid and Biochemical Molecular Biology, 118*, 304–310.

Conaglen, H. M., & Conaglen, J. V. (2009). The impact of erectile dysfunction on female partners: A qualitative investigation. *Sexual and Relationship Therapy, 23*, 147–156.

Cook-Daniels, L. (2006). Trans aging. In D. Kimmel, T. Rose, & S. David (Eds.), *Lesbian, gay, bisexual, and transgender aging—Research and clinical perspectives* (pp. 20–35). New York: Columbia University Press.

Cook-Daniels, L., & Munson, M. (2010). Sexual violence, elder abuse, and sexuality of transgender adults, age 50+: Results of three surveys. *Journal of GLBT Family Studies, 6*, 142–177.

Cooney, T. M., & Dunne, K. (2001). Intimate relationships in later life: Current realities, future prospects. *Journal of Family Issues, 22*, 838–858.

Cooper, C. A., Jadidian, A., Paggi, M., Romrell, J., Okun, M. S., Rodriguez, R. L., et al. (2009). Prevalence of hypersexual behavior in Parkinson's disease patients: Not restricted to males and dopamine agonist use. *International Journal of Geriatric Medicine, 2*, 57–61.

Cooperman, N. A., Arnsten, J. H., & Klein, R. S. (2007). Current sexual activity and risk sexual behavior in older men with or at risk for HIV infection. *AIDS Education and Prevention, 19*(4), 321–333.

Cornelison, T. L., Montz, F. J., Bristow, R. E., Chou, B., Bovicelli, A., & Zeger, S. L. (2002). Decreased incidence of cervical cancer in Midicare-eligible women. *Obstetrics and Gynecology, 100*, 79–86.

Corona, G., Lee, D. M., Forti, G., O'Connor, D. B., Maggi, M., O'Neil, T. W., et al. (2010). Age-related changes in general and sexual health in middle-aged and older men: Results from the European Male Ageing Study (EMAS). *The Journal of Sexual Medicine, 7*, 1362–1380.

Cort, E., Monroe, B., & Oliviere, D. (2004). Couples in palliative care. *Sexual and Relationship Therapy, 19*, 337–354.

Council on Social Work Education. (2011). *Rethinking aging stereotypes: "But I Don't Want to Work With Older Adults!"* Accessed 19 Oct 2011. http://www.cswe.org/CentersInitiatives/GeroEdCenter/Students/LearnMore/RethinkingAgingStereotypes.aspx

Covey, H. C. (1989). Perceptions and attitudes toward sexuality of the elderly during the Middle Ages. *The Gerontologist, 29*, 93–100.

Cranston-Cuebas, M. A., & Barlow, D. H. (1990). Cognitive and affective contributions to sexual functioning. *Annual Review of Sex Research, 1*, 119–161.

Croft, L. (1982). *Sexuality in later life: A counseling guide for physicians.* Boston: PSG Inc.

Csberg, C., & Johnson, E. (2009). Viagra selfhood: Pharmaceutical advertising and the visual formation of Swedish masculinity. *Health Care Analysis, 17*, 144–157.

Cummings, J. L. (1991). Behavioral complications of drug treatment of Parkinson's disease. *Journal of the American Geriatrics Society, 39*, 708–716.

Dail, P. W. (1988). Prime-time television portrayals of older adults in the context of family life. *The Gerontologist, 28*, 700–706.

Dasgupta, N., & Greenwald, A. G. (2001). On the malleability of automatic attitudes: Combating automatic prejudice with images of admired and disliked individuals. *Journal of Personality and Social Psychology, 81*, 800–814.

David, S., & Cernin, P. A. (2008). Psychotherapy with lesbian, gay, bisexual, and transgender older adults. *Journal of Gay and Lesbian Social Services, 20*, 31–49.

Davies, J. M., & Frawley, M. G. (1994). *Treating the adult survivor of childhood sexual abuse: A psychoanalytic perspective.* New York: Basic Books.

Deevey, S. (1990). Older lesbian women: An invisible minority. *Journal of Gerontological Nursing, 16*, 35–39.

Den Oudsten, B. L., Van Heck, G. L., Van der Steg, A. F. W., Roukema, J. A., & De Vries, J. (2010). Clinical factors are not the best predictors of quality of sexual life and sexual functioning in women with early stage breast cancer. *Psycho-Oncology, 19*, 646–656.

Department of Health and Human Services. (2010). *Medicare preventive services.* Retrieved http://www.medicare.gov/navigation/manage-your-health/preventive-services/preventive-service-overview.aspx.

Diegel, K., Schrimshaw, E. W., Brown-Bradley, C. J., & Lekas, H. (2020). Sources of emotional distress associated with diarrhea among late middle-age and older HIV-infected adults. *Journal of Pain and Symptom Management, 40*, 353–369.

Dijksterhius, A., Spears, R., & Lepinasse, V. (2001). Reflecting and deflecting stereotypes: Assimilation and contrast in impression formation and automatic behavior. *Journal of Experimental Social Psychology, 37*, 286–299.

Donaldson, R. L., & Meana, M. (2011). Early dyspareunia experience in young women: Confusion, consequences, and help seeking barriers. *The Journal of Sexual Medicine, 8*, 814–823.

Dorfman, R., Walters, K., Burke, P., Hardin, L., Karanek, R., Raphael, J., & Silverstein, E. (1995). Old, sad, and alone: The myth of the aging homosexual. *Journal of Gerontological Social Work, 24*, 29–44.

Drabble, L., Keatley, J., & Marcelle, G. (2003). Progress and opportunities in lesbian, gay, bisexual, and transgender health communications. *Clinical Research and Regulatory Affairs, 20*, 205–227.

Duffy, L. M. (1995). Sexual behavior and marital intimacy in Alzheimer's couples: A family theory perspective. *Sexuality and Disability, 13*, 239–254.

Dworkin, S. (2006). The aging bisexual: The invisible of the invisible minority. In D. Kimmel, T. Rose, & S. David (Eds.), *Lesbian, gay, bisexual, and transgender aging* (pp. 36–52). New York: Columbia University Press.

Eckert, J. K., Carder, P. C., Morgan, L. A., Frankowski, A. C., & Roth, E. G. (2009). *Inside assisted living: The search for home*. Baltimore, MD: Johns Hopkins University Press.

Ehrenfeld, M., Bronner, G., Tabak, N., Alpert, R., & Bergman, R. (1999). Sexuality among institutionalized elderly patients with dementia. *Nursing Ethics, 6*, 144–149.

Emlet, C. (2006a, August 18). Graying of the HIV/AIDs epidemic. *Seattle post-intelligencer*. Retrieved from http://seattlepi.nwsource.com/opinion/28166_aids18.html

Emlet, C. (2006b). An examination of the social networks and social isolation in older and younger adults living with HIV/AIDS. *Journal of Health and Social Work, 31*, 299–308.

Emlet, C. (2006c). "You're awfully old to have this disease": Experiences of stigma and ageism in adults 50 years and older living with HIV/AIDS. *The Gerontologist, 46*, 781–790.

Engber, D. (2007, September 27). Is it time to let the elderly have more sex? *Slate on-line magazine, 9/27*. Retrieved from www.slate.com/articles/life/the_sex_issue/2007/09/naughty_nursing_homes.html.

Eyada, M., & Atwa, M. (2007). Sexual function in female patients with unstable angina or non-ST-elevation myocardial infarction. *The Journal of Sexual Medicine, 4*, 1373–1380.

Feil, N. (1995). When feelings become incontinent: Sexual behaviors in the resolution stage of life. *Sexuality and Disability, 13*, 271–282.

Feiring, C., Simon, V. A., & Cleland, C. (2009). Childhood sexual abuse, stigmatization, internalizing symptoms, and the development of sexual difficulties and dating aggression. *Journal of Consulting and Clinical Psychology, 77*, 127–137.

Feldman, H. A., Goldstein, I., Hatzichristou, D. G., Krane, R. J., & McKinlay, J. B. (1994). Impotence and its medical and psychosocial correlates: Results of the Massachusetts Male Aging Study. *The Journal of Urology, 151*, 54–61.

Ferri, C. P., Prince, M., Brayne, C., Brodaty, H., Fratiglioni, L., Ganguli, M., et al. (2005). Global prevalence of dementia: a Delphi consensus study. *The Lancet, 366*, 2112–2117.

Finger, B. (2006). *Sexuality and chronic illness*. San Francisco, CA: Annual Conference of the American Academy of Physician Assistants.

Fisher, L. L. (2010). *Sex, romance, and relationships: AARP survey of midlife and older adults*. Washington, DC: AARP.

Fisher, D. A., Hill, D. L., Grube, J. W., & Gruber, E. L. (2007). Gay, lesbian, and bisexual content on television: A quantitative analysis across two seasons. *Journal of Homosexuality, 52*, 167–188.

Fleming, D. T., McQuillian, G. M., Johnson, R. E., Nahmias, A. J., Sevgi, O. A., Lee, F. K., et al. (1997). Herpes simplex virus type 2 in the United States, 1976-1994. *The New England Journal of Medicine, 337*, 1105–1111.

Florida Department of Health. (2007). *Florida HIV/AIDS Annual Report/Epidemiologic Profile*. Bureau of HIV/AIDS. Retrieved from http://www.doh.state.fl.us/disease_ctrl/aids/trends/epiprof/aids07.pdf

Folstein, M., Folstein, S., & McHugh, P. (1975). Mini-Mental State: A practical method for grading the cognitive status of patients for the clinician. *Journal of Psychiatric Research, 12*, 189–198.

Foster, M. (2011). *National HIV/AIDS and Aging Awareness Day, Sept. 18*. Accessed 16 Sep 2011. http://www.oaadvisors.com/2011/09/06/national-hivaids-aging-awareness-day-sept-18/.

Fox, S., & Rainie, L. (2002). *Vital decisions: How Internet users decide what information to trust when they or their loved ones are sick*. Washington, DC: Pew Internet & American Life Project. Retrieved from www.pewinternet.org/Reports/2002.

Frankowski, A. C., & Clark, L. J. (2009). Sexuality and intimacy in assisted living: Residents' perspectives and experiences. *Sexuality Research & Social Policy, 6*(4), 25–37.

Franzoi, S., & Shields, S. (1984). The body esteem scale: Multidimensional structure and sex differences in a college population. *Journal of Personality Assessment, 48*, 173–178.

Fredriksen-Goldsen, K. I., & Muraco, A. (2010). Aging and sexual orientation: A 25-year review of the literature. *Research on Aging, 32*, 372–413.

Freedman, M. R., Rosenberg, S. J., & Schmaling, K. B. (1991). Childhood sexual abuse in patients with paradoxical vocal cord dysfunction. *The Journal of Nervous and Mental Disease, 179*, 295–298.

Freud, S. (1946). *Totem and taboo: Resemblances between the psychic lives of savages and neurotics (A. A. Brill, Trans.)*. New York: Vintage (Original work published 1913).

Fry, R. P., Beard, R. W., Crisp, A. H., & McGuigan, S. (1997). Sociopsychological factors in women with chronic pelvic pain with and without pelvic venous congestion. *Journal of Psychosomatic Research, 42*, 71–85.

Gadalla, T. M. (2008). Eating disorders and associated psychiatric comorbidity in elderly Canadian women. *Archives of Women's Mental Health, 11*, 357–362.

Galindo, D., & Kaiser, F. E. (1995). Sexual health after 60. *Patient Care, 29*, 25–35.

Gamel, C., Davis, B., & Hengeveld, M. (1993). Nurses' provision of teaching and counseling on sexuality: A review of the literature. *Journal of Advanced Nursing, 18*, 1219–1227.

Garnets, L., Hancock, K. A., Cochran, S. D., Goodchilds, J., & Peplau, L. A. (1991). Issues in psychotherapy with lesbians and gay men: A survey of psychologists. *The American Psychologist, 46*, 964–972.

Gearon, C. J. (2008, February). HIV/AIDS prevention pushed for the 50+. *AARP*. Retrieved from http://www.aarp.org/relationships/love-sex/info-02-008/aids_prevention_for_50plus_pushed.html.

Geertzen, J. H. B., Van Es, C. G., & Dijkstra, P. U. (2009). Sexuality and amputation: A systematic literature review. *Disability and Rehabilitation, 31*, 522–527.

Gelo, F. (2008). *Invisible individuals: LGBT elders* (pp. 36–40). Summer: Aging Well.

George, L. K., & Weiler, S. J. (1981). Sexuality in middle and late life. *Archives of General Psychiatry, 38*, 919–923.

Gibbs, L. (2005). Applications of masculinity theories in a chronic illness context. *International Journal of Men's Health, 4*, 287–300.

Gibson, V. (2001). *Cougar: A guide for older women dating younger men*. Buffalo, NY: Firefly Books.

Ginsberg, T. B., Pomerantz, S. C., & Kramer-Feeley, V. (2005). Sexuality in older adults: Behaviours and preferences. *Age and Ageing, 34*, 475–480.

Giraldi, A., & Kristensen, E. (2010). Sexual dysfunction in women with diabetes mellitus. *Journal of Sex Research, 47*, 199–211.

Gist, Y. J., & Hetzel, L. I. (2004). *We the people: aging in the United States. Census 2000 special reports*. Washington, DC: U.S. Census Bureau.

Gladwell, M. (2005). *Blink*. New York: Little, Brown.

Glass, J. C., Mustian, R. D., & Carter, L. R. (1986). Knowledge and attitudes of health care providers toward sexuality in the institutionalized elderly. *Educational Gerontology, 12*, 465–475.

Glass, J. C., & Webb, M. L. (1995). Health care educators' knowledge and attitudes regarding sexuality in the aged. *Educational Gerontology, 21*, 713–733.

Glazer, H. I., & Laine, C. D. (2006). Pelvic floor muscle biofeedback in the treatment of urinary incontinence: A literature review. *Applied Psychophysiology and Biofeedback, 31*, 187–201.

Golan, O., & Chong, B. (1992). Sexuality and ageing: Some physical aspects. *Geriatrician, 11*, 10–11.

Goldberg, S., Sickler, J., & Dibble, S. L. (2005). Lesbians over sixty: The consistency of findings from twenty years of survey data. *Journal of Lesbian Studies, 9*, 195–213.

Goldstein, I., Lue, T. F., Padma-Nathan, H., Rosen, R. C., Steers, W. D., & Wicker, P. A. (1998). Oral sildenafil in the treatment of erectile dysfunction. *The New England Journal of Medicine, 338*, 1397–1404.

Golub, S. A., Botsko, M., Gamarel, K. E., Parsons, J. T., Brennan, M., & Karpiak, S. E. (2011). Dimensions of psychological well-being predict consistent condom use among older adults living with HIV. *Ageing International, 36*, 346–360.

Golub, S. A., Tomassilli, J., Pantalone, D., Brennan, D., Karpiak, S., & Parsons, J. (2010). Prevalence and correlates of sexual behavior and risk management among HIV-positive adults over 50. *Sexually Transmitted Diseases, 37*(10), 615–620.

Gonzalez, J. S., Hendriksen, E. S., Collins, E. M., Duran, R. E., & Safren, S. A. (2009). Latinos and HIV/AIDS: Examining factors related to disparity and identifying opportunities for psychosocial intervention research. *AIDS and Behavior, 13*, 582–602.

Goodnough, L. T., Shander, A., & Brecher, M. E. (2003). Transfusion medicine: Looking to the future. *The Lancet, 361*, 161–169.

Gorman, C. (2006). The graying of AIDS. *Time, 168*(7), 54–56.

Gott, M., & Hinchliff, S. (2003). Barriers to seeking treatment for sexual problems in primary care: A qualitative study with older people. *Family Practice, 20*, 690–695.

Gott, M., Hinchliff, S., & Galena, E. (2004). General practitioner attitudes to discussing sexual health issues with older people. *Social Science & Medicine, 58*, 2093–2103.

Greenwald, A. G., McGhee, D. E., & Schwartz, J. L. K. (1998). Measuring individual differences in implicit cognition: the Implicit Association Test. *Journal of Personality and Social Psychology, 74*, 1464–1480.

Greenwald, A. G., Nosek, B. A., & Banaji, M. R. (2003). Understanding and Using the Implicit Association Test: I. *An Improved Scoring Algorithm. Journal of Personality and Social Psychology, 85*, 197–216.

Gregg, A. P., Seibt, B., & Banaji, M. R. (2006). Easier done than undone: Asymmetry in the malleability of implicit preferences. *Journal of Personality and Social Psychology, 90*, 1–20.

Grisso, T. (2003). *Evaluating competences* (2nd ed.). New York: Plenum.

Grov, C., Golub, S., Parsons, J. T., Brennan, M., & Karpiak, S. E. (2010). Loneliness and HIV-related stigma explain depression among older HIV-positive adults. *AIDS Care, 22*, 630–639.

Gua, Y. D. (2008). Inappropriate sexual behaviours in cognitively impaired older individuals. *The American Journal of Geriatric Pharmacotherapy, 6*(5), 269–288.

Guan, J. (2004). Correlates of spouse relationship with sexual attitude, interest, and activity among Chinese elderly. *Sexuality and Culture, 8*, 104–131.

Gutmann, D. (1994). *Reclaimed powers: Men and women in later life*. Evanston, IL: Northwestern University Press.

Haber, D. (2009). Gay aging. *Gerontology and Geriatrics Education, 30*, 267–280.

Hackett, M. L., Yapa, C., Parag, V., & Anderson, C. S. (2005). Frequency of depression after stroke: A systematic review of observational studies. *Stroke, 36*, 1330–1340.

Haile, R., Padilla, M. B., & Parker, E. A. (2011). 'Stuck in the quagmire of an HIV ghetto': The meaning of stigma in the lives of older black gay and bisexual men living with HIV in New York City. *Culture, Health and Sexuality, 13*(4), 429–442.

Hammond, D. B. (1987). *My parents never had sex*. New York: Prometheus.

Hardy, D. J., & Vance, D. E. (2009). The neuropsychology of HIV/AIDS in older adults. *Neuropsychology Review, 19*, 263–272.

Harries, C., Forrest, D., Harvey, N., McClelland, A., & Bowling, A. (2007). Which doctors are influenced by a patient's age? A multi-method study of angina treatment in general practice, cardiology and gerontology. *Quality & Safety in Health Care, 16*, 23–27.

Harrington, J. M., Jones, E. G., & Badger, T. (2009). Body image perceptions in men with prostate cancer. *Oncology Nursing Forum, 36*, 167–172.

Harris, G. (2003). After 5 years, rivals emerge ready to give Viagra a fight. *The New York Times, 21*, C1.

Harrison, P. M., & Beck, A. J. (2004). *Prisoners in 2003 (Bureau of Justice Statistics Bulletin NCJ 205335)*. Washington, DC: U.S. Department of Justice.

Hash, K. M., & Netting, F. E. (2007). Long-term planning and decision-making among midlife and older gay men. *Journal of Social Work in End-of-Life & Palliative Care, 3*(2), 57–77.

Hay, A., & Johnson, S. (2001). Fundamental skills and knowledge for hospice and palliative care social workers. In Competency-based education for social workers (pp. 12–13). Arlington, VA: National Hospice and Palliative Care Organization.

Healey, T., & Ross, K. (2002). Growing old invisibly: Older viewers talk television. *Media, Culture, & Society, 24*, 105–120.

Heaphy, B. (2007). Sexualities, gender and ageing: Resources and social change. *Current Sociology, 55*, 193–210.

Hebl, M. R., & Heatherton, T. F. (1998). The stigma of obesity in women: The difference is black and white. *Personality and Social Psychology Bulletin, 24*, 417–426.

Hebrew Home. (2010). *Sexual expression policy at the home*. Retrieved from http://www.hebre-whome/org.

Heckman, T. G., Sikkema, K. J., Hansen, N., Kochman, A., Heh, V., Neufeld, S., & the AIDS & Aging Research Group. (2011). A randomized clinical trial of a coping improvement group intervention for HIV-infected older adults. *Journal of Behavioral Medicine, 34*, 102–111.

Helgeson, V. S., Lepore, S. J., & Eton, D. T. (2006). Moderators of the benefits of psychoeducational interventions for men with prostate cancer. *Health Psychology, 25*, 348–354.

Heller, L., Keren, O., Aloni, R., & Davidoff, G. (1992). An open trial of vacuum penile tumescence: Constriction therapy for neurological impotence. *Paraplegia, 30*, 550–553.

Hellerstein, H. K., & Friedman, E. H. (1970). Sexual activity and the post coronary patient. *Archives of Internal Medicine, 125*, 987–999.

Hellstrom, W. J. G., Montague, D. K., Moncada, I., Carson, C., Minhaus, S., Faria, G., et al. (2010). Implants, mechanical devices, and vascular surgery for erectile dysfunction. *The Journal of Sexual Medicine, 7*, 501–523.

Hillman, J. L. (1998). Health care providers' knowledge about HIV induced dementia among older adults. *Sexuality and Disability, 16*, 181–192.

Hillman, J. (2007). Knowledge and attitudes about HIV/AIDS among community-living older women: Reexaming issues of age and gender. *Journal of Women & Aging, 19*, 53–67.

Hillman, J. (2008a). Knowledge, attitudes, and experience regarding HIV/AIDS among older adult inner-city Latinos. *International Journal of Aging & Human Development, 66*, 243–257.

Hillman, J. (2008b). Sexual issues and aging within the context of work with older adult patients. *Professional Psychology: Research and Practice, 39*, 290–297.

Hillman, J. (2011). A call for an integrated biopsychosocial model to address fundamental disconnects in an emergent field: An introduction to the special issue on "sexuality and aging. *Ageing International, 36*, 303–312.

Hillman, J., & Stricker, G. (1994). Linkage of knowledge and attitudes toward elderly sexuality: Not necessarily a uniform relationship. *The Gerontologist, 34*, 256–260.

Hillman, J. L., & Stricker, G. (1996). Predictors of college students' knowledge and attitudes toward elderly sexuality: The relevance of grandparental contact. *Educational Gerontology: An International Journal, 22*, 539–555.

Hillman, J. L., & Stricker, G. (1998). Some issues in the assessment of HIV among older adults. *Psychotherapy, 35*, 483–489.

Hillman, J., & Stricker, G. (2001). The management of sexualized transference and countertranference with older adult patients: Implications for practice. *Professional Psychology: Research and Practice, 32*, 272–277.

Hillman, J. L., Stricker, G., & Zweig, R. A. (1997). Clinical psychologists' judgments of older adult patients with character pathology: Implications for practice. *Professional Psychology: Research and Practice, 28*, 179–183.

Hinchliff, S., & Gott, M. (2008). Challenging social myths and stereotypes of women and aging: Heterosexual women talk about sex. *Journal of Women & Aging, 20*, 65–81.

Hinrichs, K. L. M., & Vacha-Haase, T. (2010). Staff perceptions of same-gender sexual contacts in long-term care facilities. *Journal of Homosexuality, 57*, 776–789.

Hinrichsen, G. A. (2006). Why multicultural issues matter for practitioners working with older adults. *Professional Psychology: Research and Practice, 37*, 29–35.

Hinrichsen, G. A., & Clougherty, K. F. (2006). *Interpersonal psychotherapy for depressed older adults*. Washinginton, DC: American Psychological Association.

Ho, J., & Rabi, D. (2006). Sexual dysfunction in women with diabetes. *Canadian Journal of Diabetes, 19*, 5–7.

Hodson, D. S., & Skeen, P. (1994). Sexuality and aging: The hammerlock of myths. *Journal of Applied Gerontology, 13*, 219–235.

Holaday, M., Smith, D. A., & Sherry, A. (2000). Sentence completion tests: A Review of the Literature and results of a survey of members of the Society for Personality Assessment. *Journal of Personality Assessment, 74*, 371–383.

Hong, K. E. (2003). Categorical exclusions: Exploring legal responses to health care discrimination against transsexuals. *Columbia Journal of Gender and Law, 14*(1), 88.

Hooker, E. (1957). The adjustment of the male overt homosexual. *Journal of Projective Techniques, 21*, 18–31.

Hordern, A. J., & Street, A. F. (2007). Let's talk about sex: Risky business for cancer and palliative care clinicians. *Contemporary Nurse, 27*, 49–60.

Horgan, O., & MacLachlan, M. (2004). Psychosocial adjustment to lower-limb amputation: A review. *Disability and Rehabilitation, 26*, 837–850.

Horner, M. J., Ries, L. A. G., Krapcho, M., Neyman, N., Aminou, R., Howlader, N., et al. (2006). *SEER Cancer Statistics Review, 1975-2006, National Cancer Institute*. Retrieved from http://seer.cancer.gov/csr/1975_2006/.

Howard, J. R., O'Neill, S., & Travers, C. (2006). Factors affecting sexuality in older Australian women: Sexual interest, sexual arousal, relationships, and sexual distress in older Australian women. *Climacteric, 9*, 355–367.

Howlader, N., Noone, A. M., Krapcho, M., Neyman, N., Aminou, R., Waldron, W, et al. (Eds.) (2011). *SEER Cancer Statistics Review, 1975-2008*. Bethesda, MD: National Cancer Institute. Retrieved http://seer.cancer.gov/csr/1975_2008.

Hsiao, W., Bennett, N., Guhring, P., Narus, J., & Mulhall, J. P. (2011). Satisfaction profiles in men using intracavernosal injection therapy. *The Journal of Sexual Medicine, 8*, 512–517.

Hsu, L. G., & Zimmer, B. (1988). Eating disorders in old age. *The International Journal of Eating Disorders, 7*, 133–138.

Hughes, M. (2009). Lesbian and gay people's concerns about ageing and accessing services. *Australian Social Work, 62*, 186–201.

Hughes, D., & Evans, A. (2003). Health needs of women who have sex with women. *British Medical Journal, 327*, 1739–1758.

Hummert, M. L., Garstka, T. A., O'Brien, L. T., Greenwald, A. G., & Mellott, D. S. (2002). Using the implicit association test to measure age differences in implicit social cognitions. *Psychology and Aging, 17*, 482–495.

Hummert, M. L., Garstka, T. A., Shaner, J. L., & Strahm, S. (1995). Judgments about stereotypes of the elderly: Attitudes, age associations, and typicality ratings of young, middle-aged, and elderly adults. *Research on Aging, 17*, 168–189.

Hybels, C. F., & Blazer, D. G. (2003). Epidemiology of late-life mental disorders. *Clinics in Geriatric Medicine, 19*, 663–696.

Hyde, J. (2001). Understanding the context of assisted living. In K. H. Namazi & P. K. Chafetz (Eds.), *Assisted living: Current issues in facility management and resident care* (pp. 15–28). Westport, CT: Auburn House.

Hyer, L., Kramer, D., & Sohnle, S. (2004). CBT with older people: Alterations and the value of the therapeutic alliance. *Psychotherapy: Theory, Research, Practice, Training, 41*, 276–291.

Illa, L., Brickman, A., Saint-Jean, G., Echenique, M., Metsch, L., Eisdorfer, C., et al. (2008). Sexual risk behaviors in late middle age and older HIV seropositive adults. *AIDS and Behavior, 12*(6), 935–942.

Illa, L., Echenique, M., Jean, G. S., Vustamante-Avellaneda, V., Metsch, L., Mendez-Mulet, L., et al. (2010). Project ROADMAP: Reeducating older adults in maintaining AIDS prevention: A secondary intervention for older HIV-positive adults. *AIDS Education and Prevention, 22*, 138–147.

Institute of Medicine (2010, January 11). *Hepatitis and liver cancer: A national strategy for prevention and control of Hepatitis B and C.* Retrieved from http://www.iom.edu/Reports/2010/Hepatitis-and-Liver-Cancer-A-National-Strategy-for-Prevention-and-Control-of-Hepatitis-B-and-C.aspx

Ives, D. G., Lave, J. R., Traven, N. D., Schulz, R., & Kuller, L. H. (1996). Mammography and pap smear use by older rural women. *Public Health Reports, 111*, 244–250.

Jackett, M. L., Yapa, C., Parag, V., & Anderson, C. S. (2005). Frequency of depression after stroke: A systematic review of observational studies. *Stroke, 36*, 1330–1340.

Jackson, G., Arver, S., Banks, I., & Stecher, V. J. (2010). Counterfeit phosphodiesterase type 5 inhibitors pose significant safety risks. *International Journal of Clinical Practice, 64*, 497–504.

Jackson, N., Johnson, M., & Roberts, R. (2008). The potential impact of discrimination fears of older gays, lesbians, bisexuals and transender individuals living in small-to moderate-sized cities on long-term health care. *Journal of Homosexuality, 54*, 325–339.

James, J. W., & Haley, W. E. (1995). Age and health bias in practicing clinical psychologists. *Psychology and Aging, 10*, 610–616.

Janus, S. S., & Janus, C. L. (1993). *The Janus report on sexual behavior*. New York: Wiley.

Jena, A. B., Goldman, D. P., Kamdar, A., Lakdawalla, D. N., & Lu, Y. (2010). *Sexually Transmitted Diseases Among Users of Erectile Dysfunction Drugs: Analysis of Claims Data Annals of Internal Medicine, 153*, 1–7.

Jones, J. C., & Mystrom, N. M. (2003). Looking back…looking forward: Addressing the lives of lesbians 55 and older. *Journal of Women & Aging, 14*, 59–76.

Jonsson, A., Aus, G., & Bertero, C. (2009). Living with a prostate cancer diagnosis: A qualitative 2-year follow-up. *The Aging Male, 13*, 25–31.

Juraskova, I., Butow, P., Robertson, R., Sharpe, L., McLeod, C., & Hacker, N. (2003). Post-treatment sexual adjustment following cervical and endometrial cancer: A qualitative insight. *Psycho-Oncology, 12*, 267–279.

Kaas, M. J. (1981). Geriatric sexuality breakdown syndrome. *International Journal of Aging & Human Development, 13*, 71–77.

Kaiser, F. E. (1996). Sexuality in the elderly. *Geriatric Urology, 1*, 99–109.

Kaiser, F. E., & Morley, J. E. (1994). Gonadotropins, testosterone, and the aging male. *Neurobiology of Aging, 15*, 559–563.

Kaiser, F. E., Viosca, S. P., Morley, J. E., Mooradian, A. D., Davis, S. S., & Korenman, S. G. (1988). Impotence and aging: Clinical and hormonal factors. *Journal of the American Geriatrics Society, 36*, 511–519.

Karel, M. J., Knight, B. G., Duffy, M., Hinrichsen, G. A., & Zeiss, A. M. (2010). Attitude, knowledge, and skill competencies for practice in professional geropsychology: Implications for training and building a geropsychological workforce. *Training and Education in Professional Psychology, 4*, 75–84.

Kasper, J., & O'Malley, M. (2007, October 16). *Changes in characteristics, needs, and payment for care of elderly nursing home residents: 1999 to 2004.* Kaiser Commission on Medicaid and the Uninsured. Retrieved from http://www.kff.org/medicaid/upload/7663.pdf

Katz, A. (2005). Do ask, do tell: Why do so many nurses avoid the topic of sexuality? *The American Journal of Nursing, 105*(7), 66–68.

Katz, A. (2011). Sexuality at the end of life. The exchange: Canadian virtual hospice. Retrieved from www.virtualhospice.ca

Kaufman, D. W., Kelly, J. P., Rosenberg, L., Anderson, T. E., & Mitchell, A. A. (2002). Recent patterns of medication use in the ambulatory adult population of the United States: The Sloan Survey. *Journal of the American Medical Association, 287*, 337–344.

Kautz, D. D. (2007). Hope for love: Practical advice for intimacy and sex after stroke. *Rehabilitation Nursing, 32*, 95–132.

Kawakami, K., Young, H., & Dovidio, J. F. (2002). Automatic stereotyping: Category, trait and behavioral activations. *Personality and Social Psychology Bulletin, 28*, 3–15.

Kaye, R. A., & Markus, T. (1997). AIDS teaching should not be limited to the young. *USA Today Magazine, 126*, 50.

Kazemi-Saleh, D., Pishgou, B., Assari, S., & Tavallaii, S. A. (2007). Fear of sexual intercourse in patients with coronary artery disease: A pilot study of associated morbidity. *The Journal of Sexual Medicine, 4*, 1619–1625.

Kennelly, S., O'Neill, D., & O'Brien, J. G. (2011). Elder abuse in residential care. *The Lancet, 377*, 1076.

Kelly, J. (1977). The aging male homosexual: Myth and reality. *The Gerontologist, 17*, 328–332.

Kemper, E. (2003). Graying of American prisons; Addressing the continued increase in geriatric inmates. *Corrections Compendium, 28*, 22–26.

Kendig, N., & Adler, W. (1990). The implications of the acquired immunodeficiency syndrome for gerontology research and geriatric medicine. *Journal of Gerontology, 45*, M77–M81.

Kennedy, G. J., Hague, M., & Zarankow, B. (1997). Human sexuality in late life. *International Journal of Mental Health, 26*, 35–46.

Kernberg, O. F. (1991). Aggression and love in the relationship of the couple. In G. I. Fogel & W. A. Myers (Eds.), *Perversions and near-perversions in clinical practice: New psychoanalytic perspectives* (pp. 153–175). New Haven: Yale University Press.

Kessell, B. (2001). Sexuality in the older person. *Age and Ageing, 30*, 121–124.

Khoo, E. M., Tan, H. M., & Low, W. Y. (2008). Erectile dysfunction and comorbidities in aging men: An urban cross-sectional study in Malaysia. *The Journal of Sexual Medicine, 5*, 2925–2934.

Kingston, T. (2002). The challenges and rewards of life as an outspoken bisexual elder. *Outworld, 8*, 1–6.

Kinsey, A. C., Pomeroy, W. B., & Martin, C. E. (1948). *Sexual behavior in the human male*. Philadelphia: Saunders.

Kinsey, A. C., Pomeroy, W. B., Martin, C. E., & Gebhard, P. H. (1953). *Sexual behavior in the human female*. Philadelphia: Saunders.

Knauer, N. (2009). LGBT elder law: Toward equity in aging. *Harvard Journal of Law & Gender, 32*, 1–58.

Knight, B. G. (1996). *Psychotherapy with older adults* (2nd ed.). Thousand Oaks, CA: Sage.

Knoepp, L. R., Shippey, S. H., Chen, C. C. G., Cundiff, G. W., Derogatis, L. R., & Handa, V. L. (2010). Sexual complaints, pelvic floor symptoms, and sexual distress in women over forty. *The Journal of Sexual Medicine, 7*, 3675–3682.

Koerner, B. I. (2000). A lust for profits. *U.S. News & World Report, 128*(12), 36–42.

Komisaruk, B. R., & Whipple, B. (1995). Elevation of pain threshold by vaginal stimulation in women. *Pain, 21*, 357–367.

Kontula, O., & Haavio-Mannila, E. (2009). The impact of aging on human sexual activity and sexual desire. *Journal of Sex Research, 46*, 46–56.

Krychman, M. L. (2011). Vaginal estrogens for the treatment of dyspareunia. *The Journal of Sexual Medicine, 8*, 666–674.

Kudadjie-Gyamfi, E., Consedine, N. S., & Magai, C. (2006). On the importance of being ethnic: Coping with the threat of prostate cancer in relation to prostate screening. *Cultural Diversity and Ethnic Minority Psychology, 12*, 509–526.

Kulkarni, S. C., Levin-Rector, A., Ezzati, M., Murray, C. J. L. (2011). Falling behind: Life expectancy in U.S. counties from 2000 to 2007 in an international context. *Population Health Metrics, 9*, 1-12. Retrieved from http://www.pophealthmetrics.com/content/pdf/1478-7954-9-16.pdf

Kutner, M., Greenberg, E., Jin, Y, & Paulsen, C. (2006). *The health literacy of America's adults: Results from the 2003 national assessment of adult literacy* (NCES 2006-2483). U.S. Department of Education. Washington, DC: National Center for Education Statistics.

Kutner, J. S., Kassner, C. T., & Nowels, D. E. (2001). Symptom burden at the end of life: Hospice providers' perceptions. *Journal of Pain and Symptom Management, 21*, 473–480.

Kuypers, J. A., & Bengtson, V. L. (1973). Social breakdown and competence: A model of normal aging. *Human Development, 16*, 181–201.

Kuzmarov, I. W., & Bain, J. (2009). Sexuality in the aging couple, Part II: The aging male. *Geriatrics & Aging, 12*, 53–57.

Laidlaw, K., Thompson, L. W., Dick-Siskin, L., & Gallagher-Thompson, D. (2009). *Cognitive Behaviour Therapy with Older People*. New York: Wiley.

Landers, S., Mimiaga, M. J., & Krinsky, L. (2010). The open door project task force: A qualitative study of LGBT aging. *Journal of Gay & Lesbian Social Services, 22*, 316–336.

Landreville, P., Desrosiers, J., Vincent, C., Verreault, R., Boudreault, V., & The BRAD Group. (2009). The role of activity restriction in poststroke depressive symptoms. *Rehabilitation Psychology, 54*, 315–322.

Langer-Most, O., & Langer, N. (2010). Aging and sexuality: How much do gynecologists know and care? *Journal of Women & Aging, 22*, 283–289.

Lapid, M. I., Prom, M. C., Burton, M. C., McAlpine, D. E., Sutor, B., & Rummans, T. A. (2010). Eating disorders in the elderly. *International Psychogeriatrics, 22*, 523–536.

Laumann, E. O., Paik, A., Glasser, D. B., Kang, J.-H., Wang, T., Levinson, B., et al. (2006). A cross-national study of subjective sexual well-being among older women and men: Findings from the global study of sexual attitudes and behaviors. *Archives of Sexual Behavior, 35*, 145–161.

Laumann, E. O., Paik, A., & Rosen, R. C. (1999). Sexual dysfunction in the United States. *Journal of the American Medical Association, 281*, 537–544.

Lawton, M. P., & Brody, E. M. (1969). Assessment of older people: Self-maintaining and instrumental activities of daily living. *The Gerontologist, 9*, 179–186.

Lee, C. T., & Oesterling, J. E. (1995). Diagnostic markers of prostate cancer: Utility of prostate-specific antigen in diagnosis and staging. *Seminars in Surgical Oncology, 11*, 23–35.

Lee, Y., & Sontag, M. S. (2010). An assessment of the proximity of clothing to self scale for older persons. *International Journal of Consumer Studies, 34*, 443–448.

Lehabot, K., Walters, K. L., & Simoni, J. M. (2010). Abuse, mastery, and health among lesbian, bisexual, and two-spirit American Indian and Alaska Native Women. *Psychology of Violence, 1(S)*, 53–67.

Leiblum, S. R., & Rosen, R. C. (1989). *Principles and practice of sex therapy: Update for the 1990's*. New York: Guilford Press.

Lemieux, L., Cohen-Schneider, R., & Holzapfel, S. (2001). Aphasia and sexuality. *Sexuality and Disability, 19*, 253–266.

Lemieux, L., Kaiser, S., Pereira, J., & Meadows, L. M. (2004). Sexuality in palliative care: Patient perspectives. *Palliative Medicine, 18*, 630–637.

Lerner, A. J., Hedera, P., Koss, E., & Stuckey, J. (1997). Delirium in Alzheimer disease. *Alzheimer's Disease and Associated Disorders, 11*, 16–20.

Lester, E. P. (1985). The female analyst and the eroticized transference. *The International Journal of Psycho-Analysis, 66*, 283–293.

Leviton, D. (1978). Intimacy/sexual needs of the terminally ill. *Death Education, 2*, 261–280.

Levitra. (2011, November 1). *Did you know ED usually has a medical cause?* Retrieved from www.levitra.com

Levy, B. R., & Banaji, M. R. (2002). Implicit ageism. In T. Nelson (Ed.), *Ageism: Stereotyping and prejudice against older persons* (pp. 49–75). Cambridge: MIT Press.

Lewis, L. J. (2004). Examining sexual health discourses in a racial/ethnic content. *Archives of Sexual Behavior, 33*, 223–234.

Lewis, C. L., Kistler, C. E., Amick, H. R., Watson, L. C., Bynum, D. L., Walter, L. C., et al. (2006). Older adults' attitudes about continuing cancer screening later in life: A pilot study interviewing residents of two continuing care communities. *BMC Geriatrics, 3*, 6–10.

Lewittes, H. J. (1988). Just being friendly means a lot-Women, friendship, and aging. *Women & Health, 14*, 139–159.

Lichtenberg, P. A., & Strzepek, D. M. (1990). Assessments of institutionalized dementia patients' competencies to participate in intimate relationships. *The Gerontologist, 30*, 117–120.

Lindau, S., Gosch, K., Abramsohn, E., Chan, P., Krumholz, Spatz, E., et al., (2010). *Gender differences in loss of sexual activity 1-year after an acute myocardial infarction.* Presented at the American Heart Association's Quality of Care and Outcomes Research Conference.

Lindau, S. T., Leitsch, S. A., Lundberg, K. L., & Jerome, J. (2006). Older women's attitudes, behavior, and communication about sex and HIV: A community-based study. *Journal of Women's Health (2002), 15*, 747–753.

Lindau, S. T., Schumm, L. P., Laumann, E. O., Levinson, W., O'Muircheartaigh, C. A., & Waite, L. J. (2007). A study of sexuality and health among older adults in the United States. *The New England Journal of Medicine, 357*, 762–774.

Linley, L., Hall, H., & An, Q. (December, 2007). *HIV/AIDS diagnoses 2. Among persons fifty years and older in 33 states, 2001-20005.* Atlanta, GA: Centers for Disease Control and Prevention National HIV Prevention Conference.

Linneman, T. J. (2008). How do you solve a problem like Will Truman?: The feminization of gay masculinities of *Will & Grace*. *Men and Masculinities, 10*, 583–603.

Linsk, N. L. (1994). HIV and the elderly. *Families in Society, 75*, 362–372.

Lopez, D. A., Mathers, C. D., Ezzti, M., Jamison, D. T., & Marray, C. J. L. (2006). Measuring the global burden of disease and risk factors, 1990–2001. In D. a. Lopez, C. D. Mathers, M. Ezzati, E. T. Jamison, & C. J. L. Murray (Eds.), *Global burden of disease and risk factors* (pp. 1–13). Washington, DC: World Bank.

Los Angeles Times (2004, May 22). *Elderly turn to internet to find love*. Retrieved from http://www.globalaging.org/elderrights/us/2004/Elderinternet.htm

Lovejoy, J. C., Champagne, C. M., de Jonge, L., Xie, H., & Smith, S. R. (2008). Increased visceral fat and decreased energy expenditure during the menopausal transition. *International Journal of Obesity, 32*, 949–958.

Lovejoy, T. I., Heckman, T. G., Sikkema, K. J., Hansen, N. B., Kochman, A., Suhr, J. A., et al. (2008). Patterns and correlates of sexual activity and condom use behavior in persons 50-plus years of age living with HIV/AIDS. *AIDS and Behavior, 12*(6), 943–956.

Lown, B. (2000). Market health care: The commodification of health care. *Philosophy and Social Action, 26*, 57–71.

Lyden, M. (2007). Assessment of sexual consent capacity. *Sexuality and Disability, 25*, 3–20.

Malamud, W. I. (1996). Countertransference issues with elderly patients. *Journal of Geriatric Psychiatry, 29*, 33–41.

Malatesta, V., Chambless, D., Pollack, M., & Cantor, A. (1988). Widowhood, sexuality, and aging: A life span analysis. *Journal of Sex & Marital Therapy, 14*, 49–62.

Malavige, L. S., & Levy, J. C. (2009). Erectile dysfunction in diabetes mellitus. *The Journal of Sexual Medicine, 6*, 1232–1247.

Manderson, L. (2005). Boundary breaches: The body, sex and sexuality after stoma surgery. *Social Science & Medicine, 61*, 405–415.

Mandras, S. A., Uber, P. A., & Mehra, M. R. (2007). Sexual activity and chronic heart failure. *Mayo Clinic Proceedings, 82*, 1203–1210.

Marcus, M. D., Bromberger, J. T., Wei, H., Brown, C., & Kravitz, H. M. (2007). Prevalence and selected correlates of eating disorder symptoms among a multiethnic community sample of midlife women. *The Society of Behavioral Medicine, 33*, 269–277.

Marks, L. S., Duda, C., Dorey, F. J., Macairan, M. L., & Santos, P. B. (2006). Treatment of erectile dysfunction with sildenafil. *Urology, 53*, 19–24.

Marshall, B. L. (2010). Science, medicine and virility surveillance: 'Sexy seniors' in the pharmaceutical imagination. *Sociology of Health & Illness, 32*, 211–224.

Martinez, R. (2005). Prostate cancer and sex. *Journal of Gay & Lesbian Psychotherapy, 9*, 91–99.

Marwill, S. L., Freund, K. M., & Barry, P. P. (1996). Patient factors associated with breast cancer screening among older women. *Journal of the American Geriatrics Society, 44*, 1210–1214.

Masters, W. H., & Johnson, V. E. (1966). *Human sexual response*. Boston: Little, Brown.

Matthias, R. E., Lubben, J. E., Atchison, K. A., & Schweitzer, S. O. (1997). Sexual activity and satisfaction among very old adults: Results from a community-dwelling Medicare population survey. *The Gerontologist, 37*, 6–14.

Mattis, S. (1973). *Dementia Rating Scale*. Odessa, FL: Psychological Assessment Resources, Inc.

Mattson-DiCecca, A. A., Farag, N. R., & Burns, R. B. (2009). Update: A 60-year-old woman with sexual difficulties. *Journal of the American Medical Association, 301*, 94.

Mayo Clinic. (2010a, November 1). *Prostate cancer*. Retrieved from http://www.mayoclinic.com/health/prostate-cancer/DS00043

Mayo Clinic. (2010b, October 31). *Vaginal dryness*. Retrieved from http://www.mayoclinic.com/health/vaginal-dryness

Mayo Clinic. (2010c, November 1). *Vaginal atrophy*. Retrieved from http://www.mayoclinic.com/health/vaginal-atrophy

Mayo Clinic. (2011a, November 1). *Chronic pain can interfere with sexuality*. Retrieved from http://www.mayoclinic.com/health/chronic-pain

Mayo Clinic. (2011b, November 1). *Erectile dysfunction*. Retrieved from http://www.mayoclinic.com/erectiledysfunction

Mayo Clinic. (2011c, October 31). *Heart attack: Coping and support*. Retrieved from http://www.mayoclinc.com/health/heart-attack.

Mayo Clinic. (2011d, November 1). *Menopause*. Retrieved from http://www.mayoclinic.com/health/menopause.

Mayo Clinic. (2011e, October 31). *Urinary incontinence*. Retrieved from http://www.mayoclinic.com/health/incontinence.

McCarthy, B. W. (2001). Relapse prevention strategies and techniques with erectile dysfunction. *American Journal of Sex and Marital Therapy, 27*, 1–8.

McCartney, J., Izeman, H., Rogers, D., & Cohen, N. (1987). Sexuality and the institutionalized elderly. *Journal of the American Geriatrics Society, 35*, 331–333.

McCormack, K. A., Newman, D. K., Colling, J., & Pearson, B. D. (1992). A practice guideline for urinary incontinence: The challenge for nurses. *Urologic Nursing, 12*, 40–45.

McCurry, J. (2008, March 29). No sex, thank-you…we're Japanese. *The Guardian: The Observer*. Retrieved from http://www.guardian.co.uk/world/2008.

McDougall, G. J. (1993). Therapeutic issues with gay and lesbian elders. *Clinical Gerontologist, 14*, 45–57.

McElhaney, L. J. (1992). Dating and courtship in the later years. *Generations, 16*, 21–23.

McKinlay, J. B. (2000). The worldwide prevalence and epidemiology of erectile dysfunction. *International Journal of Impotence Research, 12*(4), S6–S11.

McVary, K. T. (2007). Clinical practice: Erectile dysfunction. *The New England Journal of Medicine, 357*, 2472–2481.

Medicare. (2008). *Nursing homes: resident rights*. Retrieved on October 16, 2011 from http://www.medicare.gov/nursing/residentrights.asp.

Mellor, D., Fuller-Tyszkiewicz, M., McCabe, M. P., & Ricciardelli, L. A. (2010). Body image and self-esteem across age and gender: A short-term longitudinal study. *Sex Roles, 63*, 672–681.

Meschede, T., Shapiro, T. M., Sullivan, L., & Wheary, J. (2010). *Severe financial insecurity among African-American and Latino seniors*. The Institute on Assets and Social Policy, The Heller School for Social Policy and Management at Brandeis University, and Demos. Retrieved from http://iasp.brandeis.edu/pdfs/SFSI.pdf

Meston, C. M. (1997). Aging and sexuality in successful aging. *The Western Journal of Medicine, 167*, 285–290.

Metlife Mature Market Institute. (2010). Demographic profile: America's middle boomers. Retrieved from http://www.metlife.com/assets/cao/mmi/publications/Profiles/mmi-middle-boomer-demographic-profile.pdf

Metlife Mature Market Institute, & The Lesbian and Gay Aging Issues Network of the American Society of Aging. (2010). Out and aging: The Metlife study of lesbian and gay baby boomers. *Journal of GLBT Family Studies, 6*, 40–57.

Meyer, B. J., Russo, C., & Talbot, A. (1995). Discourse comprehension and problem solving: Decisions about the treatment of breast cancer by women across the life span. *Psychology and Aging, 10*, 84–103.

Meyerowitz, J. (2002). *How sex changed: A history of transsexuality in the United States.* Cambridge: Harvard University Press.

Miller, M. (2000). Surfs up, IQs down: Top 50 web searches. *Kansas City Business Journal, 18*(29), 16.

Minichiello, V., Browne, J., & Kendig, H. (2000). Perceptions and consequences of ageism: Views of older people. *Ageing and Society, 20*, 253–278.

Ministry of Health and Long Term Care. (2011). *Seniors' care: Long term care homes.* Retrieved from http://www.health.gov.on.ca/english/public/program/ltc/15_facilities. html#4.

Molassiotis, A., Chan, C., Yam, B., Chan, E., & Lam, C. (2002). Life after cancer: Adaptation issues faced by Chinese gynecological cancer survivors in Hong Kong. *Psycho-Oncology, 11*, 114–123.

Montejo, Á. L., Majadas, S., Rico-Villademoros, F., Llorca, G., de la Gándara, J., Franco, M., et al. (2010). Frequency of sexual dysfunction in patients with a psychotic disorder receiving antipsychotics. *The Journal of Sexual Medicine, 7*, 3404–3413.

Moore, K. L. (2010). Sexuality and sense of self in later life: Japanese men's and women's reflections on sex and aging. *Journal of Cross-Cultural Gerontology, 25*, 149–163.

Moore, T. E., & Cadeau, L. (1985). The representation of women, the elderly and minorities in Canadian television commercials. *Canadian Journal of Behavioral Science, 17*, 215–225.

Moreira, E. D., Sae-Chul, K., Glasser, D., & Gingell, C. (2006). Sexual activity, prevalence of sexual problems, and associated help-seeking patterns in men and women aged 40-80 years in Korea: Data from the global study of sexual attitudes and behaviors (GSSAB). *The Journal of Sexual Medicine, 3*, 201–211.

Morgan, L. A. (2009). Balancing safety and privacy: The case of room locks in assisted living. *Journal of Housing for the Elderly, 23*, 185–203.

Morgan, E. E., Woods, S. P., Delano-Wood, L., Bondi, M. W., Grant, I., & the HIV Neurobehavioral Research Program (HNRP) Group. (2011). Intraindividual variability in HIV infection: Evidence for greater neurocognitive dispersion in older HIV seropositive adults. *Neuropsychology, 25*, 645–654.

Morley, J. E. (1996). Update on men's health: Progress in geriatrics. *Generations, 4*, 13–19.

Moye, J. (2003). Guardianship and conservatorship. In T. Grisso (Ed.), *Evaluating competencies* (pp. 309–390). New York: Plenum.

Mugavero, M. J., Castellano, C., Edelman, D., & Hicks, C. (2007). Late diagnosis of HIV infection: The role of age and sex. *The American Journal of Medicine, 120*(4), 370–373.

Muram, D., Miller, K., & Cutler, A. (1992). Sexual assault of the elderly victim. *Journal of Interpersonal Violence, 7*(1), 70–76.

Murphy, J. A., Rawlings, E. I., & Howe, S. R. (2002). A survey of clinical and psychologists on treatment lesbian, gay, and bisexual clients. *Professional Psychology: Research and practice, 33*, 183–189.

Murray, J., & Adam, B. D. (2001). Aging, sexuality and HIV issues among older gay men. *The Canadian Journal of Human Sexuality, 10*(3–4), 75–91.

Nappi, R. E., & Kokot-Kierepa, M. (2010). Women's voices in the menopause: Results from an international survey on vaginal atrophy. *Maturitas, 67*, 233–238.

National Cancer Institute. (2009). *Study questions value of prostate cancer screening.* Bethesda, MD: Author. Retrieved from http://www.cancer.gov/aboutnci/ncicancerbulletin/archive/2009/090809/page2

National Center for Assisted Living. (2008). *General principles for assisted living.* Retrieved October 16, 2011 from http://www.ahcancal.org/ncal/about/Documents/GPAssistedLiving.pdf.

National Center on Elder Abuse (2011). *Major types of elder abuse*. Retrieved from http://www.ncea.aoa.gov/Main_Site/FAQ/Basics/Types_Of_Abuse.aspx

National Center for Assisted Living. (2009). *Resident profile*. Retrieved October 16, 2011 from http://www.ahcancal.org/ncal/resources/Pages/ResidentProfile.aspx.

National Center for Assisted Living. (2011). *Better serving the lesbian, gay, bisexual and transgender populations in assisted living communities*. Retrieved October 23, 2011 from http://www.achca.org/content/pdf/LGBT-%20Final%20Draft-%20Slides%20Only-Formatted.pdf.

National Institutes of Health. (1992). NIH Consensus Statement. *Impotence, 10*(4), 1–31.

Netzley, S. B. (2010). Visibility that demystifies: Gays, gender, and sex on television. *Journal of Homosexuality, 57*, 968–986.

New York Department of Health and Hygiene. (2011). *Sexually transmitted diseases (STD): The NYC DOHMH Bureau of STD Control literature data base*. Retrieved from http://www.nyc.gov/html/doh/html/std/std-lit-clearing-house.shtml.

Nielson. (2009, December 10). *Six million more seniors using the web than five years ago*. Nielsen Wire. Retrieved from http://blog.nielsen.com/nielsenwire/online_mobile/

Niemela-Nyrhinen, J. (2007). Baby boom consumers and technology: Shooting down stereotypes. *Journal of Consumer Marketing, 24*(4), 305–312.

Nimrod, G. (2009). Seniors' online communities: A quantitative content analysis. *The Gerontologist, 50*, 382–392.

Nosek, M., Kennedy, H. P., Beyene, Y., Taylor, D., Gillis, C., & Lee, K. (2010). The effects of perceived stress and attitudes toward menopause and aging on symptoms of menopause. *Journal of Midwifery & Women's Health, 44*, 328–334.

Nowosielski, K., Drosdzol, A., Sipinski, A., Kowalczyk, R., & Skrzypulec, V. (2009). Diabetes mellitus and sexuality—does it really matter? *The Journal of Sexual Medicine, 7*, 723–735.

Nurmberg, H. G., Hensley, P. G., Gelenberg, A. J., Fava, M., Lauriello, J., & Paine, S. (2003). Treatment of antidepressant-associated sexual dysfunction with sildenafil: A randomized controlled trial. *Journal of the American Medical Association, 289*, 56–64.

Nusbaum, M. R. H., Singh, A. R., & Pyles, A. A. (2004). Sexual healthcare needs of women aged 65 and older. *Journal of the American Geriatrics Society, 52*, 117–122.

Nusbaum, E. (2009). The culture pages: The cougar movement. *New Yorker, 42*(37), 65–67.

Nusbaum, M. R. H., Hamilton, C., & Lenahan, P. (2003). Chronic illness and sexual functioning. *American Family Practice, 67*, 347–354.

Nusbaum, J. F., & Robinson, J. D. (1984). Attitudes toward aging. *Communication Research Reports, 1*, 21–27.

Nyanzi, S. (2011). Ambivalence surrounding elderly widows' sexuality in urban Uganda. *Ageing International, 36*, 378–400.

NYC Health Department. (2010). HIV epidemiology and field services semiannual report. *NYC Department of Health and Mental Hygiene, 5*(1), 1–4.

O'Connor, K. M., & Fitzpatrick, J. M. (2006). Side-effects of treatments for locally advanced prostate cancer. *British Journal of Urology International, 97*, 22–28.

Oh, S. J., Ku, J. H., Choo, M. S., Yun, J. M., Kim, D. Y., & Park, W. H. (2008). Health-related quality of life and sexual functioning women with stress urinary incontinence and overactive bladder. *International Journal of Urology, 15*, 62–67.

Onishi, J., Suzuki, Y., Umegaki, H., Endo, H., Kawamura, T., Imaizumi, M., et al. (2006). Behavior, psychological and physical symptoms in group homes for older adults with dementia. *International Psychogeriatrics, 18*, 75–86.

Orel, N. A., Stelle, C., Watson, W. K., & Bunner, B. L. (2010). No one is immune: A community education partnership addressing HIV/AIDS and older adults. *Journal of Applied Gerontology, 29*, 352–370.

Orel, N. A., Wright, J., & Wagner, J. (2004). The scarcity of HIV/AIDS risk reduction materials targeting the needs of older adults among State Departments of Public Health. *The Gerontologist, 44*, 693–696.

Oswald, R., & Masciadrelli, B. (2008). Generative ritual among nonmetropolitan lesbians and gay men: Promoting social inclusion. *Journal of Marriage & Family, 70*, 1060–1073.

Pancholy, A. B., Goldenhar, L., Fellner, A. N., Crisp, C., Kleeman, S., & Pauls, R. (2011). Resident education and training in female sexuality: Results of a national survey. *The Journal of Sexual Medicine, 8*, 361–366.

Paras, M. L., Murad, M. H., Chen, L. P., Goranson, E. N., Sattler, A. L., Colbenson, K. M., et al. (2009). Sexual abuse and lifetime diagnosis of somatic disorders: A systematic review and meta-analysis. *Journal of the American Medical Association, 302*, 550–561.

Parker, S. (2006). What barriers to sexual expression are experienced by older people in 24-hour care facilities. *Reviews in Clinical Gerontology, 16*, 275–279.

Patnaik, B. K., & Barik, M. (2005). Is there a male menopause or andropause? A review of aging male reproductive system. *Indian Journal of Gerontology, 19*, 237–242.

Peat, C. M., Peyerl, N. L., Ferraro, F. R., & Butler, M. (2011). Age and body image in Caucasian men. *Psychology of Men & Masculinity, 12*, 195–200.

Pennix, B. W., van Tilburg, T., Kriegsman, D. M., Boeke, A. J., Deeg, D. J., & van Eijk, J. T. (1999). Social network, social support, and loneliness in older persons with different chronic diseases. *Journal of Aging and Health, 11*, 151–168.

Peterson, A. C., Levin, R., & Zweig, R. (1989). The erotized transference: An adaptive point of view. *Psychoanalysis and Psychotherapy, 7*, 129–141.

Pfeiffer, E. (1977). Sexual behavior in old age. In E. W. Busse & E. Pfeiffer (Eds.), *Behavior and adaptation in later life* (pp. 130–141). Boston: Little Brown.

Pfeiffer, E., & Davis, G. C. (1972). Determinants of sexual behavior in middle and old age. *Journal of the American Geriatric Society, 33*, 635–643.

Pfizer. (2011). Viagra. Retrieved from www.Pfizer.com.

Phelps, J. S., Jain, A., & Monga, M. (2004). The PsychoedPlusMed approach to erectile dysfunction treatment: The impact of combining a psychoeducational intervention with sildenafil. *Journal of Sex and Martial Therapy, 30*, 30–314.

Plassman, B., Langa, K. M., Fisher, G., Heeringa, S. G., Weir, D. R., Ofstedal, M. B., et al. (2007). Prevalence of Dementia in the United States: The Aging, Demographics, and Memory Study. *Neuroepidemiology, 29*, 125–132.

Poggi, R. G., & Berland, D. I. (1985). The therapists' reactions to the elderly. *The Gerontologist, 25*, 508–513.

Pommerville, P. (2006). Erectile dysfunction in older males: Why not investigate and treat it? *Geriatrics and Aging, 9*, 119–123.

Pope, H. G., Phillips, K. A., & Olivardia, R. (2000). *The Adonis complex: The secret crisis of male body obsession*. New York, NY: The Free Press.

Porst, H., Padma-Nathan, H., Giuliano, F., Anglin, G., Varanese, L., & Rosen, R. (2003). Efficacy of Tadalafil for the treatment of erectile dysfunction at 24 and 36 hours after dosing: A randomized controlled trial. *Urology, 62*, 121–126.

Pratt, R. R. (2004). Art, dance, and music therapy. *Physical Medicine and Rehabilitation Clinics of North America, 15*, 827–841.

Pugh, S. (2005). Assessing the cultural needs of older lesbians and gay men: Implications for practice. *Practice: Social Work in Action, 17*, 207–218.

Putterbaugh, D. T. (2010). On the prowl. *USA Today, 139*, 54–56.

Quam, J. K., & Whitford, G. S. (1992). Adaptation and age related expectation of older gay and lesbian adults. *The Gerontologist, 32*, 367–374.

Rabow, M. W., Hauser, J. M., & Adams, J. (2004). Supporting family caregivers at the end of life: "They don't know what they don't know. *Journal of the American Medical Association, 291*, 483–490.

Rackley, J. V., Warren, S. A., & Bird, G. W. (1988). Determinants of body image in women at midlife. *Psychological Reports, 62*, 9–10.

Rakowski, W., Breen, N., Meissner, H., Rimer, B. K., Vernon, S. W., Clark, M. A., et al. (2004). Prevalence and correlates of repeat mammography among women aged 55-79 in the Year 2000 national health Interview Survey. *Preventive Medicine, 39*, 1–10.

Ramsey-Klawsnik, H. (1996). Assessing physical and sexual abuse in health care settings. In L. A. Baumhover & S. C. Beall (Eds.), *Abuse, neglect, and exploitation of older persons: strategies for assessment and intervention* (pp. 67–87). Baltimore, MD: Health Professions Press.

Ramsey-Klawsnik, H., Teaster, P. B., Mendiondo, M. S., Abner, E. L., Cecil, K. A., & Tooms, M. R. (2007). Sexual abuse of vulnerable adults in care facilities: Clinical findings and a research initiative. *Journal of the American psychiatric Nurses Association, 12,* 332–339.

Ramsey-Klawsnik, H., Teaster, P. B., Mendiondo, M. S., Marcum, J. L., & Abner, E. L. (2008). Sexual predators who target elders: Findings from the first national study of sexual abuse in care facilities. *Journal of Elder Abuse & Neglect, 20,* 353–376.

Rappaport, E. A. (1956). The management of an eroticized transference. *Psychoanalytic Quarterly, 25,* 515–529.

Redelman, M. J. (2008). Is there a place for sexuality in the holistic care of patients in the palliative care phase of life? *American Journal of Hospice & Palliative Medicine, 25,* 366–371.

Reyes, M. (2010). Field guide to the cougar. *Psychology Today, 43*(6), 34–35.

Richards, M. A., & Ramirez, A. J. (1997). Quality of life: The main outcome measure of palliative care. *Palliative Medicine, 11,* 89–92.

Riggle, E. D. B., Whitman, J. S., Olson, A., Rostosky, S. S., & Strong, S. (2008). The positive aspects of being a lesbian or gay man. *Professional Psychology: Research and Practice, 39,* 210–217.

Rios-Ellis, B., et al. (2005). Critical disparities in Latino mental health: Transforming research into action. National Council of La Raza. Retrieved from http://mhcaucus.napolitano.house.gov/reports/Critical_Disparities_in_Latino_Mental_Health.pdf.

Rivera-Romas, Z. A., & Buki, L. P. (2011). I will no longer be a man! Manliness and prostate cancer screenings among Latino men. *Psychology of Men & Masculinity, 12,* 13–25.

Roach, S. M. (2004). Sexual behavior of nursing home residents: staff perceptions and responses. *Journal of Advanced Nursing, 48*(4), 371–379.

Ronch, J. L. (1985). Suspected anorexia nervosa in a 75 year old institutionalized male: Issues in diagnosis and intervention. *Clinical Gerontologist, 4,* 31–38.

Ropelato, J. (2011). *Internet pornography statistics.* Retrieved from http://internet-filter-review.toptenreviews.com/internet-pornography-statistics.html

Rosen, R. C. (2000). Medical and psychological interventions for erectile dysfunction: Toward a combined treatment approach. In S. R. Leiblum & R. C. Rosen (Eds.), *Principles and practice of sex therapy* (3rd ed., pp. 276–304). New York: Guilford.

Rosen, T., Lachs, M. S., & Pillemer, K. (2010). Sexual aggression between residents in nursing homes: Literature synthesis of an underrecognized problem. *Journal of the American Geriatrics Society, 58,* 1070–1079.

Rosen, R. C., & Leiblum, S. R. (1993). Treatment of male erectile disorders: Current options and dilemmas. *Journal of Sexual and Marital Therapy, 8,* 5–9.

Rosen, R. C., & Leiblum, S. R. (1995). Treatment of sexual disorders in the 1990s: An integrated approach. *Journal of Consulting and Clinical Psychology, 63,* 877–890.

Rosen, R. C., Wing, R. R., Schneider, S., Wadden, T. A., Foster, G. D., West, D. W., et al. (2009). Erectile dysfunction in type 2 diabetic men: Relationship to exercise fitness and cardiovascular risk factors in the Look AHEAD Trial. *The Journal of Sexual Medicine, 6,* 141–1422.

Rosenbaum, T. Y. (2010). Musculosketetal pain and sexual function in women. *The Journal of Sexual Medicine, 7,* 645–653.

Rosenblatt, A., Samus, Q. M., Steele, C. D., Baker, A. S., Harper, M. G., Brandt, J., et al. (2004). The Maryland Assisted Living Study: Prevalence, recognition, and treatment of dementia and other psychiatric disorders in the assisted living population of central Maryland. *Journal of the American Geriatrics Society, 42,* 1618–1625.

Rosenzweig, R., & Fillit, H. (1992). Probable heterosexual transmission of AIDS in an aged woman. *Journal of the American Geriatrics Society, 40,* 1261–1264.

Rossouw, J. E., Anderson, G. L., Prentice, R. L., LaCroix, A. Z., Kooperberg, C., Stefanick, M. L., Jackson, R. D., Beresford, S. A., Howard, B. V., & Johnson, K. C., Kotchen, J. M., Ockene, J., & Writing Group for the Women's Health Initiative Investigators. (2002). Risks and benefits of estrogen plus progestin in healthy postmenopausal women. *Journal of the American Medical Association, 288,* 321–333.

Roughan, P. A., Kaiser, F. E., & Morley, J. E. (1993). Sexuality and the older woman. *Care of the Older Woman, 1,* 87–106.

Rubin, R. (2004). Men talking about Viagra. *Men and Masculinities, 7*, 22–30.

Rybarczyk, B., DeMarco, G., DeLaCruz, M., Lapidos, S., & Fortner, B. (2001). A classroom mind/body wellness intervention for older adults with chronic illness: comparing immediate and 1-year benefits. *Behavioral Medicine, 27*, 15–27.

Sacerdoti, R. C., Lagana, L., & Koopman, C. (2010). Altered sexuality and body image after gynecological cancer treatment: How can psychologists help? *Professional Psychology: Research and Practice, 41*, 533–540.

Safarinejad, M. R. (2010). The effects of the adjunctive bupropion on male sexual dysfunction induced by a selective serotonin reuptake inhibitor: A double-blind placebo-controlled and randomized study. *British Journal of Urology International, 106*, 840–847.

Safarinejad, M. R. (2011). Reversal of SSRI-induced female sexual dysfunction by adjunctive bupropion in menstruating women: A double-blind placebo-controlled and randomized study. *Journal of Psychopharmacology, 25*, 370–378.

Sanchez, T. H., & Gallagher, K. M. (2006). Factors associated with recent sildenafil (Viagra) use among men who have sex with men in the United States. *Epidemiology and Social Science, 42*, 95–100.

Sand, M. S., Fisher, W., Rosen, R., Heiman, J., & Eardley, I. (2008). Erectile dysfunction and constructs of masculinity and quality of life in the Multinational Men's Attitudes to Life Events and Sexuality (MALES) study. *The Journal of Sexual Medicine, 5*, 583–594.

Santoro, N., & Komi, J. (2009). Prevalence and impact of vaginal symptoms among postmenopausal women. *The Journal of Sexual Medicine, 6*, 2133–2142.

Sbrocco, T., Weisberg, R. B., & Barlow, D. H. (1995). Sexual dysfunction in the older adult: Assessment of psychosocial factors. *Sexuality and Disability, 13*, 201–218.

Schewe, C. D., & Balazs, A. L. (1992). Role transitions in older adults: A marketing opportunity. *Psychology and Marketing, 9*, 85–99.

Schiavi, R. C., & Rehman, J. (1995). Sexuality and aging. *The Urologic Clinics of North America, 22*, 711–726.

Schick, V., Herbenick, D., Reece, M., Sanders, S. A., Dodge, B., Middlestadt, S. E., et al. (2010). Sexual Behaviors, Condom Use, and Sexual Health of Americans Over 50: Implications for Sexual Health Promotion for Older Adults. *The Journal of Sexual Medicine, 7*, 315–329.

Schmitz, M. A., & Finkelstein, M. (2010). Perspectives on poststroke sexual issues and rehabilitation needs. *Topics in Stroke Rehabilitation, 17*(3), 204–213.

Schmucker, D. L. (2005). Age related changes in liver structure and function: Implications for disease? *Experimental Gerontology, 40*, 650–659.

Schnatz, P. F., Whitehurst, S. K., & O'Sullivan, D. M. (2010). Sexual dysfunction, depression, and anxiety among patients of an inner-city menopause clinic. *Journal of Women's Health, 19*, 1843–1849.

Schoedl, A. F., Costa, M. C. P., Mari, J. J., Mello, M. F., Tyrka, A. R., Carpenter, L. L., et al. (2010). The clinical correlates of reported childhood sexual abuse: An association between age at trauma onset and severity of depression and PTSD in adults. *Journal of Child Sexual Abuse, 19*, 156–170.

Schonberg, M. A., McCarthy, E. P., York, M., Davis, R. B., & Marcantonio, E. R. (2007). Factors influencing elderly women's mammography screenings decisions: Implications for counseling. *BMC Geriatrics, 16*, 7–26.

Schover, L. R., Fouladi, R. T., Warneke, C. L., Neese, L., Klein, E. A., Zippe, C., et al. (2004). Seeking help for erectile dysfunction after treatment for prostate cancer. *Archives of Sexual Behavior, 33*, 443–454.

Schulz, R., Williamson, G. M., & Bridges, M. (1991). *Limb amputation among the elderly: Psychosocial factors influencing treatment.* Washington, DC: AARP Andrus Foundation.

Scogin, E., & McElreath, L. (1994). Efficacy of psychosocial treatments for geriatric depression: A quantitative review. *Journal of Consulting and Clinical Psychology, 62*, 69–73.

Segraves, R. T., & Segraves, K. B. (1995). Human sexuality and aging. *Journal of Sex Education and Therapy, 21*, 88–102.

Serati, M., Salvatore, S., Uccella, S., Nappi, R. E., & Bolis, P. (2009). Female urinary incontinence during intercourse: A review on an understudied problem for women's sexuality. *The Journal of Sexual Medicine, 6*, 40–80.

Serrano, J. P., Latorre, J. M., Gatz, M., & Montanes, J. (2004). Life review therapy using autobiographical retrieval practice for older adults with depressive symptomatology. *Psychology and Aging, 19*, 270–277.

Services and Advocacy for Gay, Lesbian, Bisexual, and Transgender Elders; SAGE. (2010). *SAGE resources*. Retrieved October 23, 2011, http://sageusa.org/resources/resource_view. cfm?resource=224.

Sexual Advice Association. (2011). *Sexual Advice Association Fact Sheets*. www.sda.uk.net. Retrieved June 11, 2011.

Shabsigh, R., Kaufman, J., Magee, M., Creanga, D., Russell, D., & Budhwani, M. (2010). Lack of awareness of erectile dysfunction in many men with risk factors for erectile dysfunction. *BMC Urology, 10*, 18.

Shaeer, O., & Shaeer, K. (2011). The global on-line sexuality survey (GOSS): Erectile dysfunction among Arabic-speaking internet users in the Middle East. *The Journal of Sexual Medicine, 8*, 2152–2163.

Shea, J. L. (2011). Older women, martial relationships, and sexuality in China. *Ageing International, 36*, 361–377.

Sheikh, J. I., & Yesavage, J. A. (1986). Geriatric Depression Scale (GDS): Recent evidence and development of a shorter version. *Clinical Gerontologist: The Journal of Aging and Mental Health, 5*, 165–173.

Shippy, R. M., Cantor, M. H., & Brennan, M. (2004). Social networks of aging gay men. *Journal of Men's Studies, 13*, 107–120.

Shrestha, L. B., & Heisler, E. J. (2011). *The changing demographic profile of the United States*. Washington, DC: Congressional Research Service. Retrieved from http://www.fas.org/sgp/crs/misc/RL32701.pdf.

Siegel, K., Schrimshaw, E. W., Brown-Bradley, C. J., & Lekas, H. (2010). Sources of emotional distress associated with diarrhea among late middle-age and older HIV-infected adults. *Journal of Pain and Symptom Management, 40*, 353–369.

Sigmund, J. A. (2002). Sildenafil advertising and the realities of sildenafil treatment. *The Journal of Clinical Psychiatry, 63*, 1183.

Silverberg, M. J., Keyden, W., Horberg, M. A., DeLorenzo, G. N., Klein, D., & Quesenberry, C. P., Jr. (2007). Older age and the response to and tolerability of antiretroviral therapy. *Archives of Internal Medicine, 167*, 684–691.

Simon, J. (2010). Low sexual desire-Is it all in her head? Pathology, diagnosis and treatment of hypoactive sexual desire disorder. *Postgraduate Medicine, 122*, 128–136.

Singer, P. A., Martin, D. K., & Kelner, M. (1999). Quality end-of-life care: Patient's perspectives. *Journal of the American Medical Association, 281*, 163–168.

Slusher, M. P., Mayer, C. J., & Dunkle, R. E. (1996). Gays and lesbians older and wiser (GLOW): A support group for older gay people. *The Gerontologist, 36*, 118–123.

Smith, K. P., & Christakis, N. A. (2009). Association between widowhood and risk of diagnosis with a sexually transmitted infection in older adults. *American Journal of Public Health, 99*, 2055–2062.

Smith, M. M., & Schmall, V. I. (1983). Knowledge and attitudes toward sexuality and sex education of a select group of older people. *Gerontology & Geriatrics Education, 3*(4), 259–269.

Snyder, R. J., & Zweig, R. A. (2010). Medical and psychology students' knowledge and attitudes regarding aging and sexuality. *Gerontology and Geriatrics Education, 31*, 235–255.

Sowers, M., Zheng, H., Tomey, K., Karvonen-Gutierrez, C., Jannausch, M., Li, X., et al. (2007). Changes in body composition in women over six years at midlife: Ovarian and chronological aging. *The Journal of Clinical Endocrinology and Metabolism, 92*, 895–901.

Sparrow, W. (2008, June 28). Sex in depth: A new wrinkle in Japan's porn boom. *Asia Times*. Retrieved from http://www.atimes.com/atimes/Japan/JF28Dh01.html

Speer, D. C., & Schneider, M. G. (2003). Mental health needs of older adults and primary care: Opportunities for interdisciplinary geriatric team practice. *Clinical Psychology: Science and Practice, 10*, 109–111.

Spillman, B. C., & Black, K. J. (2006). *The size and characteristics of the residential care population: Evidence from three national surveys.* Washington, DC: Office of the Assistant Secretary for Planning and Evaluation, U.S. Department of Health and Human Services.

Spitzer, R. (1981). The diagnostic status of homosexuality in DSM-III: A reformulation of the issues. *The American Journal of Psychiatry, 138*, 210–215.

Stannek, T., Hurny, C., Schock, O. D., Bucher, T., & Munzer, T. (2009). Factors affecting self-reported sexuality in men with obstructive sleep apnea syndrome. *The Journal of Sexual Medicine, 6*, 3415–3424.

Starr, B. D., & Weiner, M. B. (1981). *The Starr-Weiner report on sex and sexuality in the mature years.* Briarcliff Manor, NY: Stein & Day.

Starratt, C., Fields, R. B., & Fishman, E. (1992). Differential utility of the NCSE and MMSE with neuropsychiatric patients. *The Clinical Neuropsychologist, 6*, 331.

Stausmire, J. M. (2004). Sexuality at the end of life. *The American Journal of Hospice & Palliative Care, 21*, 33–39.

Stavis, P. F. & Walker-Hirsch, L. (1999). Consent to sexual activity. In R. D. Dinerstein, S. S. Herr, & J. L. O'Sullivan. (Eds.) *A guide to consent* (pp. 57–67). Washington, DC: American Association on Mental Retardation.

Stead, M. L., Brown, J. M., Fallowfield, L., & Selby, P. (2003). Lack of communication between healthcare professionals and women with ovarian cancer. *British Journal of Cancer, 88*, 666–671.

Steinmetz, K. L., Coley, K. C., & Pollock, B. G. (2005). Assessment of geriatric information on the drug label for commonly prescribed drugs in older people. *Journal of the American Geriatrics Society, 53*, 891–894.

Stone, L. (1977). *The family, sex, and marriage in England, 1500–1800.* New York: Harper & Row.

Swartz, M. N., Healy, B. P., & Musher, D. M. (1999). *In K. K. Holmes, P. F. Sparling, P. Mardh, S. M. Lemon, W. E. Stamm, P. Piot, & J. N. Wasserheit. (Eds.). pp. 487-509. Sexually transmitted diseases* (3rd ed.). New York: McGraw-Hill.

Tally, M. (2006). She doesn't let age define her": Sexuality and motherhood in recent "middle-aged chick flicks. *Sexuality and Culture, 10*, 33–55.

Tamam, Y., Tamam, L., Akil, E., Yasan, A., & Tamam, B. (2008). Post-stroke sexual functioning in first stroke patients. *European Journal of Neurology, 15*, 660–666.

Tannenbaum, C., Corcos, J., & Assalian, P. (2006). The relationship between sexual activity and urinary incontinence in older women. *Journal of the American Geriatrics Society, 4*, 1220–1224.

Teaster, P. B., Otto, J. M., Dugar, T. D., Mendiondo, M. S., Abner, E. L., Cecil, K. A. (2006). *The 2004 survey of state Adult Protective Services: Abuse of adults 60 years of age and older* [Report to the National Center on Elder Abuse, Administration on Aging]. Washington, DC: National Center on Elder Abuse.

Teaster, P. B., Ramsey-Klawsnik, H., Mendiondo, M. S., Abner, E., Cecil, K., & Tooms, S. (2007). From behind the shadows: A profile of the sexual abuse of older men residing in nursing homes. *Journal of Elder Abuse & Neglect, 19*, 29–45.

Teaster, P. B., & Roberto, K. A. (2003). Sexual abuse of older women living in nursing homes. *Journal of Gerontological Social Work, 40*, 105–119.

Teaster, P. B., & Roberto, K. A. (2004). Sexual abuse of older adults: APS cases and outcomes. *The Gerontologist, 44*, 788–796.

Ter Kuile, M. M., Both, S., & van Lankveld, J. J. D. M. (2010). Cognitive behavioral therapy for sexual dysfunctions in women. *The Psychiatric Clinics of North America, 33*, 595–610.

Thompson, H. S., & Ryan, A. (2009). The impact of stroke consequences on spousal relationships from the perspective of the person with the stroke. *Journal of Clinical Nursing, 18,* 1803–1811.

Tiggemann, M., Martins, Y., & Kirkbride, A. (2007). Oh to be lean and muscular: Body image ideals in gay and heterosexual men. *Psychology of Men and Masculinity, 8,* 15–24.

Toh, S., Hernandez-Diaz, S., Logan, R., Rossouw, J. E., & Hernan, M. A. (2010). Coronary health disease in postmenopausal recipients of estrogen plus progestin therapy: Does the increased risk ever disappear?: A randomized trial. *Annual of Internal Medicine, 152,* 211–217.

Toyama, M. (2008, June 17). Japan's booming sex niche: Elder Porn. *Time Magazine.* Retrieved from http://www.time.com/time/magazine/article/0,9171,1818203,00.html

Tsai, C., & Pan, S. (2010). The impact of irritative lower urinary tract symptoms on erectile dysfunction in aging Taiwanese males. *The Aging Male, 13,* 179–183.

Tsatali, M. S., Tsolaki, M. N., Christodoulou, T. P., & Papaliagkas, V. T. (2011). The complex nature of inappropriate sexual behaviors in patients with dementia: Can we put it into a frame? *Sexuality and Disability, 29,* 143–156.

United States Bureau of the Census. (2011). *Population estimates.* Retrieved http://www.census.gov/popest/estimates.html.

United States Department of Health and Human Services. (2001). *Fact sheet: The many faces of aging.* Washington, DC: Author.

United States Preventive Services Task Force. (2008). Screening for prostate cancer: U.S. Preventive Services Task Force recommendation statement. *Annals of Internal Medicine, 149,* 185–191.

United States Public Health Service. (1999). *The Surgeon General's call to action to prevent suicide.* Washington, DC: Author.

Van Deusen, J. (1997). Body image of non-clinical and clinical populations of men: A literature review. *Occupational Therapy in Mental Health, 13,* 37–57.

van Houtum, W. H., Rauwerda, J. A., Ruwaard, D., Schaper, N. C., & Bakker, K. (2004). Reduction in diabetes-related lower-extremity amputations in the Netherlands: 1991-2000. *Diabetes Care, 27,* 1042–1046.

van Lankveld, J. J. D. M., Granot, M., Weljmar, S., Willibrord, C. M., Binik, Y. M., Wesselmann, U., et al. (2010). Women's sexual pain disorders. *The Journal of Sexual Medicine, 7,* 615–631.

Vares, T., & Braun, V. (2006). Spreading the word, but what word is that? Viagra and male sexuality in popular culture. *Sexualities, 9,* 315–322.

von Sydow, K. (1995). Unconventional sexual relationships: Data about German women ages 50 to 91 years. *Archives of Sexual Behavior, 24,* 271–283.

Waite, L. J., Laumann, E. O., Das, A., & Schumm, L. P. (2009). Sexuality: Measures of partnerships, practices, attitudes, and problems in National Social Life, Health, and Aging Study. *The Journals of Gerontology: Series B: Psychological Sciences and Social Sciences, 64*(S1), 56–66.

Walker, B. L., & Harrington, D. (2002). Effects of staff training on staff knowledge and attitudes about sexuality. *Educational Gerontology, 28,* 639–654.

Wallace, M. (1992). Management of sexual relationships among elderly residents of long term care facilities. *Geriatric Nursing, 13,* 308–311.

Wallace, M., & Safer, M. (2009). Hypersexuality among cognitively impaired older adults. *Geriatric Nursing, 30,* 230–237.

Walz, T. (2002). Crones, dirty old men, sexy seniors: Representations of the sexuality of older persons. *Journal of Aging and Identity, 7,* 99–112.

Walz, T. H., & Blum, N. S. (1987). *Sexual health in later life.* Lexington, MA: Lexington Books.

Waslow, M., & Loeb, M. B. (1979). Sexuality in nursing homes. *Journal of the American Geriatric Society, 27,* 73–79.

Watters, Y., & Boyd, T. V. (2009). Sexuality in later life: opportunity for reflections for healthcare providers. *Sexual and Relationship therapy, 24,* 307–315.

Weeks, D. J. (2002). Sex for the mature adult: Health, self-esteem and countering ageist stereotypes. *Sexual and Relationship Therapy, 17*, 231–240.

Weeks, D. J., & James, J. (1998). *Secrets of the superyoung.* New York: Villard.

Weitz, R. (2010). Changing the scripts: Midlife women's sexuality in contemporary US film. *Sexuality & Culture, 14*, 17–32.

Welch, H. G., Schwartz, L. M., & Woloshin, S. (2005). Prostate-specific antigen levels in the United States: Implications of various definitions for abnormal. *Journal of the National Cancer Institute, 97*, 1132–1137.

Wentzell, E., & Salmeron, J. (2009). You'll "get Viagraed:" Mexican men's preference for alternative erectile dysfunction treatment. *Social Science & Medicine, 68*, 1759–1765.

Wertheimer, A. (2003). *Consent to sexual relations.* Cambridge, UK: Cambridge University Press.

West, S. L., D'Aloisio, A. A., Agans, R. P., Kalsbeek, W. D., Borisov, N. N., & Thorp, J. M. (2008). Prevalence of low sexual desire and hypoactive sexual desire disorder in a nationally representative sample of US women. *Archives of Internal Medicine, 168*, 1441–1449.

White, C. B. (1982). A scale for the assessment of attitudes and knowledge regarding sexuality in the aged. *Archives of Sexual Behavior, 11*, 491–502.

Wienberg, C. (2008, April 15). Call girls at nursing home fuel debate in Denmark (Update 1). Bloomberg. Retrieved from www.Bloomberg.com.

Wigfall, L. T., Williams, E. M., Sebastian, N., & Glover, S. H. (2010). HIV testing among deep South residents 40 to 64 years old with cardiovascular disease and/or diabetes. *Journal of the National Medical Association, 102*, 1150–1157.

Wild, S., Roglic, G., Green, A., Sicree, R., & King, H. (2004). Global prevalence of diabetes: Estimates for the year 2000 and projections for 2030. *Diabetes Care, 27*, 1047–1053.

Wiley, D., & Bortz, W. M. (1996). Sexuality and aging: Usual and successful. *The Journals of Gerontology, 51*, M142–M150.

Willert, A., & Semans, M. (2000). Knowledge and attitudes about later life sexuality: What clinicians need to know about helping the elderly. *Contemporary Family Therapy, 22*, 415–435.

Williams, M. E., & Freeman, P. A. (2007). Transgender health: Implications for aging and caregiving. *Journal of Gay & Lesbian Services, 18*, 93–108.

Winn, R. L., & Newton, N. (1982). Sexuality in aging: A study of 106 cultures. *Archives of Sexual Behavior, 11*, 283–298.

Wiwanitkit, V. (2008). Sexuality in Parkinsonism. *Sexuality and Disability, 26*, 105–108.

Wolf, D. G. (1982). *Growing older: Lesbians and gay men.* Berkeley: University of California Press.

Wyatt, S., Henwood, F., Hart, A., & Smith, J. (2005). The digital divide, health information and everyday life. *New Media & Society, 7*, 199–218.

Wyman, J. F., Burgio, K. L., & Newman, D. K. (2009). Practical aspects of lifestyle modifications and behavioural interventions in the treatment of overactive bladder and urgency urinary incontinence. *International Journal of Clinical Practice, 63*, 1177–1191.

Xavier, J., Hitchcock, D., Hollinshead, S., Keisling, M., Lewis, Y., Lombardi, E., et al. (2004). *An overview of U.S. trans health priorities: A report by the Eliminating Disparities Working Group.* Washington, D.C.: National Coalition for LGBT Health.

Xu, J., Dochanek, K. D., Murphy, S. L., & Tejada-Vera, B. (2010). Deaths: Final data for 2007. *National Vital Statistics Reports, 58*(19), 1–135.

Yang, K., & Jolly, L. D. (2008). Age cohort analysis in adoption of mobile data services: Gen Xers versus baby boomers. *Journal of Consumer Marketing, 25*, 272–280.

Youn, G. (2009). Marital and sexual conflicts in elderly Korean people. *Journal of Sex and Marital Therapy, 35*(3), 230–238.

Zablotksy, D. (1998). Overlooked, ignored, and forgotten: Older women at risk for HIV infection and AIDS. *Research on Aging, 20*, 760–775.

Zang, Y., Chung, L. Y. F., & Wong, T. K. S. (2008). A review of the psychosocial issues for nurses in male genitalia-related care. *Journal of Clinical Nursing, 17*, 983–998.

Zeiss, R. A., Delmonico, R. L., Zeiss, A. M., & Dornbrand, L. (1991). Psychologic disorder and sexual dysfunction in elders. *Clinics in Geriatric Medicine, 7*, 133–151.

Zeranski, L., & Halgin, R. P. (2011). Ethical issues in elder abuse reporting: a professional psychologist's guide. *Professional Psychology: Research and Practice, 42*, 294–300.

Zerninke, K. (2007, November 18). Love in the time of dementia. *New York Times*. Retrieved from http://www.nytimes.com/2007/11/18/weekinreview/18zernike.html?pagewanted=all

Zilbergeld, B. (1992). *The new male sexuality*. New York: Bantam Books.

Index*

Printed by Books on Demand, Germany